The First Gene

The First Gene

The Birth of Programming, Messaging and Formal Control

David L. Abel, Director, The Gene Emergence Project,

Department of ProtoBioCybernetics and ProtoBioSemiotics,
The Origin of Life Science Foundation, Inc.

LongView Press™–Academic Biological Research Division

New York, N.Y.

Author: David L. Abel
Authors Chapter 5:   Kirk K. Durston & David K. Y. Chiu
Author Chapter 10   Donald E. Johnson
Editor: David L. Abel
Cover Illustration, design © 2011 Jessie Nilo
Interior photos, tables, charts and graphics: As noted in book

Library of Congress Control Number: 2011940333

Library of Congress Subject Heading Suggestions:
1. QH325 Life – Origin
2. Biogenesis
3. Molecular evolution
4. Information theory in biology
5. Evolutionary genetics
6. Genetic code

BISAC / BASIC Classification Suggestions:
1 SCI029000   SCIENCE / Life Sciences / Genetics & Genomics
2. SCI049000   SCIENCE / Life Sciences / Molecular Biology
3. SCI009000   SCIENCE / Life Sciences / Biophysics
4. SCI017000   SCIENCE / Life Sciences / Cell Biology

ISBN-13: ISBN: 978-0-9657988-9-1   V 1.0

Abel, David L. 1946 —

The First Gene: The Birth of Programming, Messaging and Formal Control. An Anthology
David L. Abel, Editor. Includes bibliographic references, glossary and index

LongView Press™—Academic, 2011, Biological Research Division
244 5th Avenue, Suite # G228, New York, NY 10001-7604
Printed in the United States of America on acid-free paper

Dedication

This anthology is dedicated to

all those challengers of Kuhnian Paradigm Ruts

who risk their careers and reputations

raising an eyebrow of skepticism

over theories that are pontificated to be scientific fact

by a thoroughly entrenched hierarchy and majority,

but which are in fact unfalsifiable,

completely unsubstantiated empirically,

lacking a single prediction fulfillment,

and not even logically defensible.

The First Gene

David L. Abel, Editor

Table of Contents

Preface

Few specialists exist in the world in either of two new scientific disciplines: ProtoBioCybernetics and ProtoBioSemiotics. Because of the paucity of researchers in these two fields, the Editor of this anthology was forced to introduce both scientific disciplines by contributing most of the chapters in this initial anthology. Hopefully, this work will seed a long list of contributors to the next ProtoBioCybernetic and ProtoBioSemiotic anthologies. No two fields of science are more fascinating and challenging.

Donald E. Johnson is uniquely qualified to contribute to both of these fields by virtue of having multiple PhD's in chemistry, information theory and computer science. His book *Programming of Life* is a national best seller, and contains much material directly relevant to this anthology. His book *Probability's Nature and Nature's Probability (A call to scientific integrity)* also takes an honest look at the constraints of probabilistic resources on what could possibly have occurred randomly over large periods of time.

David Chiu is a mathematician and Kirk Durston is a biophysicist, both at the University of Guelph. Their peer-reviewed science journal publications have centered on the difficult problem of *measuring* the change in Functional Sequence Complexity (FSC) of proteins during evolution. Change in the FSC of proteins as they evolve can be measured in "Fits"— Functional bits. The ability to quantify changes in biofunctionality during evolutionary transition represents one of the most important advances in biological research in recent decades. See especially, Durston, K.K.; Chiu, D.K.; Abel, D.L.; Trevors, J.T. 2007, Measuring the functional sequence complexity of proteins, Theor Biol Med Model, 4, 47 (Free on-line access at http://www.tbiomed.com/content/4/1/47).

Because this book is also being made available in e-book format (e.g., for Kindles), many awkward internet links have been deliberately left in the text and reference lists.

A glossary is included to define and expound on many technical terms used within this anthology.

David L. Abel, Director
The Gene Emergence Project
Department of ProtoBioCybernetics & ProtoBioSemiotics
The Origin of Life Science Foundation, Inc.

Acknowledgements

I would like to thank the many selfless anonymous peer reviewers of each paper in this anthology, sometimes as many as six per paper, and especially Donald E. Johnson for his management of the anonymity of reviewers of Dr. Abel's manuscripts. Don Johnson also anonymously peer-reviewed all chapters but his own. James Harding and Morris Hedge provided extensive high-quality critiques, copy-editing, and proof-reading of all manuscripts even after the peer review process was completed. LongView Press—Academic provided crucial professional technical assistance in preparing the book manuscript for publication. Jessie Nilo did a nice job on cover design and graphic art. John Gagliardi suggested additional glossary terms be added. Jed Macosko edited some of the promotional material.

Finally, the Editor would like to acknowledge the financial support for this work coming in the form of a grant from The Origin of Life Science Foundation, Inc. An anonymous multi-millionaire donor has generously provided the funding over many years for every Gene-Emergence Project the Board of Directors has ever approved. This includes voluntarily becoming legally bound and subject to the regular audits needed to guarantee the underwriting of The Origin of Life Prize[TM] (www.lifeorigin.info). The Origin of Life Prize[TM] focuses research attention specifically on the problem of gene emergence, code-origin, and the birth of programming, messaging and formal controls in a prebiotic naturalistic world. The Prize is judged by the scientific community at large, particularly by those investigators most involved with life-origin research.

Introduction

Evolution theory is quick to offer an explanatory model for how existing Prescriptive Information (PI) in genomes could have been progressively *modified*:

> Extremely rare beneficial genetic mutations can occur and be preserved through differential survival and reproduction of the fittest phenotypic organisms.

Painfully lacking within evolution theory, however, is an explanation for how *any* Prescriptive Information (PI) got written or programmed into the genome *in the first place*. The origin of initial genetic/genomic instructions and epigenomic regulation is typically side-stepped by life-origin science. The whole point of The Origin of Life Prize^{TM} was to stimulate research specifically into this neglected area. After over a decade of submissions to the Origin of Life Prize^{TM}, no submission has made it past the first tier of naturalistic scientists screening the submissions and judging their relevance.

The origin of control and regulation in nature is the subject of this anthology. It is also the focus of a new scientific discipline known as ProtoBioCybernetics. "Cybernetics" is the study of various means of steering and organizing *controls*, not just the effects of mere physicodynamic constraints. Constraints are blind and indifferent to formal success. Controls steer events toward pragmatic goals.

"Bio" of course refers to life. "Proto" refers to "first" or "earliest" [life]. Thus ProtoBioCybernetics is the focused study of how control mechanisms arose in inanimate nature (in a prebiotic environment) to organize the very first protocells and protometabolism.

A closely related new field is ProtoBioSemiotics. Controlling messages would have had to be sent, received and understood at their destination in any early protocellular metabolic scheme or system. Constructive controlling messages are meaningful and functional. Such messages steer events toward computational success or algorithmic optimization; they are not just meaningless, redundant, extremely low-informational signals such as pulsars give off.

How did a prebiotic natural environment of mere mass/energy interactions generate meaningful, functional messages? How did chance and necessity prescribe the ability of the receiver to follow arbitrary rules required for decoding? How could the laws of physics and chemistry have enabled molecules to understand linguistic-like symbol systems, and act on such messages within the first protocells? This is the subject of ProtoBioSemiotics.

This anthology specifically addresses questions that for all too long have been swept under the rug of honest scientific investigation. The result has been prolonged entrapment in the greatest Kuhnian paradigm rut in the history of science.

The First Gene, David L. Abel, Editor 2011, pp 1-18 ISBN: 978-0-9657988-9-1

1. What is ProtoBioCybernetics?

David L. Abel

Department of ProtoBioCybernetics/ProtoBioSemiotics
Director, The Gene Emergence Project
The Origin-of-Life Science Foundation, Inc.
113 Hedgewood Dr. Greenbelt, MD 20770-1610 USA

Abstract: Cybernetics addresses *control* rather than mere constraints. Cybernetics incorporates Prescriptive Information (PI) into various means of steering, programming, communication, instruction, integration, organization, optimization, computation and regulation to achieve formal function. "Bio" refers to life. "Proto" refers to "first." Thus, the scientific discipline of Proto-BioCybernetics specifically explores the often-neglected derivation through "natural process" of initial control mechanisms in the very first theoretical protocell. Whether an RNA World, Peptide World, Lipid World, or other composomal Metabolism-First model of life-origin is pursued, selection for biofunction is required prior to the existence of a living organism. For gene emergence, selection for *potential* biofunction (programming at decision nodes, logic gates and configurable switch-settings) quickly becomes the central requirement for progress.[*]

Reprint request: David L. Abel, Director, The Gene Emergence Project, Department of ProtoBioCybernetics/ProtoBioSemiotics, The Origin-of-Life Science Foundation, Inc., 113 Hedgewood Dr., Greenbelt, MD 20770-1610 USA E-mail: life@us.net

*Sections from previously published peer-reviewed science journal papers [1-9] have been incorporated with permission into this chapter.

Introduction: Exploring the birth of Control

Explaining the scientific discipline of ProtoBioCybernetics first requires defining control and differentiating it from mere constraints [7]. Controls steer events, usually toward formal function and utility. Controls are the subject of cybernetics. The spontaneous orderliness of nature described by the laws of physics and chemistry are blind to formal function and utility.

Inanimate nature constrains events; it does not control events. Nature constrains events without regard to utility. Physicodynamics (mass/energy interactions) have never been observed to foster function at the programming, computational, or linguistic level.

Establishing bona fide control first requires at least one bit of Shannon uncertainty. A bit of uncertainty corresponds to one binary *choice opportunity* at a decision node, *not a choice*. Controls must be chosen. Control requires programming freedom from fixed law. Controls are thus a product of "free-will" determinism, not cause-and-effect physicodynamic determinism. Controls integrate events and objects into a functional state known as "organization." The reader is invited to try to falsify the following null hypothesis, *"Controls alone, not constraints, cause formal function and utility to come into existence."* Without controls, self-ordering can occur in nature, but not bona fide organization. As explained later in this anthology, purposefully chosen constraints are a form of formal control, and do not serve to falsify the above null hypothesis.

Perhaps the earliest historical demonstration of control in the study of physics, thermodynamics and kinetics is the thought experiment known as Maxwell's demon [10] (see Chapter 2 and 4). The demon *chooses* when to open and close a trap door to separate hotter, faster-moving ideal gas molecules from cooler, slower moving molecules. The free-will selective operation of this trap door by the demon alone creates an energy differential and work potential out of physicodynamically inert molecules. The demon's choices constitute the first instance of formal *control*. An agent's purposeful choices control the physical state. Maxwell's demon was the first naturalistic "engineer," except that the demon is not naturalistic, but is a choosing agent that has no place in naturalistic physics texts. Of course, constraints limit behavior too, but without any consideration or pursuit of formal function and utilitarian goal. Physicodynamic constraints are blind to the notions of function, usefulness, formal work, utility, and pragmatism.

Controls are provided to physical systems through Prescriptive Information (PI) [2, 6, 8, 11, 12]. "Prescriptive information either tells us what choices to make, or it is a recordation of wise choices already made." [6] Carrying this idea further, "Prescriptive information either instructs or directly produces nontrivial function at its destination." [2, 8] Externally provided algorithmic processing is usually needed to realize utility from a Turing-tape-like linear digital source of PI such as DNA. But, ribozymes provide exception to this. The PI in ribonucleotide sequencing in ribozymes can act directly to catalyze without any externally applied algorithmic processing. This is the appeal of the preRNA and RNA World models of life origin.

Symbol systems and configurable switch-settings allow recordation of purposeful choices into physicality. PI instructs or produces nontrivial function through control choices originating with each decision-node selection. PI, the use of symbol systems, and the various mechanisms of instantiating control into physicality will all be explained more thoroughly in the chapters that follow in this anthology.

1. What does "formal" mean?

Most academicians readily agree that language, mathematics, programming, and logic theory are formalisms. Few realize exactly why. The word "formal" relates to Plato's forms and Aristotle's appreciation of general classes of form and function that transcend particular physical structure and shape.

Formalisms typically employ representationalism (e.g., math and letter symbol systems, words and language)—something that physicality cannot generate or participate in. Formal choices can be represented within linear digital prescriptions using sequences of "1" vs. "0" to represent each switch-setting to "on" vs. "off," "Yes" vs. "No," or "Open" vs. "Closed." Symbol systems are governed by arbitrary rules, not laws. The rule could just as easily be that "1" represents "Closed." Laws describe the invariant deterministic behavior of inanimate nature. Rules can be readily broken, and govern voluntary, choice-contingent behavior. All formalisms arise out of uncoerced choices in the pursuit of function and utility.

A "formalism" is a concept or idea like the category and generalization known as a "paper clip." (See Figure 1) There are many different kinds of paper clips. But when we say "paper clip," everyone knows what we mean without knowing the details of which specific kind of paper clip we are talking about. The general "form" of a paper clip comes to our mind as both a formal *representation* of meaning (which physics and chemistry cannot participate in) and as a *generalization* of that form that transcends the mass, energy and structure of each specific kind of paper clip. Formalisms make generalizations and categorizations possible.

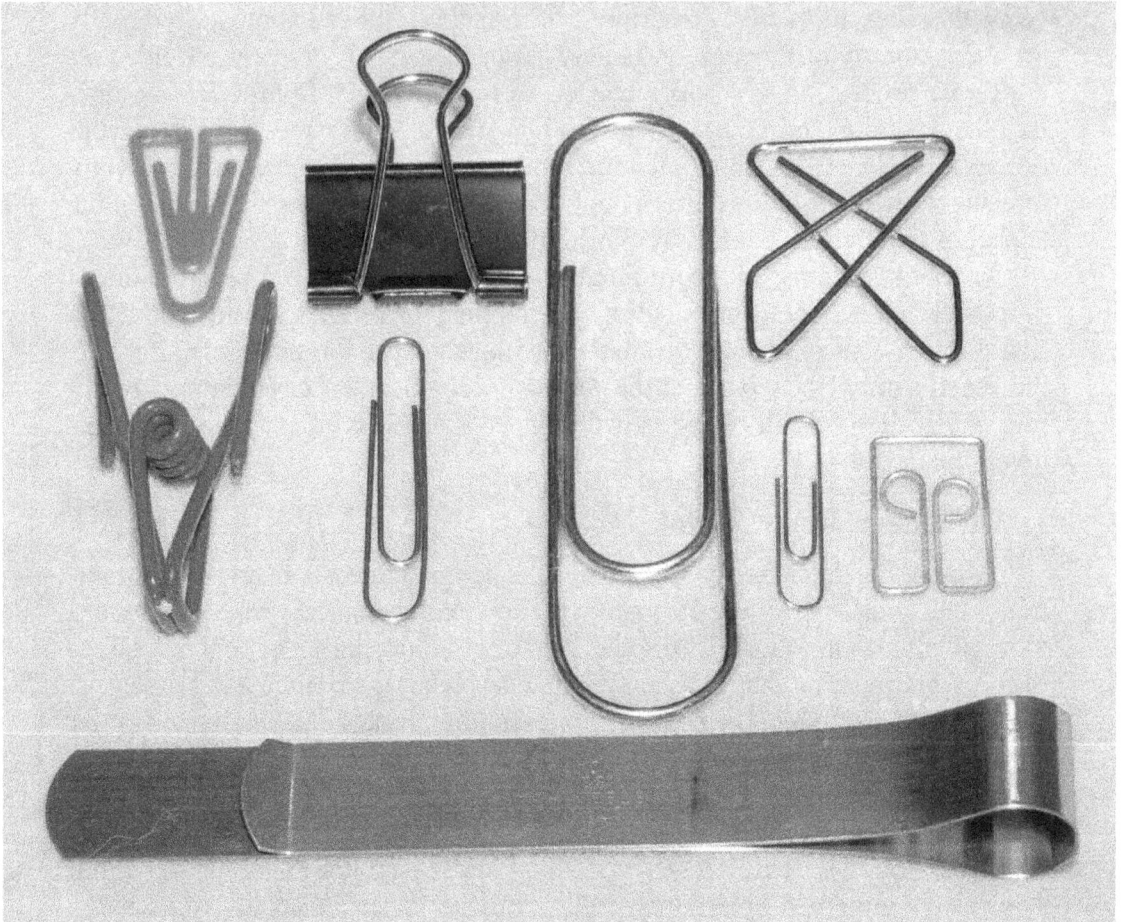

Figure 1. The idea of "paperclip" transcends any unique physical structure, shape, or scale, and it also transcends any particular type or style of paper clip. It is a conceptual category, not a particular physical entity. In this sense, "paperclip" is an abstract representational formalism rather than a physical object.

Listed below are aspects of reality that are all formalisms. None of these formalisms can be encompassed by a consistently held naturalistic worldview that seeks to reduce all things to physicodynamics:
- Mathematics
- Language
- Inferential and deductive logic theory
- The sign/symbol/token systems of semiosis

- Decision theory
- Cybernetics (including computer science)
- Computation
- Integrated circuits
- Bona fide organization (as opposed to mere self-ordering in chaos theory)
- Semantics (meaning)
- Pursuits of goals
- Pragmatic procedures and processes
- Art, literature, theatre, ethics, aesthetics
- The personhood of scientists themselves—the ultimate formalism

All of the above formalisms depend upon *choice contingency* rather than chance contingency or necessity. To hypothesize various theories of evolution, we substitute selection pressure for choice contingency. But whether we are talking about natural selection or artificial selection, the bottom line is still *selection* of the fittest or most pragmatic from among real options.

Nontrivial formal systems have never been observed to arise from "coin flips" at successive bifurcation points. Decision nodes must be true to their descriptive name. If guesses are made at decision nodes, both reason and empirical experience teach us that little or no utility will be generated. Wise choices must be made with intent to achieve logical, cybernetic, computational, and linguistic function. "Garbage in, garbage out," programmers quip. The criterion of wise choices from among real options is incorporated into the generation of any kind of nontrivial organized system.

"Paper clip" can be a single physical entity. But it can also be a formal generalization, a category, or a representation of *a class of entities*, all of which have the same basic function even though no two physical structural descriptions are the same. Three paperclips in Figure 1 vary only in scale; but different scale of the same entity represents still another kind of formalism. The collective descriptive category of "paper clip" is an idea—a formalism. This aspect of formalism was recognized by Plato when he thought about general or universal "form."

2. Physicality vs. Nonphysical Formalisms

Physicalism cannot address or deal with this kind of formal reality. It's like saying "genome" to address the entire class of all specific genetic sequences. Genes and the supposedly "non-coding" DNA sequences are responsible for *prescribing* each specific protein's amino acid sequencing and all of the microRNA controls—not mere physicodynamic constraints—that regulate genes and holistic metabolism. And of course the phenomenon of "regulation" is also formal rather than physical. Controls cannot be reduced to the chance and/or necessity of mere physical constraints.

Aristotelian "formal" and "final" causes cannot be reduced to mere mass/energy interactions. Certainly physicodynamics pursues no "final causes." Physicality has no desires, goals or aspirations. Inanimate physicality has no sensation of or motivation to pursue utility. Physicality is blind to pragmatism.

Formalisms alone generate nontrivial function and utility. Language, mathematics, inferential logic theory, computer programming, knowledge, the "aboutness" of information, and the scientific method itself are just a few examples of formalisms that cannot be reduced to physicodynamic interactions. Formalisms are nonphysical (Table 1). Formalisms always require abstract, conceptual choice contingency. Formalisms can be instantiated (recorded) into a physical medium or matrix. But the formalism itself remains nonphysical despite its instantiation into physicality.

Why is it possible to commit errors in mathematical manipulations, but not possible for gravity, electromagnetism, the strong or weak nuclear force to commit errors? The answer is that the cause-and-effect determinism of inanimate (nonliving) physicodynamic orderliness contains no decision nodes, logic gates or configurable switch-settings (Table 1). Since it is impossible to choose, it is impossible to err. Physicodynamic effects are forced by physical law. The only freedom that exists is a standard deviation bell curve stemming from the uncertainty of heat agitation and complex interactions of known physical forces. Yet-to-be-discovered forces and invariant physicodynamic laws would offer no programming freedom.

What about the "form" of a pattern in beach sand left by wave action? Isn't that formal? No! Such patterns can be explained by a chain or sequence of purely physical events. The pattern has regularity, but it does not result from "formal causation." The pattern has nothing to do with choice contingency and control. It has only to do with constraints and low-informational law. The same is true of the dissipative structures of chaos theory. They are self-ordered, not self-organized. What causes the confusion here is failing to ask whether these phenomena are controlled or merely constrained. We must differentiate between redundant, oscillating, low-informational physicodynamic patterns (similar to pulsar-like signals) vs. formal phenomena that invariably involve choice contingency, not chance contingency or law. Pulsar signals cannot generate meaningful messages because they are merely self-ordered by fixed law (necessity).

Table 1. *Comparison of Formalisms to Constrained Physicality.*

Attribute	Formalisms	Constrained Physicality
Physicodynamic	No. Utterly nonphysical	Yes. Entirely physicodynamic
Options/Possibilities	Many	Few
Uncertainty	High prior to choices	Little
Constrained	No	Yes
Controlled	Yes	No
Limited by forced, fixed laws	No	Yes
Limited by voluntary rule obedience	Yes	No
Chance contingent	No	Some
Choice contingent	Yes	No
Decision nodes	Yes	No
Logic gates	Yes	No
Configurable switch-settings	Yes	No
Abstract or tangible	Abstract	Tangible
Conceptual	Yes	No, except the mathematical nature of the laws themselves
Caused by	Choice determinism	Law-like necessity
Nontrivial function-producing	Yes	Never once observed
Goal oriented	Yes	Never
Which side of The Cybernetic Cut	Far side	Near side
Symbols/Representationalism used	Yes	Never
Meaning generated	Yes	Never
Sophisticated utility generated	Yes	Never
Useful and Pragmatic	Yes	Blind and Indifferent

When we hear the word "formal," we need to *stop* thinking "order" or "pattern." We need to think *"choice contingency."* Physicality cannot participate in formalisms such as language, mathematics, coding, translations, programming, logical inference, circuit integration, engineering, ethics, aesthetics, and the scientific method itself. All of these formalisms require "arbitrary" choices. Arbitrary does not mean random. It means choices uncoerced-by-law. Necessity would program every logic gate the same way every time, by law. If inanimate nature did the programming, it would generate a computer program consisting of all 0's, or a program consisting of all 1's.

Mere coin flips at decision nodes will not work either to explain formalisms. No computationally successful program was ever generated by a random number generator. To generate nontrivial functional Markov chains always involves behind-the-scenes steering. So-called "evolutionary algorithms" (a

Self-contradictory nonsense term), if they produce any formal function, are always artificially controlled. Optimization of genetic algorithms is always choice-contingent, and therefore formal rather than physical.

3. Intuitive, Semantic, Functional Information (FI)

The Prescriptive Information (PI) mentioned at the start of this chapter is a subset of intuitive or semantic (meaningful) information. Semantic information conveys meaningful and functional messages (semiosis) from a source that can be understood by a receiving agent at its final destination—at the far end of a Shannon channel. Adami rightly argues that information must always be *about* something [13]. "Aboutness" is a common focus of attention in trying to elucidate what makes information intuitive [14, 15]. But aboutness is always abstract, conceptual, and formal. Efforts to define aboutness in purely physical terms have frustrated bioinformationists for decades [16-19]. The difficulty of defining and understanding semantic information is especially acute in genetics [20, 21]. Oyama points to the many problems trying to relate semantic information to cellular biology [22]. Some investigators attempt to deny that genes contain meaningful information and true instructions [23-30]. Their arguments strain credibility.

Jablonka rightly argues that Shannon information is insufficient to explain biology [31]. She points to the required interaction between sender and receiver. Jablonka emphasizes both the function of bioinformation and its "aboutness," arguing that semantic information only exists in association with living or designed systems. "Only a living system can make a source into an informational input" [31, pg. 588]. Jablonka correctly senses the *formal* nature of semantic and intuitive information. Formalisms of all kinds involve abstract ideas and agent-mediated purposeful choices. Inanimate physics and chemistry have never been shown to generate life or formal choice-based systems.

Semantic information, unlike Shannon "information," is Functional Information (FI). Shannon "information" is a misnomer. What is usually called "Shannon information" is in fact either Shannon "uncertainty" or "reduced uncertainty" (poorly termed "mutual entropy"). Neither can prescribe or generate formal function. Shannon uncertainty is nothing more than a measure of combinatorial probabilism [32]. Bits of Shannon uncertainty can measure "binary choice opportunities." Under no circumstances can bits ever measure actual specific choices. *Yet specific choices at bona fide binary decision nodes alone generate semantic information*—choosing from among two real options: either "On" is picked, or "Off" is picked. A logical "excluded middle" prevails. Indecision is not allowed at programming nodes except to deliberately provide end-user programming options. In computer science this is formally represented symbolically by the programmer's choice of either a "1" or a "0" at each logic gate. At every decision node, either a "1" or a "0" must be picked to generate both Functional Information (FI) and its subset of Prescriptive Information (PI) at that node.

The reason bits cannot measure specific choices is that no standard fixed unit of measure exists to quantify specific choices made at each unique position in the programming string of decision nodes. To generate bits of "reduced uncertainty" requires a great deal of cognitive background knowledge independent of the "before" or "after" bit measurements themselves. To reduce negative uncertainty requires the accrual of positive knowledge that reduces and counterbalances that negative uncertainty. The "after" measurement of bits must be educated by gained certainty before it can be compared to the "before" measure of uncertainty to establish "reduced uncertainty." Even then, reduced uncertainty is a very limited form of knowledge. Shannon uncertainty cannot progress to becoming FI without smuggling in positive information from an external source. Only then is Shannon uncertainty (bits) reduced.

Even though bits cannot measure FI, it is important to remember that to record FI does require 1) Shannon uncertainty (contingency: events could have occurred differently despite physical constraints), 2) freedom of selection (the ability to choose), and 3) intent (goal and/or purpose). No system can contain FI that is fully physicodynamically determined. The necessity of law-like physical behavior disallows contingency and freedom of selection. What is forced by law cannot offer choice, goal or purpose. In addition, for FI to be generated, the possibility of error and symbolic misrepresentation must exist [33]. Sterelny and Griffiths also argue that the semantic content of information, including genetic information, can be stored and expressed at a later time. Immediacy of cause-and-effect is not required. Finally, choice contingency is a form of determinism. Determinism is not limited to physicodynamics. Choice contingency, when instantiated into physicality, can become a true cause of physical effects.

4. Descriptive Information (DI) vs. Prescriptive Information (PI)

The source of the external positive background information needed to reduce uncertainty (discussed above) can come in the form of Descriptive Information (DI). Intuitive semantic information (Functional Information, or FI), technically has two subsets: Descriptive (DI) and Prescriptive (PI). Unfortunately, many semantic information theorists make the mistake of thinking of functional information solely in terms of human epistemology and specifically description (DI). This in effect limits the meaning of "function." DI provides valued common-sense knowledge to human beings about the way things already are. *Being* can be described to provide one form of function. But this subset of intuitive and semantic information, while highly functional, is very limited and grossly inadequate to address many forms of *instruction and control*. Prescriptive information (PI) does far more than describe. We can thoroughly describe a new Mercedes automobile, providing a great deal of DI in the process. But this functional DI might tell us almost nothing about *how to design, engineer and build* that Mercedes. The term "functional information" as used in peer-reviewed naturalistic biological literature by Nobel laureate Jack Szostak et al in 2003 [34] [35, 36] can be a completely inadequate descriptor of the "how to" information—the instructions—required to organize and program sophisticated utility. *Potential* formal function must be *prescribed* in advance by Prescriptive Information (PI) via decision node programming, not just described after the fact. As its name implies, PI specifically conceives and prescribes utility. PI programs computational success in advance of halting. While it is true that halting must be empirically verified (the halting problem [37, 38]), *computational success still must be prescribed in advance of its realization.* PI either tells us what choices to make, or it is a recordation of wise choices already made [12]. When we install computer software, we are installing PI. Yet PI is not just limited to instruction. PI can also indirectly generate nontrivial computational success and cybernetic function in conjunction with external algorithmic processing.

PI can perform nonphysical "formal work." PI can then be instantiated into physicality to marshal physical work out of nonphysical formal work [6, 10]. Cybernetic programming is only one of many forms of PI. Ordinary language itself, various communicative symbol systems, logic theory, mathematics, rules of any kind, and all types of controlling and computational algorithms are forms of PI.

PI arises from expedient choice commitments at bona fide decision nodes [6, 9, 39]. Such decisions steer events toward pragmatic results that are valued by agents. Empirical evidence of PI arising spontaneously from inanimate nature is sorely lacking [1, 9]. Neither chance nor necessity has been shown to generate prescriptive information [1, 3, 4, 6-9, 12, 40, 41]. Choice contingency, not chance contingency, prescribes nontrivial function.

PI typically is recorded into a linear digital symbol system format. Symbols represent purposeful choices from an alphabet of symbol options. Symbol selection is made at bona fide decision nodes. Selection of particular sequences of symbols (syntax) must follow prescribed arbitrary rules. It is only when these rules are followed by both sender and receiver that a meaningful/functional message can be successfully conveyed to its destination (semiosis). A meaning*less* message (another self-contradictory nonsense term) would fulfill no purpose and provide no functionality. It would therefore not qualify definitionally as a "message." It would in fact be nothing more than a signal. Signals are not necessarily messages. A consistently repeating pulsar signal is not a meaningful message, and therefore not a message at all. Yet a pulsar signal contains high order and pattern.

It is common for non-specialists in biocybernetics and biosemiotics to try to define messages erroneously in terms of "patterns." The patterns in the sand caused by wave action of the sea, for example, convey no meaningful message or cybernetic programming. As we shall see in later chapters, neither order nor patterns are the key to meaning, regulation, control or function. *Selection for potential function* at bona fide decision nodes and logic gates is. more conceptually complex PI is needed to compute and organize metabolism and life than is needed to generate our most advanced computer systems. Life is the most sophisticated of all integrated meta-systems. Prescriptive Information is much more than intuitive semantic information. PI requires anticipation, "choice with intent," and the diligent pursuit of Aristotle's "final function" at successive bona fide decision nodes. PI either instructs or directly produces formal function at its destination through the use of controls, not mere constraints [6, 11]. Once again, PI either tells us what choices to make, or it is a recordation of wise choices already made.

Decision node choices can also be recorded or instantiated into physicality via logic-gate and configurable switch-settings. When mental symbols are recorded onto physical objects, they are called "tokens." The small blocks of wood with letters written on them in a Scrabble game are tokens. When physical tokens are *chosen* to spell words, the symbol system is called a Material Symbol System (MSS) [42, 43]. Although the tokens are physical, the selection of each token to spell meaningful syntax is not physical. Each selection is abstract, conceptual, nonphysical, choice-based, rule-guided, and formal. The same is true of the symbol meaning itself. Meaning is arbitrarily assigned to each representational symbol. The latter is a purely formal control function, not a physicodynamically constrained interaction.

The prior selection of each nucleotide and syntactical nucleotide sequencing is a form of linear digital programming of potential function. Transcription and mRNA editing must be completed by additional algorithmic processing. The "messenger molecules" are rigidly bound with covalent bonds before that biofunction is ever realized phenotypically. Protein primary structure (linear digital sequence) must be completed in the ribosome before folding into molecular machines begins. All of this linear digital prescription must take place long before any fittest living organisms can be favored by the environment (The GS Principle) [5]. Later we will look at the incredible additional layers and dimensions of PI that are instantiated into both DNA and other parallel nanocomputers systems in the cell.

Definitions of information that are limited to human epistemology are not helpful when it comes to gene and functional small RNA emergence. Genetic and genomic PI was instructing the organization of metabolism long before Homo sapiens arrived on the scene to ponder it. Humanity's knowledge and definitions of information are irrelevant to the question of how protocells could have objectively prescribed biofunction and self-organized in a prebiotic environment. We need to cultivate a less anthropocentric and less subjective understanding of PI and its MSS coding in the genome to make any progress in life-origin science. Molecular biological encryption/decryption cannot be reduced to a product of human consciousness. Linear digital prescription using a material symbol system, its coding and decoding, its error-correction mechanisms, and its noise-reducing Hamming redundancy block coding produced not only cellular metabolism, but human brain function with its consciousness. Our knowledge of these phenomena is secondary, not primary. Our consciousness and epistemology is not the center of biological and cosmic reality. No room exists for solipsism within a consistently held naturalistic and evolutionary worldview. Naturalism metaphysically presupposes an external-to-mind literal objective history of real physical transitions from objective simple one-celled organisms to primate brains. Consciousness is secreted by the brain (just as liver secretes bile) in a naturalistic worldview. Consciousness is not ultimate as envisioned by solipsism, and the derivation of consciousness from physicodynamics must be fully elucidated before equating naturalism with science.

5. The focus of ProtoBioCybernetics

ProtoBioCybernetics seeks to study the source and derivation of Prescriptive Information (PI) in inanimate nature. PI is the "how to" information that we call "instructions." Genomes were giving instructions and computing long before *Homo sapiens* existed. Belief in chemical/molecular evolution presupposes that physicodynamics alone generated formal instructions sufficient to organize a protometabolism in a structural protocell that spontaneously came to life. The sharp focus of the discipline of ProtoBioCybernetics addresses questions of how inanimate nature could have:

1. Chosen wisely from among physicodynamically indeterminate options
2. Valued and pursued formal function to which physicodynamics is blind
3. Anticipated what would "work" before that utility came into existence

4. Wrote formal rules governing behavior not forced by fixed laws
5. Programmed, measured and computed formally controlled systems
6. Generated the very first Prescriptive Information (PI)
7. Generated the first material symbol system with which to record and replicate PI.
8. Organized both protocell structure and protometabolism.
9. Programmed specific reaction sequences, pathways and cycles.
10. Integrated all those specific reactions into a cooperative protometabolic system
11. Established protometabolic control and regulatory mechanisms.
12. Selected the syntax of "alphabetical characters" (nucleotides) so as to "spell" and "encode" meaningful (biofunctional) messages that prescribe amino acid sequence
13. Devised a noise-reducing Hamming "block code" (triplet codons)
14. "Foreknew" which nucleotide sequences would prescribe (only upon later translation) the needed amino acid sequences.
15. "Foreknew" which sequences of amino acids (primary molecular structure) would only later fold according to minimum Gibbs free-energy dictates to produce thousands of needed three-dimensional molecular machines and computers.
16. Devised a formal codon table and bijective 3-to-1 translation system
17. Devised a representational heritable symbol system "independent" of phenotype that could evolve, but still retain already achieved progress so that the wheel didn't have to be completely re-invented with each new phenotypic reproduction.
18. Overcame the various well-known chicken-and-egg problems of life origin research.
19. Isolated out only homochiral nucleic acid and peptides.
20. Employed only 3'5' phosphodiester bonds in nucleic acid
21. Employed only peptide bonds in polypeptides
22. Achieved dehydration synthesis of heteropolymers, not just homopolymers, in aqueous environments
23. Synthesized exceedingly hard-to-make building blocks such as cytosine.
24. Overcome molecular instability of many key components of life over vast periods of time while life was slowly getting organized supposedly by small increments.

How did unaided physicodynamics accomplish any one of these *formal* feats of control? How did physicodynamics integrate all of these individual formal feats into a utilitarian metasystem called "metabolism"? Formalisms are abstract, conceptual, choice-contingent, organizational, nonphysical, mentally steered constructions. Physicodynamics does not practice the fine arts of purposeful choice, language, mathematics, logic theory, algorithmic optimization, programming, computation, and the pursuit of formal function and utility. What possible physicodynamic mechanisms could have existed in a prebiotic environment for inanimate physicality to generate such formalisms—such PI, organization, control and regulation?

The standard contention of naturalistic life-origin science is that almost none of these *conceptual* complexities of current life were needed or present in the first protocells. The composome and micelle to vesicle model of life-origin seems to circumvent the steep vertical cliff of Mount Improbable, providing the gradual back-side slope to that vertical cliff [44]. But the nagging problem for the scientific discipline of ProtoBioCybernetics is that a critical mass of minimal integration, cooperation and control is required even for the simplest of theoretical protocells to self-organize. No motivation, reason or mechanism seems to exist for accumulation through gradualism of any one, let alone *all*, of the above formal integrations needed even for the simplest conceivable protometabolism. A punctuated equilibrium approach to abiogenesis is statistically prohibitive by hundreds of orders of magnitude, and is definitively falsified with sound application of the Universal Plausibility Metric and Principle [45].

The very existence of bona fide "self-organization" has been called into question despite thousands of peer-reviewed papers and books that simply presuppose its reality [1, 6, 7, 9, 12, 46, 47]. Often low-informational spontaneous self-ordering phenomena (e.g., Prigogine's dissipative structures of chaos theory) are confused with imagined "self-organization." [9] The two have little in common. Self-ordering phenomena perform no formal functions that could be organized into sustainable utility, let alone a cybernetic protometabolic metasystem.

Thus the subjects of study within the discipline of ProtoBioCybernetics, and this anthology of peer-reviewed works, include:

A. The three fundamental categories of reality: chance, necessity, and selection
B. The three subsets of sequence complexity, Random Sequence Complexity (RSC), Ordered Sequence Complexity (OSC), and Functional Sequence Complexity (FSC)
C. Physicodynamics (physicochemical mass/energy interactions) vs. nonphysical

D. formalisms
E. Constraints vs. Controls
F. Law vs. Rules
G. The formal nature of function, work and utility
H. Physicodynamic determinism vs. programming determinism
I. Decision nodes, logic gates and configurable switch-settings
J. The Cybernetic Cut and one-way nature of the Configurable Switch Bridge
K. Order and Pattern vs. Complexity and Noise
L. Self-ordering vs. self-organization
M. Mere combinatorial complexity vs. bona fide organization
N. Shannon uncertainty vs. semantic information
O. Description vs. Prescription
P. Functional Information transcends mere Shannon reduced uncertainty
Q. Prescriptive Information (PI) transcends mere Descriptive Information (DI)
R. The instantiation of nonphysical formalisms into physicality
S. Moving far from equilibrium
T. Chaos theory vs. Systems theory
U. Material Symbol Systems cannot be reduced to physicality
V. Can composomes (which include ribozyme complexes) evolve?
W. The GS (Genetic Selection) Principle
X. The highly selective nature of membrane active transport
Y. What might be a protocell's minimal genome?
Z. The Formalism > Physicality (F > P) Principle

In summary, we might ask, "To what degree do the PI, biocybernetic and biosemiotic aspects of cellular metabolism conform to the cognitive and psychological criteria of these formalisms?" How did inanimate physical nature generate such nonphysical formalisms sufficient to organize life? Do the biological controls and messages that integrate metabolism have conceptual meaning? If not, how can meaning be divorced from such exquisite genomic and epigenomic instructions, metabolic integration and organization? Is anything more goal-directed than the holistic metabolism needed to be and stay alive? This is the subject matter of ProtoBioCybernetics.

References

1. Abel, D.L. 2009, The capabilities of chaos and complexity, Int. J. Mol. Sci., 10, (Special Issue on Life Origin) 247-291 Open access at http://mdpi.com/1422-0067/10/1/247
2. Abel, D.L.; Trevors, J.T. 2006, More than metaphor: Genomes are objective sign systems, Journal of BioSemiotics, 1, (2) 253-267.
3. Abel, D.L. 2007, Complexity, self-organization, and emergence at the edge of chaos in life-origin models, Journal of the Washington Academy of Sciences, 93, (4) 1-20.
4. Abel, D.L. 2008, 'The Cybernetic Cut': Progressing from description to prescription in systems theory, The Open Cybernetics and Systemics Journal, 2, 234-244 Open access at www.bentham.org/open/tocsj/articles/V002/252TOCSJ.pdf
5. Abel, D.L. 2009, The GS (Genetic Selection) Principle, Frontiers in Bioscience, 14, (January 1) 2959-2969 Open access at http://www.bioscience.org/2009/v14/af/3426/fulltext.htm.
6. Abel, D.L. 2009, The biosemiosis of prescriptive information, Semiotica, 2009, (174) 1-19.
7. Abel, D.L. 2010, Constraints vs. Controls, Open Cybernetics and Systemics Journal, 4, 14-27 Open Access at http://www.bentham.org/open/tocsj/articles/V004/14TOCSJ.pdf.
8. Abel, D.L.; Trevors, J.T. 2005, Three subsets of sequence complexity and their relevance to biopolymeric information., Theoretical Biology and Medical Modeling, 2, 29 Open access at http://www.tbiomed.com/content/2/1/29.
9. Abel, D.L.; Trevors, J.T. 2006, Self-Organization vs. Self-Ordering events in life-origin models, Physics of Life Reviews, 3, 211-228.
10. Abel, D.L. 2011, Moving 'far from equilibrium' in a prebitoic environment: The role of Maxwell's Demon in life origin. In Genesis - In the Beginning: Precursors of Life, Chemical Models and Early Biological Evolution Seckbach, J.Gordon, R., Eds. Springer: Dordrecht.
11. Abel, D.L. Prescriptive Information (PI) [Scirus Topic Page]. http://www.scitopics.com/Prescriptive_Information_PI.html (Last accessed July, 2011).
12. Abel, D.L.; Trevors, J.T. 2007, More than Metaphor: Genomes are Objective Sign Systems. In BioSemiotic Research Trends, Barbieri, M., Ed. Nova Science Publishers: New York, pp 1-15
13. Adami, C. 1998, Introduction to Artificial Life. Springer/Telos: New York.
14. Bruza, P.D.; Song, D.W.; Wong, K.F. 2000, Aboutness from a common sense perspective, JASIS, 51, (12) 1090-1105.
15. Hjorland, B. 2001, Towards a theory of aboutness, subject, topicallity, theme, domain, field, content . . ., and relevance, Journal of the American Society of Information Systems and Technology, 52, (9) 774-778.
16. Maynard Smith, J. 1999, The 1999 Crafoord Prize Lectures. The idea of information in biology, Q Rev Biol, 74, (4) 395-400.
17. Maynard Smith, J. 2000, The concept of information in biology, Philosophy of Science, 67, (June) 177-194 (entire issue is an excellent discussion).
18. Szathmary, E. 1996, From RNA to language, Curr Biol, 6, (7) 764.
19. Szathmary, E. 2001, Biological information, kin selection, and evolutionary transitions, Theoretical Population Biology, 59, 11-14.
20. Johnson, D.E. 2010, Probability's Nature and Nature's Probability (A call to scientific integrity). Booksurge Publishing: Charleston, S.C.
21. Johnson, D.E. 2010, Programming of Life. Big Mac Publishers: Sylacauga, Alabama.
22. Oyama, S. 2000, The Ontogeny of Information: Developmenal Systems and Evolution (science and cultural theory). Duke University Press: Durham, N.C.
23. Sarkar, S. 2000, Information in genetics and developmental biology: Comments on Maynard Smith, Philosophy of Science, 67, 208-213.
24. Sarkar, S. 1996, Biological information: a skeptical look at some central dogmas of molecular biology. In The Philosophy and History of Molecular Biology: New Perspectives, Sarkar, S., Ed. Kluwer Academic Publishers: Dordrecht, pp 187-231.
25. Boniolo, G. 2003, Biology without information, History and Philosophy of the Life Sciences, 25, 255-73.
26. Salthe, S.N. 2005, Meaning in nature: Placing biosemitotics within pansemiotics, Journal of Biosemiotics, 1, 287-301.
27. Salthe, S.N. 2006, What is the scope of biosemiotics? Information in living systems. . In Introduction to Biosemiotics: The New Biological Synthesis Barbieri, M., Ed. Springer-Verlag New York, Inc.: Dordrecht, The Netherlands; Secaucus, NJ, USA
28. Kurakin, A. 2006, Self-organization versus Watchmaker: molecular motors and protein translocation., Biosystems 84, (1) 15-23.
29. Mahner, M.; Bunge, M.A. 1997, Foundations of Biophilosophy. Springer Verlag: Berlin.

30. Kitcher, P. 2001, Battling the undead; how (and how not) to resist genetic determinism. In Thinking About Evolution: Historical Philosophical and Political Perspectives, Singh, R. S.; Krimbas, C. B.; Paul, D. B.Beattie, J., Eds. Cambridge University Press: Cambridge, pp 396-414.

31. Jablonka, E. 2002, Information: Its interpretation, its inheritance, and its sharing, Philosophy of Science, 69, 578-605.

32. Shannon, C. 1948, Part I and II: A mathematical theory of communication, The Bell System Technical Journal, XXVII, (3 July) 379-423.

33. Griffiths, P.E.; Sterelny, K. 1999, Sex and Death: An Introduction to Philosophy of Biology. Univ. of Chicago Press: Chicago.

34. Szostak, J.W. 2003, Functional information: Molecular messages, Nature, 423, (6941) 689.

35. Hazen, R.M.; Griffin, P.L.; Carothers, J.M.; Szostak, J.W. 2007, Functional information and the emergence of biocomplexity, Proc Natl Acad Sci U S A, 104 Suppl 1, 8574-81.

36. Sharov, A. 2009, Role of utility and inference in the evolution of functional information, Biosemiotics, 2, 101-115.

37. Davis, M. 1958, Computability and Unsolvability. McGraw-Hill: New York.

38. Turing, A.M. 1936, On computable numbers, with an application to the *entscheidungs problem*, Proc. Roy. Soc. London Mathematical Society, 42, (Ser 2) 230-265 [correction in 43, 544-546].

39. Kaplan, M. 1996, Decision Theory as Philosophy. Cambridge Univ. Press: Cambridge.

40. Trevors, J.T.; Abel, D.L. 2004, Chance and necessity do not explain the origin of life, Cell Biology International, 28, 729-739.

41. Abel, D.L. 2002, Is Life Reducible to Complexity? In Fundamentals of Life, Palyi, G.; Zucchi, C.Caglioti, L., Eds. Elsevier: Paris, pp 57-72.

42. Rocha, L.M. 2000, Syntactic autonomy: or why there is no autonomy without symbols and how self-organizing systems might evolve them, Annals of the New York Academy of Sciences, 207-223.

43. Rocha, L.M. 2001, Evolution with material symbol systems, Biosystems, 60, 95-121.

44. Dawkins, R. 1996, Climbing Mount Improbable. W.W. Norton & Co: New York.

45. Abel, D.L. 2009, The Universal Plausibility Metric (UPM) & Principle (UPP), Theor Biol Med Model, 6, (1) 27 Open access at http://www.tbiomed.com/content/6/1/27.

46. Abel, D.L. 2008 The capabilities of chaos and complexity, Society for Chaos Theory: Society for Complexity in Psychology and the Life Sciences, International Conference at Virginia Commonwealth University, Richmond, VA., Aug 8-10.

47. Overman, D.L. 1997, A Case Against Accident and Self-Organization. Rowman and Littlefield Publishers, Inc.: New York.

The First Gene, David L. Abel, Editor 2011, pp 19-54 ISBN: 978-0-9657988-9-1

2. The Three Fundamental Categories of Reality[*]

David L. Abel

Department of ProtoBioCybernetics/ProtoBioSemiotics
Director, The Gene Emergence Project
The Origin-of-Life Science Foundation, Inc.
113 Hedgewood Dr. Greenbelt, MD 20770-1610 USA

Abstract. Contingency means that events could unfold in multiple ways in the midst of, and despite, cause-and-effect determinism. But there are two kinds of contingency: Chance and Choice/Selection. Chance and Necessity cannot explain a myriad of repeatedly observable phenomena. Sophisticated formal function invariably arises from choice contingency, not from chance contingency or law. Decision nodes, logic gates and configurable switch settings can theoretically be set randomly or by invariant law, but no nontrivial formal utility has ever been observed to arise as a result of either. Language, logic theory, mathematics, programming, computation, algorithmic optimization, and the scientific method itself all require purposeful choices at bona fide decision nodes. Unconstrained purposeful choices must be made in pursuit of any nontrivial potential function at the time each logic gate selection is made. Natural selection is always post-programming. Choice Contingency (Selection *for potential (not yet existing)* function, not just selection *of the best already-existing* function) must be included among the fundamental categories of reality along with Chance and Necessity.

Correspondence/Reprint request: Dr. David L. Abel, Department of ProtoBioCybernetics/ProtoBioSemiotics, The Origin-of-Life Science Foundation, Inc., 113 Hedgewood Dr. Greenbelt, MD 20770-1610 USA E-mail: life@us.net

[*]Sections from previously published peer-reviewed science journal papers [1-9] have been incorporated with permission into this chapter.

Introduction: The big three: Chance, Necessity and Choice/Selection

Three fundamental categories of reality exist: Chance, Necessity (law-like determinism), and choice/selection [1, 3, 7, 10]. Why must Selection be included along with "Chance and Necessity" [11] as a fundamental category of reality?

First, biological science currently presupposes natural selection as its primary organizing Principle. Without *selection*, evolution is impossible.

Second, linear digital genetic instructions represent *selection-based* cybernetic programming. An essential biological process in life, transcription to translation, uses a semiotic symbol system and encryption/decryption as evidenced by the conceptual codon table. No direct physicochemical reactions take place between codons and amino acids. Symbols are formal representations of meaning, not merely physical objects. Each triplet codon not only represents, but prescribes a single codon or stop instruction. Each triplet codon is a Hamming "block code" that reduces noise pollution in the Shannon communication channel. These are abstract conceptual formalisms, not physical objects.

In computer programming we use a fixed block of 7 bits to prescribe each ASCII symbol (letter). The reason is to reduce the likelihood of noise scrambling the signal transmission, meaning and function of each ASCII symbol. This is a Hamming redundancy "block code." A constant number of bits is used to represent each "letter" of the "alphabet." The Postal Service also uses a redundancy block code (in the form of a bar code) to represent *each digit* of every zip code [12]. In the same way, life uses a redundancy block code—three nucleotides to symbolically prescribe each codon symbol That symbol not only represents, but prescribes each amino acid or stop instruction.

Two bits of Shannon uncertainty exist for each potential nucleotide selection. This is evidenced by the are four options at each locus in a nucleic acid string. Thus, a triplet codon is a 6-bit symbol (2 additive bits of uncertainty X 3 loci each in the string), similar to a 7-bit ASCII symbol representing each textual letter.

Symbols must be *chosen* from an alphabet of symbols. Nucleotides must be selected from a phase space of four token options at each locus in the nucleic-acid single positive informational strand. The single positive informational strand corresponds to the first RNAs in an "RNA World" model of life origin, except that codonic prescription of amino acid sequence would not have been a factor.

Base-pairing is not the issue when it comes to investigating the origin of the Prescriptive Information (PI) [6]. The sequencing of nucleosides provides linear digital prescription of cybernetic function. This would also have been true of ribozymes in theoretical preRNA and RNA Worlds. A single primary structure (one long particular sequence of ribonucleotides) must fold back onto itself to make the secondary structure of an RNA ribozyme.

The third reason selection must be recognized as a fundamental category of reality is that the scientific method itself presumes the reality and reliability of formal rationality, mathematics, algorithmic optimization, cybernetic programming, and successful computations. All of these operational tools depend not upon physicodynamic necessity, but upon formal decision theory [13-15]. Decision theory in turn depends upon choice contingency. The practice of science would be impossible without purposeful choices at bona fide decision nodes and logic gates. Chance and necessity are completely inadequate to describe the most important elements of what we repeatedly observe in intra-cellular life, especially. Science must acknowledge the reality and validity not only of a very indirect, post facto natural selection, but of purposeful selection for potential function as a fundamental category of reality. To disallow purposeful selection renders the practice of mathematics and science impossible.

1. Necessity

The order and regularity of nature observed in cause-and-effect determinism can be expressed in the form of parsimonious mathematical formulas, equalities and inequalities. These formulas work as compression algorithms for reams of data because of fixed force and mass/energy relationships in nature. These relationships even incorporate numerical constants. We can count on these formal, mathematical regularities to predict future physical interactions. Given initial conditions, we can calculate in advance what is going to happen physicodynamically. The effects are determined by known combinations of causes and their mathematical relationships.

Cause-and-effect determinism produces highly-ordered sequences of events containing almost no uncertainty or information. Such sequences of events can be described using a compression algorithm much shorter than the sequence of events being described. Reams of experimental data can be reduced to one small equation such as $F = ma$. The latter ability is the very definition of high order, low uncertainty, and minimal Shannon "information" content [12, 16-19].

Physicodynamic determinism is often referred to as "necessity" [11]. Some origin-of-life specialists believe that life arose from cause-and-effect physicochemical determinism—"it had to happen, it could not have been otherwise," as Pier Luigi Luisi summarizes the perspective [20]. Eors Szathmary calls this "the gospel of inevitability" [21]. Christian de Duve and Harold Morowitz are prominent advocates of this view.

2. Chance Contingency

"Chance" is the word used by Nobel laureate Jacques Monod in his famous book *Chance and Necessity* [11] to contrast "necessity" in a false dichotomy. In recent years the term "chance" has been largely replaced in scientific literature by the word "contingency." But this too has its problems, as more than one type of contingency exists. Not all contingency is random. Thus we will use the term "chance contingency" to specifically refer to the kind of "chance" referred to by Monod.

The single word "contingency" means that events could have happened other than what unfolded despite physicodynamic constraints [22] (See also OLEB journal **40** (4-5), October, 2010 for an excellent series of papers on contingency vs. determinism). Outcomes are not fully determined by prior cause-and-effect chains. Variability and degrees of freedom exist. Complex outcomes, at least, are not "necessary"—they are not mandated by natural laws working on initial conditions. More precisely, so many independent cause-and-effect chains interact that the result appears to be random, as in a very unlikely car accident with multiple interactive causative factors. Monod and Gould were prominent advocates of contingency. Luisi contrasts the two perspectives this way: contingency argues, "It could have not happened;" necessity argues, "It must happen." [23] For metaphysical naturalists viewing chance as nothing more than extremely complex or as-of-yet-unelucidated physicodynamic causation, contingency is nothing more than "the outcome of a particular set of simultaneous concomitant effects that apply in a particular point of time/space" [23].

Chance contingency is exemplified by heat agitation and Brownian Motion of molecules in gas and fluid phases. Some argue that all physical behavior is ultimately caused, and that chance contingency is only an illusion. Combinations of forces and their effects can be extremely complex. Undiscovered forces, matter and their relationships may also be at work [24]. But functionally, on both the macroscopic and microscopic quantum levels, distinct advantages obtain from regarding chance contingency as real and for quantifying possible outcomes statistically. Thus, we tentatively refer to chance contingency as "randomness." We might prefer to say, "Functionally, we treat chance-contingent events as though they were random."

A succession of "fair" coin flips provides an example of independent *chance-contingent* events with unweighted means. Physicodynamic constraints exert no bias on whether the outcome is heads or tails with a theoretical "fair" coin. Physical constraints act equally on both physicodynamic possibilities. Chance contingency allows the outcome to be statistically predictable because of the absence of both law-like necessity *and* controls (choice contingency). The coin toss is said to be "fair" because the mean is not weighted by physicodynamic influence *or* experimenter preference. The statistical outcome is not prejudiced or biased.

In a very general sense, chance contingency can be considered predictable (e.g., Gaussian curves). Relative degrees of determinism and chance contingency can also co-exist. Weighted means can be calculated for situations with seemingly incomplete determinism.

Chance is never a physical cause. Chance is a formal, mathematical and statistical mental construction. Chance can have no physical effects because chance is not a physical cause. We cannot attribute the Prescriptive Information (PI) [6] in nucleotide and codon sequence to chance because chance is a cause of nothing. We can describe— even predict—combinatorial outcomes using formal statistics. But statistical *descriptions* do not cause physical interactions, biofunctional syntax, or anything else.

Noise is closely related to chance and randomness. Noise has never been observed to cause nontrivial formal function either. Extreme measures are taken in communication engineering to minimize the deleterious effects of noise pollution on meaningful messages traveling through a Shannon channel. The redundancy coding explained in the Introduction of this chapter is designed to compensate for and overcome relentless noise intrusions into communication efforts. Noise is consistently counterproductive to meaningful and functional communication.

Probabilistic combinatorialism measures chance contingency. It cannot measure choice contingency. But even probabilistic combinatorialism has its boundaries that limit possible outcomes. These too are an indirect form of constraint.

Whatever one's perspective on chance, chance contingency at least *appears* not to be forcefully determined. Even though chance may be a combination of complex and unknown causations, chance is still considered *operationally*, at least, to be physicodynamically inert. This means that chance contingency is, from a practical standpoint, physicochemically indeterminate. It is decoupled from and incoherent with straightforward cause-and-effect chains of law-like "necessity." But even if we view chance contingency as nothing more than unelucidated interactive and complex physical causation, no naturalistic mechanistic explanation exists for the generation of formal utility from chance. Neither physicodynamic determinism nor random noise has ever been observed to generate programming, algorithmic optimization, computation, or nontrivial formal function of any kind.

3. Choice Contingency and Selection

Contingent events sometimes have an additional unexpected attribute. Not only do they appear not to be caused by physicodynamic determinism, they don't appear to be random either. They can look like a random string, but at the same time can instruct or produce (along with algorithmic processing) very sophisticated molecular machines. Protein sequences are found to be within 1% of an expected random sequence [25], yet clearly are not just stochastic ensembles. Only one 150-mer stochastic ensemble out of 10^{77} folds into a shape with *any* known biologic function [26-28]. To organize any protometabolism, scores, if not hundreds, of folds with highly specific functions would have to be marshaled at the right place and time and integrated into one holistic scheme. Given all of the available probabilistic resources since the Big Bang [29], the happenstantial spontaneous generation of even the simplest protometabolism would violate the Universal Plausibility Principle [30] (be definitively falsified with a $\xi < 1$).

This protein example highlights a key point: contingent events can manifest *evidence of formal control* even in the absence of physical constraints. When contingency is steered or controlled by purposeful selection from among real options, *choice contingency* is at work, not chance contingency. Choice contingency, like chance contingency, shares the operational property of physicodynamic inertness. But choice contingency introduces determinism back into the mix at the moment of purposeful selection. It's just a completely different kind of determinism. It is not a physicodynamic determinism. It is a formal determinism—choice or *cybernetic determinism.* Events are neither constrained nor random. They are choice contingent. Such events become the effects of formal cybernetic causation.

The effect of "pawn to King's bishop 4" is not caused by any law of physics and chemistry. The effect is caused by arbitrary choice contingency—by cybernetic determinism—not by environmental constraints or physical law. Environmental constraints (e.g., a flood or fire) may preclude or terminate the chess game altogether. But the inanimate environment will not choose specific moves that win chess games (formal function).

Choice Contingent Causation (CCC) can generate extraordinary degrees of *unique functionality* that has never been observed to arise from randomness or necessity. Highly pragmatic choice contingency is consistently associated with purposeful steering toward potential utility.

The kind of contingency associated with sophisticated cybernetic function is invariably associated with what philosophers of science call "agency." The hallmark of agency is *the ability to voluntarily pursue and choose for potential function.* Potential means "not yet existent." If anything is repeatedly observable in science, it is abundant evidence of agency's unique ability to exercise formal CCC in generating potential formal functionality. The only exception to human agency's unique ability to do this is life itself, which is of course what produces agency. Life itself is utterly dependent upon cybernetic programming—a phenomenon never observed independent of agency. Thus we are confronted with still another chicken-and-egg dilemma of life-origin science. Whatever the resolution of this riddle, one thing is for certain. We are forced to consider two kinds of contingency, 1) Chance contingency and 2) Choice contingency as fundamental categories of reality along with law-like necessity.

Sometimes both chance contingency and our choice contingency are partially constrained by environmental circumstances. An engineer, for example, has to work around physicodynamic reality in designing and engineering machinery and buildings. Choice contingency itself is exercised solely within the degrees of freedom available to it. But this does not keep choice contingency from being 100% formally determined. Choice contingency is never *determined* by constraints or physicodynamic necessity. If it were, no Shannon uncertainty or choice potential at decision nodes would exist. The sole determinate of choice contingency is deliberate, unencumbered, purposeful selections in pursuit of potential function. Agents merely make these formal free choices in the midst of whatever physical boundaries exist. We choose around those constraints, and at times even chose to make use of those constraints in our design and engineering plan.

The simplest choice contingent systems are digital binary ones with discrete Yes/No logic gates. These represent binary decision nodes providing totally free choice-contingent selection opportunities. A well-designed configurable binary switch has no middle-ground setting—it provides a logical "excluded middle." It is either on or off at all times. The number of binary decision nodes (logic gates, binary configurable switches) is measured in "bits." Note that *bits never measure binary choices.* Bits measure only the number of binary decision *nodes*. Bits are a measure of binary choice opportunities, not the specific binary choices themselves that are the essence of Prescriptive Information (PI). The ability to choose "On" or "Off" without any physicodynamic constraints provides the ideal choice-contingent opportunity. Choice contingency is the major component of any kind of nonphysical, abstract, conceptual formalism.

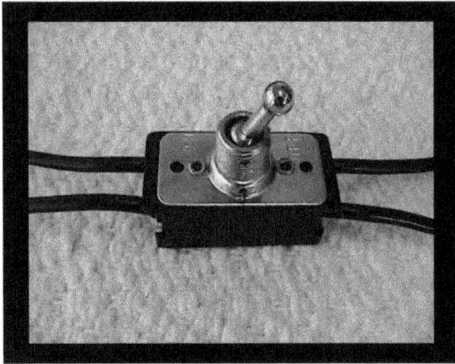

a)   b)

Figure 1. a) A binary configurable switch. Though physical, the switch-setting is nonetheless physicodynamically inert ("dynamically decoupled or incoherent"[31, 32]). No physical force field determines the direction this knob is pushed. The vector of knob push is determined by formal choice contingency alone, not by chance or necessity, and not by order or complexity.

Figure 1. b) An integrated circuit board arises only out of unified, coherent, purposefully cooperative, truly organized logic-gate switch settings. The number of permutations of voluntary (choice-contingent; configurable) switch setting combinations quickly becomes staggering. Often only one configuration achieves a certain functional computational halting.
(Used with permission, Abel, D.L. 2009, The capabilities of chaos and complexity, Int. J. Mol. Sci., 10, (Special Issue on Life Origin) 247-291)

Figure 1a shows an old-fashioned binary configurable switch. Such a switch represents the simplest decision node. Everything computational and organizational stems back to binary decision nodes. Binary decision nodes are the basis of all formal function. Even analog and index systems are ultimately based on binary choices. An analog rheostat knob, for example, must be designed to increase power when turned in one direction (e.g., clockwise) and to decrease power when turned in the opposite direction (e.g., counterclockwise).

Configurable switches are dynamically inert (dynamically incoherent; dynamically decoupled from physicodynamic causation) [31, 32]. This means that on a horizontal switch board, the force of gravity works equally on all potential switch positions. Physicodynamics plays no role in which way the switch knob is pushed. This is the very meaning of "configurable" switches. Their setting is completely decoupled from physicodynamic causation. They can only be set by formal choice contingency, not by chance or law. It is the freedom of formal choice at configurable switches that makes all forms of formal sophistication possible in any physical system. Nonphysical formalism alone determines each switch setting. The switch is a "dynamically-inert configurable switch."

Mere "bifurcation points" (forks in the road) are not synonymous with bona fide *decision* nodes. Bifurcation points can be traversed by chance contingency. A path can be taken randomly. When we come to a fork in the road, we can flip a coin to decide which way to go, but only with likely failure to reach the desired destination. The more forks in the road on our journey, the less likely chance contingency is to get us there. Organization and formal utility are achieved through rationally wise purposeful selections of what path to take at each fork in the road, not through the selection of a path based on coin tosses. When pseudo-selections are made randomly at bifurcation points, it has the same effect as noise pollution on the transmission of meaningful instructions. Rapid deterioration of programming function and computational success occurs with randomization of "selection" at bifurcation points. Logic gate settings are reduced to uneducated guesses (See Figure 2).

The existence of bifurcation points and mere "nodes" in neural nets does not account for computational success. Random selections lack purpose and goal, with predictable results. Choice with intent alone steers rats through a maze with increasing speed as their learning progresses. What exactly is learned? The best successive *choices* at true decision nodes needed to escape the maze are learned. Anticipation and planning are involved prior to each decision node commitment. The same is true of controlled openings and closings of logic gates or configurable switches. The latter requires bona fide choices made with wise steering and programming intent if any sophisticated formal function is expected to arise. Chance and necessity have nothing to do with purposeful choices in pursuit of potential function. Neither chance contingency nor law can program or be programmed into sophisticated function. Purposeful programming choices alone make possible unlimited design and engineering successes.

One of very few paths (W) that lead to algorithmic function out of 2^n branches

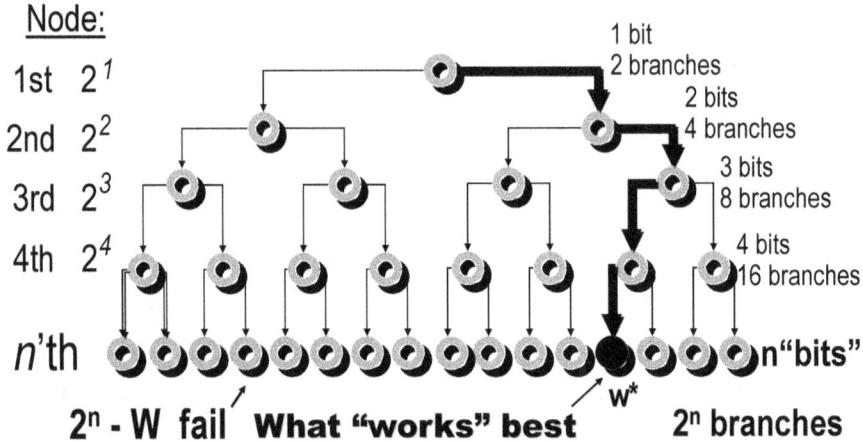

Figure 2. A dendrogram showing all possible sequences (branches or paths) of decision node options. "w*" represents the best algorithmic path to achieve maximum function. The "W" in "2^n-W" represents all paths that produce any degree of algorithmic utility. Notice that *all* paths contain equal (n) bits of Shannon so-called "information" regardless of whether the sequence of specific choice commitments accomplishes anything useful. (Used with permission, Abel DL: Complexity, self-organization, and emergence at the edge of chaos in life-origin models. *Journal of the Washington Academy of Sciences* 2007, **93**:1-20.)

Nontrivial function is only achieved through selection *for potential function* at each individual decision node (Figure 1). This is the essence of "programming." When purpose, goal, and intent are removed from the equation, "choice" becomes the equivalent of "stabs in the dark," random number generation and noise. No one has ever observed a nontrivial computational program arise from a random number generator. This is all the more significant given that not even the so-called "true random number generators" have been proven not to be technically random. Atmospheric noise and even the points in time at which a radioactive source decays continue to be subject to the critique of hard determinists. It remains to be seen whether the very recent (Sept/2010) random number generator at Max Planck Institute is truly random [33]. But either way, we can rest assured, no random number generator will be found generating any sophisticated formal function. Neither randomness nor the cause-and-effect determinism of nature has ever been demonstrated to generate nontrivial algorithmic utility. The blind belief that physicality alone can generate nonphysical formalisms is unfalsifiable, and therefore not a scientific hypothesis. It is a violation of The Cybernetic Cut [4] (See Chapter 3). It is totally without empirical support. No prediction fulfillments exist. More importantly, perhaps, is that the notion is a logically deductive impossibility.

Algorithmic optimization typically produces highly informational instructions and control. As James Harding points out [personal communication], "The process of optimizing algorithms is that of making choices to transform one set of instructions with another set for the purpose of improving along one or more axes of control (e.g., speed, size, simplicity or clarity)." Any physical matrix capable of retaining large quantities of PI must offer high degrees of Shannon uncertainty and high bit content [2, 8, 34]. High bit content refers only to combinatorial possibilities within the physical matrix. But a high number of combinatorial possibilities are an essential requirement of any physical medium if PI is to be instantiated into that medium.

No known natural process exists that spontaneously writes meaningful or functional syntax. Only agents have been known to write or program meaningful and pragmatic linear digital PI [6, 35, 36, pg 46]. Physicality cannot compute or make arbitrary symbol selections according to arbitrarily written rules. Physicality cannot compress. Physicality cannot value or pursue formal utility. Physicality is blind to pragmatic considerations, all of which are formally valued and pursued. The physicodynamics of inanimate mass and energy cannot selectively steer physical events toward algorithmic optimization. Many epigenetic factors notwithstanding, genes and genomic processes largely *program* phenotypes using a formal material symbol system (MSS) [31, 37, 38]. Neither chance nor necessity can explain this undeniable and repeatedly observable phenomenon.

3.1 Selection *OF EXISTING* Fitness (Natural Selection)

Two kinds of selection exist: 1) Selection *of existing* function (e.g., natural selection; differential survival and reproduction of already-programmed, already-living organisms) versus: 2) Selection *for potential* function (e.g., artificial selection at decision nodes in pursuit of formal function that does not yet exist for the environment to favor).

Not all selection by agents is for potential function. Agents can also select the best existing function. We pick the best commercial software package off the shelf for purchase. Based on word of mouth recommendations, we judge it to be the "fittest" software because we have heard it is the most helpful, most reliable and least expensive.   Less fit software, and the companies that produce it, tend to die out.   We may not know anything at all about programming or how the software came into existence.   The market just favors the best software that already occupies the shelf.

Selection *of existing* fitness, but not *for potential* fitness, can also be accomplished by "selection pressure."   Natural selection consists of differential survival and reproduction of the fittest already-computed, already-living phenotypes. But it occurs only at the organismic level of already-living small populations of organisms. "Survival of the fittest" is the very indirect environmental "selection" of the best existing genera of organisms.   It is a stretch to call evolution "selection" in that all that really happens is that inferior organisms tend to die off quicker.   It is an even bigger stretch to call evolution "selection pressure."   Environmental stresses challenge all living organisms to survive.   Less fit organisms (poorly programmed) tend to fail more often than the fittest organisms (well programmed for all environmental challenges).   Evolution is nothing more than differential survival and reproduction of the fittest already-programmed, already-living organisms. Thus natural selection is a unique case of after-the-fact, very indirect "selection" by default.   Selection is not intended; it just happens secondarily. No purpose guides natural selection events. No true decision nodes are involved because evolution has no goal. In this sense, selection "pressure" is a complete misnomer.   No pressure exists to choose anything.   Except for environmental stress, evolution occurs more in a vacuum than under any directional pressure. Differential survival is more happenstantial than pushed. No selection occurs at the genetic programming level where biofunction must be integrated and life organized [5, 39]. Differential survival and reproduction of already-programmed, already-living phenotypic organisms is purely eliminative [36]. It plays no role in the programming of new organisms[5, 39]. Differential survival and reproduction is always after-the-fact, never pursued.

Natural selection is selection only of existing living phenotypic fitness. Natural selection cannot select for potential fitness.   Environmental selection favors only the best already programmed, already living organisms and small groups of organisms.

We must also remember that *natural selection does not favor isolated biofunction.* Selection pressure favors only the survival-of-the-fittest holistic, already-living organisms. No organism would be alive without thousands of cooperating molecular machines, integrated biochemical pathways and cycles, and the formal goal of maintaining a homeostatic metabolism. All of these algorithmic processes must be optimized and in place before any organism can live, let alone constitute the fittest selectable life. Chang *et al.* [18] state:

Chemical evolution' should not be confused with Darwinian evolution with its requirements for reproduction, mutation and natural selection. These did not occur before the development of the first living organism, and so chemical evolution and Darwinian evolution are quite different processes.

3.2 Selection *FOR POTENTIAL* Fitness (Artificial Selection: Choice)

Selection for potential fitness is always artificial rather than natural. Selection for potential fitness is a formal, not a physical enterprise. Selection for potential fitness occurs at decision nodes. Symbols systems and configurable switch settings are used to represent those decisions. Examples of formal selection include language, cybernetic programming, logic, math, computation, algorithmic optimization, design and engineering function, organization of any kind.

Artificial selection is the essence of formalism. Artificial selection always involves purposeful choices at true decision nodes, logic gates and configurable switch settings. As is the case with Maxwell's Demon's trap door operation [40], the door must be opened and closed with intent, not randomly, and not by fixed law, if a heat-energy gradient (work potential) is expected to arise from an ideal gas distribution. An agent invariably exercises choice to program nontrivial formal function. The same is true in any form of semiosis—messaging. To generate a message requires purposeful selection of symbols from an alphabet of symbols according to formal rules to spell meaningful words and sentences. Semiosis is impossible without choices for *potential* function made at individual decision nodes in a string of decision nodes. Despite decades of concentrated research on consciousness and artificial intelligence, choice contingency remains elusive when approached from the direction of physicodynamics and naturalism alone. The mind/body problem is alive and well in the philosophy of biology [4, 7, 41-45]. Ultimately, the mind/body problem boils down to the fact that chance and necessity cannot generate choice contingency—the essence of any formalism.

Natural selection lies in the Selection of Existing (phenotypic) Fitness category. The fittest already-programmed, already-living organisms. differentially survive and reproduce better than less fit living organisms. Evolution in the end is nothing more than differential survival of the fittest already existing organisms. Evolution tells us nothing about how any organism came into existence. Organisms have to be programmed to exist and be alive. They consist of operating systems, software, and millions of nanocomputers running constantly in every living cell [36, 46, pg 47].

Scientifically Addressable Presupposed Objective Reality

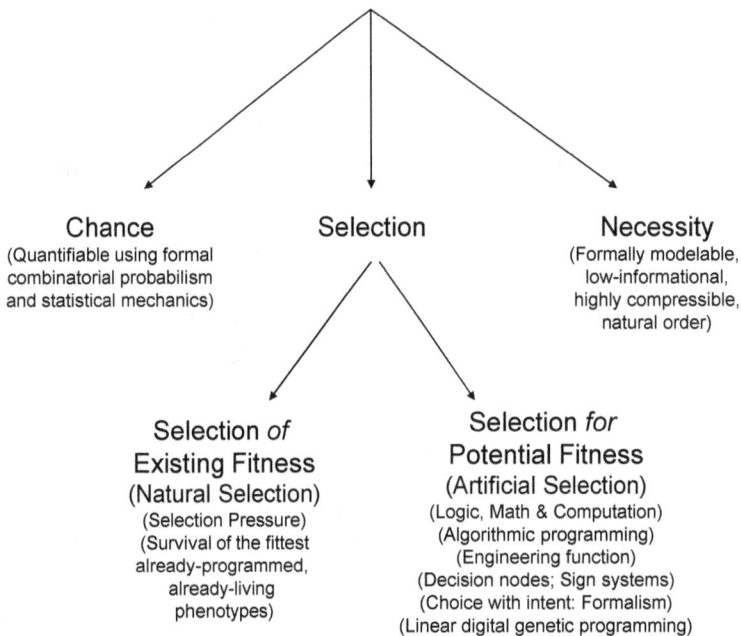

Chance
(Quantifiable using formal
combinatorial probabilism
and statistical mechanics)

Selection

Necessity
(Formally modelable,
low-informational,
highly compressible,
natural order)

Selection *of* Existing Fitness
(Natural Selection)
(Selection Pressure)
(Survival of the fittest
already-programmed,
already-living
phenotypes)

Selection *for* Potential Fitness
(Artificial Selection)
(Logic, Math & Computation)
(Algorithmic programming)
(Engineering function)
(Decision nodes; Sign systems)
(Choice with intent: Formalism)
(Linear digital genetic programming)

Figure 3. The scientific method itself presumes the reality and reliability of choice-contingent language, formal rationality, mathematics, cybernetic programming, and predictive computations. In addition, biological science presupposes natural *selection* as its most fundamental paradigm. Science, therefore, must acknowledge the validity of Selection as a foundational category of reality along with Chance and Necessity.

(Modified from: Abel DL: The biosemiosis of Prescriptive Information (PI). *Semiotica* 2009, **2009**:1-19)

It is well known that Chance Contingency alone cannot generate the programming or computation needed to organize any organism, let alone the fittest organisms. Chance contingency plus selection of the best already living organisms cannot generate life's programming and computation either. Programming and computation have never been observed to arise from any source other than Choice Contingency, never Chance Contingency or Necessity.

The sign/symbol/token systems employed by language, logic theory, mathematics, cybernetics, engineering function, and linear digital genetics all reside in the category of Selection *for* Potential Fitness. Nucleotides must be selected at the molecular/genetic level prior to the realization of any function or life. The differential survival known as "natural selection" and "evolution" is not operational when nucleotide sequencing must be programmed into nucleic acid to prescribe amino acid sequence in proteins, and simultaneously mRNA regulatory control in its complementary strand.

Linear digital genetic programming using a Hamming block code of 3 nucleotide selections to represent and prescribe each amino acid selection is a form of selection for potential fitness, not selection of existing fitness [2, 6, 8]. Genetic programming cannot be explained by natural selection [5]. The environment cannot select for potential function. Evolution has no goal or programming ability at the genetic level.

As discussed above, the selection of each nucleotide corresponds to the setting of a four-way quaternary configurable switch. Three quaternary switch settings in a row prescribe each amino acid "letter" of a very long protein "word." No fitness exists for the environment to favor or select at the level of 3'5' phosphodiester bond formation between nucleotides. These informational biopolymers must be sequenced prior to the realization of any prescriptive, enzymatic, or regulatory function. Selection at the level of nucleotide sequencing clearly falls within the category of "Selection for potential function" rather than the category of "Selection of existing function." This is called the GS (Genetic Selection) Principle[5]. The GS Principle states that selection must occur at the decision-node level of rigid covalent bond linkage of specific monomers to form functional syntax. After-the-fact selection of already-computed phenotypic fitness is not sufficient to explain genetic programming or the metabolism it organizes.

4. Constraints vs. Controls

Great confusion has resulted from sloppy interchangeable use of the terms "constraints" and "controls" [1, 4, 47, 48]. Science emphasizes precise definitions for good reason. In the case of constraints vs. controls, however, contributors to scientific literature have often been grossly negligent. As a result, numerous fallacious inferences have been propagated. Sloppy definitions often cause "category errors" in particular. Varying contexts, hierarchical levels of application, and subjective word connotations have further blurred the dichotomy. Proper definitions of these two terms hold the key to understanding whether life is truly unique from inanimate physics and chemistry.

The orderliness of nature exists in fixed mass/energy relationships and constants described by "laws." These best-thus-far generalizations describe highly probable cause-and-effect chains of behavior. Despite our quantum world enlightenment, determinism in the macroscopic world is still a highly useful and reliable concept. "Necessity" refers to this highly predictable determinism. Since the probability of law-like cause-and-effect chains approaches 1.0, the uncertainty of outcome is therefore very low. Under conditions of such low uncertainty (low Shannon bits), the prescription of sophisticated organization becomes impossible [6, 8, 49]. Uncertainty and freedom are first required before PI can be generated [6, 7]

The laws of physics and chemistry are basically compression algorithms for reams of experimental data. The laws themselves contain very little information (e.g., $F = ma$). We celebrate the parsimony and universality of these low-informational laws. Life, on the other hand, is highly informational. Metabolic organization and control is highly programmed. Life is marked by the integration of large numbers of computational solutions into one holistic metasystem. No as-of-yet undiscovered law will ever be able to explain the highly informational organization of living organisms. The latter would be a mathematical/logical (deductive) impossibility that cannot be overturned by any amount of future observation, abduction or induction. There are simply not enough bits of uncertainty in any law, nor enough "information" (reduced uncertainty, "mutual entropy" in applying a law to the data) to *prescribe* the integration of so many complex pathways, cycles and regulation schemes into a holistic metabolism.

Whereas chance contingency cannot cause any physical effects, *choice contingency* can. But choice contingency, like chance contingency, is formal, not physical. So how could nonphysical choice contingency possibly become a cause of physical effects? The answer lies in our ability to *instantiate* formal choices into physical media. As we shall see below, formal choices can be represented and recorded into physicality using purposefully chosen physical symbol vehicles in an arbitrarily assigned material symbol system. Choices can also be recorded through the setting of configurable switches. Configurable switches are physicodynamically indeterminate (inert; decoupled from and incoherent with physicodynamic causation) [31, 32]. This means that physicodynamics plays no role in how the switch is set. Physicodynamic factors are equal in the flipping of a binary switch regardless of which option is formally chosen. Configurable switches represent decision nodes and logic gates. They are set according to arbitrary rules, not laws. Here arbitrary does not mean random. Arbitrary means "not physicodynamically determined, but freely chosen" [12, 50, 51]. Arbitrary means "freely selectable"—choice contingent.

Below are listed the necessary and sufficient criteria for differentiating constraints from controls.

4.1 What are constraints?

Constraints consist of 1) initial conditions (when not chosen by experimenters), 2) the orderliness of nature itself (the "laws" of physics and chemistry), and 3) the bounds of statistical variation (e.g., standard deviation) stemming from factors such as heat agitation, complex interaction of forces, quantum indeterminacy, etc.).

Initial conditions are usually viewed as the result of prior cause-and-effect physicodynamic chains. Initial conditions in combination with the high dependability of precise physical interactions severely constrain outcome space. No local intent or purpose is involved in these constraints. The constraints just ontologically exist. Our various epistemological and metaphysical slants of interpretation are irrelevant to the fact of these objective constraints.

Constraints manifest no deliberate directionality or purpose. Constraints occur as the result of prior cause-and-effect determinism. Such cause-and-effect chains are oblivious to pragmatic goals. Even evolution has no goal [11, 52-54]. Constraints limit potential freedom indiscriminately with regard to function. Constraints exist in the form of unselected initial conditions and fixed low-informational laws. Constraints are thus utterly indifferent to utility.

Forces act physicodynamically with great regularity upon initial state conditions. Quantum indeterminism at the microscopic level does not prevent the reliable mathematical prediction of nature's macroscopic orderliness. Events are said to be caused by physical forces and their resulting mass/energy interactions. The force constants and the regularity of natural force interactions constitute a form of constraint. Thus, not only are the local initial conditions viewed as constraints, but also the high dependability (orderliness; regularity) of physicodynamic interactions.

Finally, statistical curves describing random variation allow us to predict the relative limits of variation. While statistical descriptions do not constitute a cause of physical effects, they do provide an indirect practical sense of constraint on possible outcomes.

Inanimate nature has no goals. This includes evolution. The use of the term "constraints" to refer to any formally *steered* utilitarian process is therefore erroneous. Likewise, referring to a pragmatically blind physicodynamic causal chain or to the spontaneously self-ordering dissipative structures of chaos theory [55] as a "process" is technically incorrect.

"Natural process" proceeds without regard to formal function or any goal of pragmatic outcome. This raises the question of the legitimacy of using the term "process" in the commonly used phrase "natural process." A certain wish fulfillment emerges from our naturalistic metaphysical presuppositions that uncontrolled physicodynamic phenomena will spontaneously self-organize into extraordinary degrees of formal ingenuity. Empirical support, logic, and prediction fulfillment evidence is sorely lacking for this blind, unfalsifiable belief.

The etymology of "process" traces back to "processus" and relates to "procedure." A procedure is a *formal* undertaking involving decision nodes, directionality, purpose, and goal. Processes and procedures are undertaken to achieve Aristotelian "final" function. Processes and procedures require wise anticipatory programming decisions. Utility is desired and sought after in any bona fide process.

Mere physicodynamic constraints and cause-and-effect deterministic chains cannot prescribe formal goals or generate cybernetic processes and procedures. They can only generate ordered sequences of physicochemical cause-and-effect chains with no orientation toward utility. Mere cause-and-effect chains may lead to self-ordering phenomena such as bathtub drain vortexes and the shapes of a candle flame. But unselected constraints and physicodynamic cause-and-effect chains have never been observed to steer events toward, let alone through, formal utilitarian processes, procedures, algorithmic optimizations, circuit integration, or computational solutions.

Unfortunately, it has become all too common to refer to mere physicodynamic causal chains like star formation as a "process." General scientific concepts and terms were sometimes poorly defined originally (e.g., "work," "system," "constraints" used erroneously to refer to "controls"). Fundamental confusion resulted. Over the last 100 years this same confusion has extended into multiple specialized fields (e.g. solid state physics, weather forecasting, astronomy, information theory, cybernetics). Once incorporated into the many branching specialized fields of science, the linguistic confusion only evolves independently into ever worsening varieties of nonsense in each specific field. Even when fundamental definitional errors are finally corrected, it becomes almost impossible to undo the damage in each specialized field. Astronomers are not going to stop using the word "process" to refer to the uncontrolled, merely constrained chain of deterministic physicodynamic events that cause star formation. But this does not change the fact that star formation is not a cybernetic process. It is just a cause-and-effect physicodynamic chain with some degree of statistical variation. All we can do is to call attention to some of the errors in fundamental scientific thought and terminology, and hope that the correction eventually filters down to each scientific specialty. Until then, the terminology advocated in this paper will seem idiosyncratic and at odds with long established use in multiple fields of science. During the long reign of Ptolemaic astronomy, Copernican concepts and terminology were also initially idiosyncratic.

In the mean time, we must remain clear that bona fide processes are technically controlled, cybernetically guided (programmed), goal-oriented, and organized. They are not merely ordered by the fixed, low-informational, unimaginative orderliness and cause-and-effect chains of nature.

The determinism arising from prior cause-and-effect deterministic chains has nothing to do with pragmatic goals. Constraints are limited in effect to slight statistical variation. Unchosen constraints provide no nontrivial formal function. Only our purely metaphysical commitment to philosophic naturalism sustains our religious faith in a spontaneous physical generation of formalism. This belief is utterly without scientific support.

The roles of quantum indeterminacy and the statistical variations of complex causation are often hotly debated. Even the strictest metaphysical naturalism and cause-and-effect determinism never seem able to totally obliterate chance contingency [24]. Again, both chance and choice contingencies mean that events could unfold with multiple outcomes despite constraining initial conditions and the law-like regularities of nature.

4.2 What are controls?

Constraints can permit some degree of chance-contingency freedom. But controls always manifest the exercise of purposeful selection for function from within that freedom. Controls involve *steering* events toward some *useful* end. Controls are exercised in the pursuit of formal goals such as computational halting, logically sound syllogisms, and linguistic communication.

Cybernetic function requires freedom of selection which law-like determinism precludes. Wise programming always involves choice contingency exercised at bona fide *decision nodes*, not mere "bifurcation points." Bifurcation points do represent the larger category of contingency. Bifurcation points can be traversed with nothing but chance contingency (e.g., coin flips to determine which branch of the fork to take). But no nontrivial formal function has ever been engineered with mere chance contingency at bifurcation points. Only wise choice contingency produces sophisticated utility. And any attempt to reduce decision nodes to mere bifurcation points results in rapid deterioration of any potential nontrivial formal function. The existence of mere bifurcation points, neural net nodes, or "buttons and strings" does not account for computational success. Organization and formal utility are achieved through the controlled opening and closing of logic gates. The latter requires bona fide choices made with steering and programming intent.

Table (1) The contrast between physicodynamic constraints and formal controls

Constraints	Controls
Physical / Dynamic	Nonphysical / Formal / Conceptual / Abstract
Naturally-occurring initial conditions	Agent-chosen initial conditions
The fixed orderliness of nature itself constrains	Dynamically-inert configurable switch settings control
Necessity / Chance contingency statistical bounds	Choice contingency
No goal, directionality, or intent	Purpose-driven
Non-pragmatic; *any* cause-and-effect chain prevails	Pragmatic intent and results
Bifurcation points only; No bona fide decision nodes	Decision-node choice commitments
State-based	Deliberately engineered
A string of dissipative structures momentarily occur on a unidirectional physicodynamic time vector	Time-independent programming choices can be symbolically represented and instantiated into switch settings at any time
Simple / highly-ordered / regular Monotonous / redundantly structured	Cybernetically Complex Algorithmically optimized and conceptually organized
Unimaginative	Imaginative
A natural state in physical state space	Choice contingency engineers formal function
Blindly constrains fixed law-like behavior. Deterministic without regard to formal pragmatic benefit.	Deliberately steers toward sophisticated utility through particular settings of configurable switches that are decoupled from deterministic laws.
Constraints are not capable of measuring initial conditions or manipulating formal equations	Formalism measures (*represents*) initial conditions and controls manipulate mathematical equations (e.g., F = ma)
Can not compute	Can compute
Cannot steer toward or pursue pragmatic goals	Steers, integrates circuits, and pursues formal goals
Blind to formal function	Formally prescribes function into physicodynamic reality
Differential survival/reproduction of the fittest organisms *only secondarily constrains* the population.	Linear digital prescription/regulation computes into existence all organisms *prior to* natural selection of the fittest phenotypes.

Nonphysical formal choices made with intent at decision nodes can determine the course of physicodynamic events. Such decisions instantiate purposeful choices (e.g., programming choices) into physicality. Such instantiation of controls into physicality should never be confused with mere constraints. Constraints are circumstantial elements of prior cause-and-effect chains, not programming decisions. Controls circumvent, "outsmart," and even make use of constraints in order to achieve formal (choice-based) utility. Constraints are purely physicodynamic—physical. Controls are formal and nonphysical.

5. Laws vs. Rules

The physical "Laws" are viewed as "deterministic" of invariant "necessity." They are seen as universally applicable to macroscopic physicodynamic interactions. Physicochemical phenomena unfold according to the dictates of fixed mathematical relationships. We can reliably predict physicodynamic outcomes specifically because they are constrained by invariant laws, not rules. Reliance on the necessity described by laws is what allowed us to put a man on the moon.

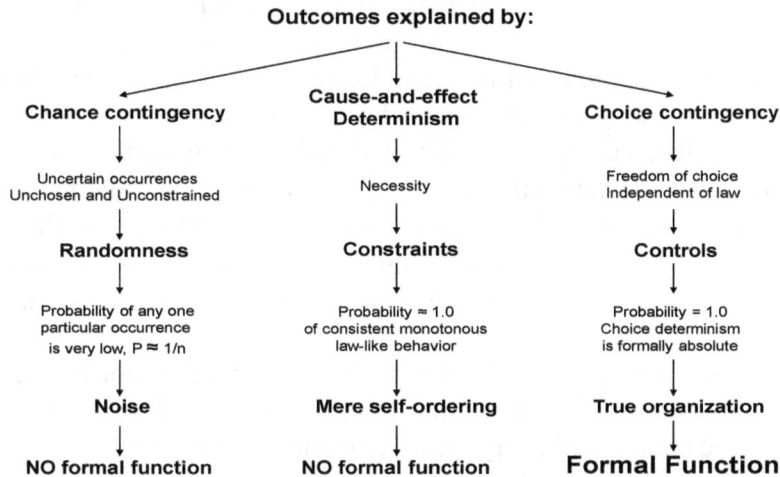

Outcomes explained by:

Chance contingency	Cause-and-effect Determinism	Choice contingency
Uncertain occurrences Unchosen and Unconstrained	Necessity	Freedom of choice Independent of law
Randomness	**Constraints**	**Controls**
Probability of any one particular occurrence is very low, $P \approx 1/n$	Probability ≈ 1.0 of consistent monotonous law-like behavior	Probability $= 1.0$ Choice determinism is formally absolute
Noise	**Mere self-ordering**	**True organization**
NO formal function	**NO formal function**	**Formal Function**

Figure 4. **The three major categories of outcome/behavior.**

Used with permission Abel DL, Constraints vs. Controls, Open Cybernetics and Systemics Journal, 2010, 4:14-27

Rules, on the other hand, govern only voluntary behavior and help guide in establishing pragmatic controls. We can break the rules any time we want, though often at the expense of math errors, programming bugs, inefficient function, or punishment. Rules apply only to choice contingency, not physicodynamic determinism.

Physical interactions do not and cannot arbitrarily choose whether to obey the Laws of motion. Invariant laws dictate the outcomes of mass/energy interactions in inanimate nature. Rules apply to formalisms. Great care must be taken in the proper use of the terms "rules" and "laws." Much confusion has resulted from sloppy interchangeable use of the two terms. The so-called "Laws of Logic," for example, are not laws! They should be called "the *Rules* of inference." They can clearly be broken resulting in disastrous fallacious inferences. The limitless functional benefits of choice-contingent freedom (e.g., mathematical manipulations; computation; computer programming) is fraught with the curse of the possibility of loss of that formal function (e.g., math errors and fatal program bugs). What we call "the Law of the Land" (legislative law, even "the Ten Commandments") is technically not law, but *rules* governing voluntary behavior.

In language and operating systems, choices of alphanumeric characters are controlled by the arbitrary rule conventions of that language. An example would be the high frequency of occurrence of the letter "u" after the letter "q" in English. Such arbitrary rule controls must never be confused with the physicodynamic law constraints of physicality. No law of nature forces u's to follow q's. The sequencing of letters in language is arbitrary. The formal rule could be broken if desired, but only at the expense of efficient communication of meaning in that language. Utility and efficiency would be compromised due to loss of communication. But no law of motion would be violated if we changed our arbitrary linguistic convention (rule). The letters on this page are physical. But their sequencing and function are formal, not physical. They function as physical symbol vehicles in a formally generated material symbol system [56, pg. 262].

Formalisms are governed by arbitrarily written rules, not by inescapable physicodynamic laws. The word "arbitrary" is often confused with "random." In a cybernetic context, arbitrary refers to choice contingency in the sense that no selection is constrained by cause-and-effect determinism. Neither is it forced by external formal controls. The choice at any decision node is uncoerced by necessity. But it is not just contingent (could occur in multiple ways despite the orderliness described by the laws of physics). Any of the switch options, or any member of a finite alphabet, can be purposefully selected. The chooser has complete freedom of choice with intent without constraint. The weighted means of Shannon uncertainty cannot explain the purposeful choices required for semiosis, for example. The door is opened to formalism because the mind is free to choose any physical option with purpose.

No such freedom exists in any law-determined system. *Laws constrain; they do not control.* To control is to steer. Where there is no freedom of choice, steering is not possible. Laws describe an orderliness that forces outcomes. This is the very reason we are able to predict outcomes in physics. *Laws produce order, not organization.* Organization is formal and choice-based. Little flexibility other than heat agitation and the complexity of interacting causes exist to produce chance contingency in inanimate nature. But such contingency never generates choice with intent, formal computational success, engineering prowess, or true organization. *The laws and constraints of inanimate nature operate without regard to pragmatic goals* [11, 57-59]. To look to laws (especially to "yet-to-be discovered" imagined laws) as an explanation for the derivation of formal controls of physicality is not only empirically unfounded, it is logically fallacious (a category error). No law can produce algorithmic organization or computational success.

6. The instantiation of Controls into physicality

We have seen that the opportunity to *choose with intent* from among real options (choice contingency) is an essential ingredient of all formalisms. But how can this freedom of purposeful selection get instantiated into a physical world of mass/energy and cause-and-effect determinism?

Controls and formalisms are nonphysical. But "instantiation" allows them to be recorded into and transmitted through physical media. A unique situation must exist within any physical system to allow the introduction of nonphysical formal controls. Controls can steer physical events toward formal goals and can generate utilitarian physical constructions via wise design and engineering decisions. The easiest way to instantiate controls into physicality is to purposefully select the constraints, such as when an experimenter deliberately chooses the initial conditions of an experiment. We can also incorporate choice contingency into physicality using especially designed and engineered physical devices with unique properties. We call these devices *configurable switches* and *logic gates*. Configurable switches and logic gates are physical devices that can register into physicality, and physically utilize, the nonphysical formal choices of mind. These configurable switches and logic gates must first be physicodynamically indeterminate ("dynamically-inert," dynamically incoherent) [31, 32]. But the invention of such switches alone is not enough. We must also have the formal wise choice contingency to set each switch and to coordinate and integrate the succession of all switch settings so as to integrate circuits.

Although statistical differences and patterns distinguish one linear digital prescriptive string from another, no Prescriptive Information exists because of probabilistic combinatorialism [60]. PI only exists at the moment a particular choice for potential function is made [8]. When a nucleotide is rigidly (covalently) bound to the single-stranded string, the four-way configurable switch knob is actually pushed in one of four possible directions. At that moment all Shannon uncertainty is replaced with formal causation. The vector of the four-way switch knob is determined by choice contingency, not by physicodynamics. It is only when one of the four options is actually selected for potential function that PI comes into existence. It is only when that choice initiates movement of the physical switch knob in one of the four directions that formalism is instantiated into physicality.

Programmed events and processes leading to sophisticated function are steered by decision-node choice commitments. Even analog and index systems require formal choices to implement. Choices made with intent can become causes of physical effects [61, 62]. These causes originate in a purely formal world, but enter into the physical world via specific configurable switch settings to become physicodynamic causes. We call this realization of formal control over physicodynamic causation the *instantiation* of formalism into physicality. Configurable switches must be specifically designed and engineered to open or close purely by formal choice, independent of any physicodynamic determinants. Of course a force must be applied to set the switch. But the question is "Which particular setting?" Whether the binary switch knob on a horizontal switch board is pushed to the right or to the left cannot be addressed by physicodynamics. The law of gravity, for example, acts equally on either option.

How are abstract control choices recorded into mass/energy? How can nonphysical choice contingency wind up controlling physicality? We shall examine 5 different ways.

6.1 Control through choice of constraints

Initial conditions can be chosen by investigators as the starting point of their experimentation. Under these circumstances, the *chosen* constraints rightly can be considered controls [51, 63, 64]. But these constraints become controls only because those constraints were purposefully selected to steer events toward the experimenters' desired results. The constraints themselves do no steering toward any formal utilitarian goal. Choice contingency alone achieves nontrivial integration, organization, and function.

No matter how well a bridge is designed, the river bottom must have adequate physical conditions at the foundation of the main bridge supports. Thus controls cannot be divorced from physicodynamic reality. But no matter how ideal the physical rock bed at these points is, no bridge will spontaneously form from physicodynamics alone. The engineers must either work around or make use of existing physical constraints when they make their design choices according to the formal *rules* (not laws) of safe bridge-building. Engineers must even make choices in view of their anticipation of future circumstantial constraints. An example is the requirement to design a bridge to survive a 100-year flood. But the dichotomy between anticipated environmental constraints (infrequent floods) and controls remains intact.

The choice of constraints, the selection of particular configurable switch settings, the choice of tokens, the choice of which iterative product to proceed with in an optimization process, formal organization schemes, the integration of physical components into a formally functional machine, and the selection of logic gate settings to achieve computation and potential integrated circuits—all of these are functions of choice contingency even though they utilize physical entities. Such formal choice contingency allows us to make use of physical objects to design and engineer physical manifestations of formalisms.

Purely physicodynamic air flow, force and friction cause airplane lift. But the chosen aspects of airplane wing design and engineering are alone what harness those physical factors into airplane flight (formal utility). Lift is ultimately prescribed and produced through physicodynamically indeterminate configurable switch settings. Every individual design enhancement in the wing and fuselage comes in the form of a formal decision-node choice. Flight is not optimized by "bifurcation points" (mere choice *opportunities*). Flight is optimized by wisely choosing which path at each bifurcation point to take. Logic-gate settings must be ideally programmed to optimize the formal utility of desired flight.

In the same way, Maxwell's Demon [65-67] is only able to dichotomize faster moving ideal gas molecules from slower moving ones through *formal choices* of when to open and close the trap door between compartments [68, 69]. Why else would such a ridiculous cartoon personage ever have been introduced into the scientific literature of physics? The reason is that no physicodynamic explanation could be found to explain the sustained nontrivial journey away from equilibrium and disorganization. Only a choosing agent could generate a sophisticated utilitarian heat engine. The Second Law can only be locally and temporarily circumvented to accomplish *useful* work through formal controls, not through spontaneous physicodynamic constraints. Without choice contingency and its controls and regulations, no locally sustained circumvention of the 2nd Law would be possible. Physicodynamic behavior would always quickly revert to obeying the 2nd Law in the absence of formal interventions. In one sense, the 2nd law is always obeyed physicodynamically, even in open systems. But in another sense, the instantiation of formalisms into physicality allows local systems to temporarily overcome the overall natural trend towards disorganization. The ability to purposefully dichotomize physical objects into formal categories and to organize and integrate pragmatic processes and procedures are formal accomplishments, not physicodynamic chains.

Constraints alone simply do not integrate, organize, or optimize algorithmic function. Constraints, including spontaneous initial conditions, forces, and the deterministic cause-and-effect chains of nature, cannot compute, program, or craft nontrivial formally creative machines. Constraints cannot generate representational symbol systems or linear digital prescription of any kind, including the genetic instructions that prescribe regulatory proteins and micro-RNAs. Without formal controls that select and organize physicodynamic constraints, we would have no complex machines, no computers, no buildings, no bridges or any other kind of engineering marvel.

6.2 Configurable switch settings

Configurable switches are physical devices that are designed and engineered to be set by choice contingency alone. Configurable switches are unique physical entities that are specifically designed to record nonphysical formal decisions into physical reality. The switches are themselves physical. Physicodynamic action is required to flip the switch. But with respect to *which switch option is chosen,* they are physicodynamically inert. Controls, therefore, can be instantiated into physicality using physicodynamically indeterminate configurable switch settings [31, 32]. They are not set by chance or cause-and-effect chains. They are physical logic gates with an excluded middle. They provide a means whereby mental choices—formal choices—can be instantiated into physicality.

No more energy is required to flip a quaternary (four-way) switch knob to the right than to the left, or away from than towards the choosing agent. Initial conditions, physical forces, energy requirements, and rate constants are equal for all options afforded by a well-designed quaternary configurable switch. Physicodynamics offers no help in elucidating why a quaternary switch knob was set to one of four possible positions, or why a combination of successive switch settings achieved such impressive correlated formal function.

Can we describe any gradual "degrees of organization" that are possible in the flipping of each binary switch knob? Note that the pictured switch knob cannot be found in a neutral position. The switch is designed with a logical "excluded middle." It will always be found in either the on or off position. Such configurable switches are designed to record yes/no, on/off, 1/0 purposeful programming choices. There is no gradation of selection at each individual binary decision node. The switch knob will be found in either the right or left position.

Configurable switches are physicodynamically inert (physicodynamically indeterminate, dynamically incoherent; dynamically decoupled from physical causation)[31, 32]. Rocha sometimes calls this "dynamic discontinuity." The very reason configurable switches are configurable is that their settings are not determined by physicodynamic cause-and-effect. Switch settings are set only by free-will selections from among real options. No laws are broken. But the laws of physics cannot explain what configurable switch settings accomplish (e.g., integrated circuits, formal computations by physical computers). This means that on an old-fashioned horizontal switch board, the force of gravity works equally on all potential switch positions. Physicodynamics plays no role in which way the switch knob is pushed. This is the very meaning of "configurable" switches. Their settings are completely decoupled from physicodynamic causation. They can only be set by formal choice contingency, not by chance or law. It is the freedom of formal choice at configurable switches that makes all forms of formal sophistication possible in any physical system. Nonphysical formalism alone determines each switch setting. The switch is a "dynamically-inert configurable switch."

The switch in Figure 1a happens to be a binary switch. We could have just as easily photographed a quaternary switch. With a quaternary switch, the knob could be pushed away from you, pulled toward you, pushed to the right, or pushed to the left. A quaternary configurable switch represents 2 bits of uncertainty. The option space of four possible equally available nucleotides also represents 2 bits of uncertainty. Each potential add-on locus in a forming single-stranded oligoribonucleotide in an imagined primordial soup adds an additional 2 bits of uncertainty to the strand. The same is true of a single-stranded (positive, instructional) DNA polymer. Each locus corresponds to a four-way (quaternary) configurable switch. The high degree of uncertainty in a potential single-stranded DNA physical matrix is what allows DNA to retain such tremendous amounts of information. Spinelli & Mayer-Foulkes[70] found specific statistical differences between exon and intron DNA sequences, referring to them as "linguistic DNA features." Large numbers of other researchers have found linguistic like properties in DNA PI as summarized by Searls[71].

No laws of physics are violated in the programming of configurable switches. Yet the effects of the particular functional settings of these configurable switches cannot be reduced to laws and constraints. Their functionality stems directly from their formally chosen settings. This constitutes the only known mechanism of bona fide controls. Configurable switches are the key to escaping the bounds of low-informational (highly constrained and ordered) physicodynamics to soar into unlimited formal creativity. Programmatically set configurable switches are also the key to exceeding the relative pragmatic uselessness of chance contingency.

The formally determined course of the flow of energy through these physical devices produces an *organized* (not merely physicodynamically ordered or constrained) physical output. This formal organization is alone what makes possible local pockets of temporary entropy evasion and seeming entropy reversal. The highly ordered dissipative structures of Prigogine achieve no such local evasions of the Second Law. But by formal programming and design, otherwise useless energy can be transduced by engineered mechanisms into usable energy. Entropy is shifted from the local to the larger peripheral environment. The algorithmic organization that achieves this is not physically derived. Such organization is always formal and decision-node based. Nonphysical PI is required.

6.3 The selection of tokens from an alphabet of physical symbol vehicles

We spell formal words in our minds through the arbitrary choice of letters from an alphabet of letters. These choices of letters are completely uncoerced by physicodynamic determinism. We then transmit those words through instantiation into a physical medium—sound waves, emails, morse code, hard-copy letters, smoke signals. We also record nonphysical words into physicality by picking physical Scrabble *tokens* and arranging them in an order that corresponds to the formal letter sequence of words in our minds. These are all forms of instantiation of formalisms, in this case language, into physicality.

6.4 Formal prescription and integration of physical components into machines

A cake is physical. Yet it comes into existence only through organizing a list of needed ingredients and following the (PI) found in a formal recipe explaining the process of how to bake that cake.

Not only does life not spontaneously generate (The First Law of Biology [Rudolf Virchow, 1858], "All life must come from previously existing life," has never been falsified), sophisticated machines do not spontaneously generate. Individual parts must be crafted and manufactured to particular specifications. Then those parts must be assembled in a very specific way so as to generate a device that can perform some desired task. The level of sophistication required to eliminate chance and/or necessity as a plausible hypothesis for generation of a machine is pretty minimal. Consider the relatively simple machine of a paper clip visualized in Fig 1 of Chapter 1. The most common form of paper clip is nothing more than a long cylinder of malleable metal alloy and constant diameter bent back on itself. How many paper clips in the history of human observation have been observed to spontaneous self-organize from iron ore in the ground?

Science is about repeated observation and predictability. If a simple paper clip has never been observed to spontaneously generate from inanimate nature, what gives us permission to declare as scientific fact that the lowest order conceptually complex machines spontaneously formed in inanimate nature? The supposedly simplest archaea or eubacteria contains hundreds of thousands of nanocomputers, operating systems, softwares, biochemical pathways and cycles leading to exquisite organization and the relentless pursuit of the goal of staying alive. Inanimate physicodynamics can achieve none of the above. It can't even perceive function, let alone pursue the goal of and program a potential computational function that doesn't even exist yet.

7. Physicodynamic Determinism vs. Programming Determinism

We are accustomed to using the term "determinism" to refer to inanimate physicodynamic cause-and-effect chains. Physicodynamic determinism is the source of the term "necessity." Events are necessary because they are invariantly caused by the orderliness of fixed relationships in physical inanimate nature. But in reality, two kinds of determinism exist—1) physicodynamic and 2) cybernetic:

7.1 Physicodynamic determinism

The forced exerted on an object is equal to its mass times its acceleration ($F = ma$). This mathematical relationship that we call a "law" is fixed—invariant in classical physics.

Physicodynamic determinism arises out of cause-and-effect chains. Physicodynamic interactions are governed by the orderly relationships described by mathematical constants and laws. Except for the effects of seemingly random heat agitation, complex causation interactions, and mild statistical bell curves of variation, the orderliness of macroscopic physical interactions is largely fixed. Contingency, including both chance contingency and especially choice contingency, is precluded. No freedom exists in these fixed physicodynamic relationships that would permit selection of some paths and rejection of other paths. Thus, we refer to such largely inescapable cause-and-effect chains as "necessity.' To try to extract any type of programming freedom at decision nodes, logic gates, or configurable switch settings from "necessity" is ludicrous. No yet-to-be discovered "law of self-organization" could logically exist. Laws describe incontrovertible regularity, not selectable bifurcation points. For a configurable switch to serve as a logic gate requires that this switch be specifically designed to provide a unique purpose. The setting of the switch must be independent from the physicodynamics of the switch itself. The switch setting must be "physicodynamically inert" or "physicodynamically indeterminate." Although a physical device, it can only be set by nonphysical choice contingency. None of the four fundamental forces of nature determine *how* the switch is set—which direction the binary switch knob is pushed.

Why do we call physicodynamic relationships "natural?" We define these relationships and predict with them using nonphysical mathematical formulae and equations. All mathematics is formal. Mathematics is abstract, conceptual, choice-based, and requires use of a representational symbol system. Mathematics requires rules, not laws, to govern the voluntary behavior of mathematicians. Mathematicians are free to err. Their calculations are not forced by deterministic laws. But if they expect to derive utility from their mathematical manipulations, they had better obey the purely formal rules of mathematical deduction. If they want their scientific conclusions to be valid, they had better obey the rules (not laws) of logical inference. And they must make wise choices at every decision node. Yet there is no opportunity for choice within physicodynamic determinism.

In addition, all mathematics in physics flows from unproven mental pre-assumptions called axioms. There is nothing about axioms that is physical or "natural" in a physicodynamic sense. The same is true of the rules of equation manipulation.

Most of physics consists of manipulating these mathematical expressions and predicting based on their formal calculations. We can know what the physical outcome will be before anything physical happens. There is nothing "natural" about such purely mathematical prediction fulfillments. To try to call physics and the subject of physics "natural" in the sense of insisting both are derived from a mass/energy explosion is laughable. Einstein rightly pointed out the impossibility of ridding science of metaphysics. He tried his best to minimize metaphysics, with poor success.

7.2 Cybernetic (programming) determinism

Cybernetics is the study of control. Cybernetic determinism programs PI (see Chapter 1, section 4). PI either instructs or indirectly produces (e.g., via already-programmed computational robots) formal function and organization. Cybernetic determinism arises only out of choice contingency. Purposeful choices have to be made at bona fide decision nodes, logic gates, and configurable switch settings. Programming choices, a form of control, determine computational outcomes and the pragmatic value of operating systems and softwares.

Choice contingency and its effect can be a lot more dramatically determinative than the fixed and boring orderliness of nature. Choices matter. Programming and computational capabilities are endless. The utility that can be generated by cybernetic determinism is mind-boggling. Cybernetic determinism alone integrates elements into a holistic cooperative scheme. Bona fide organization is always the result of cybernetic determinism, not mindless inexorable physicodynamic regularity.

The determinism of choice contingency is realized only after the choice is made. At that point, cybernetic determinism becomes theoretically formally absolute with a probability of 1.0. At the moment a binary switch is *reset*, however, the probability returns for an instant to 0.5 with one bit of Shannon uncertainty. Once reset, it immediately returns to p = 1.0. Cybernetic determinism, being formal rather than physical, is by far more definitive and absolute than the so-called "necessity" of physicodynamic causal chains. Because of heat agitation, quantum and other stochastic factors, the so-called necessity of physicodynamics must be described with a certain bell-curve relativism. Once absolute cybernetic determinism is instantiated into a physical medium, however, then the physical medium's relative "necessity" takes over (e.g., a physical configurable switch could malfunction in accord with the 2nd Law).

Physical tokens can also be arbitrarily, yet purposefully, selected from an alphabet of tokens within a material symbol system. Deliberate selections for potential meaning or function are made from an alphabet of symbols and a lexicon of word-like short symbol sequences. The same is true of nucleotide selections and amino acid selections during biopolymer formation. For nucleic acid and sRNAs to wind up functional, their sequencing must be proper before they prescribe or fold. The rigid covalent bonds between monomers provides cybernetic determinism of potential formal function at the level of molecular and genetic construction. Only later is it realized that the sequencing is cybernetically determinative of a certain folding, three-dimensional shape, highly sophisticated biofunction and contribution to system integration.

Choices acting into the physical world don't violate physical laws. But such control choices can temporarily circumvent or make use of physical laws to achieve formal function.

Although statistical differences and patterns distinguish one linear digital prescriptive string from another, no PI exists because of probabilistic combinatorialism [60]. PI only exists at the moment a particular choice for potential function is made [8]. When a nucleotide is rigidly (covalently) bound to the single-stranded string, the four-way configurable switch knob is figuratively pushed in one of four possible directions. At that moment all Shannon uncertainty is replaced with *formal CCC (Choice-Contingent Causation)—cybernetic determinism*. The vector of the four-way switch knob is determined by choice, not by physicodynamics or chance. It is only when one of the four options is actually selected so as to prescribe potential function that PI comes into existence. It is only when that choice initiates movement of the physical switch knob in one of the four directions that formalism is instantiated into physicality."

8. Conclusion: Prescription, regulation and control require Choice Contingency

Constraints should never be confused with controls. Constraints refer to the cause-and-effect deterministic orderliness of nature, to local initial conditions, and to the stochastic combinatorial boundaries that limit possible outcomes. Bits, bifurcation points and nodes represent "choice opportunities," not choices. Controls require uncoerced purposeful selections from among real options. Controls alone steer events toward formal pragmatic ends. Inanimacy is blind to and does not pursue utility. Constraints produce no integrative or organizational effects. Only the purposeful choice of constraints, not the constraints themselves, can generate bona fide controls. Configurable switch settings allow the instantiation of formal choice contingency into physicality. While configurable switches are themselves physical, the setting of these switches to achieve formal function is nonphysical and physicodynamically indeterminate—decoupled from and incoherent with physicodynamic causation [31, 32]. The mental choice of tokens (physical symbol vehicles) in a material symbol system (MSS) also instantiates nonphysical formal PI into physicality. The essence of any formalism is the exercise of purposeful choice contingency.

In a formal process, bifurcation points become true decision nodes when choice with intent determines the selected path. Anticipation and planning are involved prior to the commitment. Deliberate choice of path makes possible unlimited design and engineering successes. Nontrivial function is only achieved through selection *for potential function*. When purpose, goal, and intent are removed from the equation, "choice" becomes the equivalent of random number generation.

Three pressing questions are of immediate interest to ProtoBioCybernetics:

> 1) What are the necessary and sufficient conditions for turning physicodynamics into controls, regulation, organization, engineering, and computational feats?

> 2) How did inanimate nature give rise to a formally-directed, linear, digital, semiotic and cybernetic life?

3) How does nonphysical volition arise out of physicality to then establish control over that physicality?

References

1. Abel, D.L. 2009, The capabilities of chaos and complexity, Int. J. Mol. Sci., 10, (Special Issue on Life Origin) 247-291 Open access at http://mdpi.com/1422-0067/10/1/247
2. Abel, D.L.; Trevors, J.T. 2006, More than metaphor: Genomes are objective sign systems, Journal of BioSemiotics, 1, (2) 253-267.
3. Abel, D.L. 2007, Complexity, self-organization, and emergence at the edge of chaos in life-origin models, Journal of the Washington Academy of Sciences, 93, (4) 1-20.
4. Abel, D.L. 2008, 'The Cybernetic Cut': Progressing from description to prescription in systems theory, The Open Cybernetics and Systemics Journal, 2, 234-244 Open access at www.bentham.org/open/tocsj/articles/V002/252TOCSJ.pdf
5. Abel, D.L. 2009, The GS (Genetic Selection) Principle, Frontiers in Bioscience, 14, (January 1) 2959-2969 Open access at http://www.bioscience.org/2009/v14/af/3426/fulltext.htm.
6. Abel, D.L. 2009, The biosemiosis of prescriptive information, Semiotica, 2009, (174) 1-19.
7. Abel, D.L. 2010, Constraints vs. Controls, Open Cybernetics and Systemics Journal, 4, 14-27 Open Access at http://www.bentham.org/open/tocsj/articles/V004/14TOCSJ.pdf.
8. Abel, D.L.; Trevors, J.T. 2005, Three subsets of sequence complexity and their relevance to biopolymeric information., Theoretical Biology and Medical Modeling, 2, 29 Open access at http://www.tbiomed.com/content/2/1/29.
9. Abel, D.L.; Trevors, J.T. 2006, Self-Organization vs. Self-Ordering events in life-origin models, Physics of Life Reviews, 3, 211-228.
10. Abel, D.L. The Cybernetic Cut [Scirus Topic Page]. http://www.scitopics.com/The_Cybernetic_Cut.html (Last accessed May, 2011).
11. Monod, J. 1972, Chance and Necessity. Knopf: New York.
12. Yockey, H.P. 1992, Information Theory and Molecular Biology. Cambridge University Press: Cambridge.
13. Kaplan, M. 1996, Decision Theory as Philosophy. Cambridge Univ. Press: Cambridge.
14. Chernoff, H.; Moses, L.E. 1986, Elementary Decision Theory. 2 ed., Dover Publications: Mineola, N.Y.
15. Resnik, M.D. 1987, Choices: An Introduction to Decision Theory. University of Minnesota Press: Minneapolis, Minn.
16. Kolmogorov, A.N. 1965, Three approaches to the quantitative definition of the concept "quantity of information", Problems Inform. Transmission, 1, 1-7.
17. Chaitin, G.J. 1988, Algorithmic Information Theory. Revised Second Printing ed., Cambridge University Press: Cambridge.
18. Li, M.; Vitanyi, P. 1997, An Introduction to Kolmogorov Complexity and Its Applications. 2 ed., Springer-Verlag: New York.
19. Yockey, H.P. 2002, Information theory, evolution and the origin of life, Information Sciences, 141, 219-225.
20. Luisi, P.L. 2010, Open questions on the origins of life: Introduction to the Special Issue, OLEB, 40, 353-355.
21. Szathmary, E. 2002, The gospel of inevitability NATURE 419, (24 October) 779-780.
22. Luisi, P.L. 2003, Contingency and determinism, Philos Transact A Math Phys Eng Sci, 361, (1807) 1141-7.
23. Luisi, P.L. 2010, Contingency and Determinism in the origin of life, and elsewhere, OLEB, 40, (4-5 October) 356-361.
24. Pearle, J. 2000, Causation. Cambridge University Press: Cambridge.
25. Weiss, O.; Jimenez-Montano, M.A.; Herzel, H. 2000, Information content of protein sequences, J Theor Biol, 206, (3) 379-86.
26. Axe, D.D. 2000, Extreme functional sensitivity to conservative amino acid changes on enzyme exteriors, J Mol Biol, 301, (3) 585-95.
27. Axe, D.D. 2004, Estimating the prevalence of protein sequences adopting functional enzyme folds, J Mol Biol, 341, (5) 1295-315.
28. Axe, D.D. 2010, The case against a Darwinian origin of protein folds, BIO-complexity, 1, 1-12.
29. Dembski, W. 1998, The Design Inference: Eliminating Chance Through Small Probabilities. Cambridge University Press: Cambridge.
30. Abel, D.L. 2009, The Universal Plausibility Metric (UPM) & Principle (UPP), Theor Biol Med Model, 6, (1) 27 Open access at http://www.tbiomed.com/content/6/1/27.
31. Rocha, L.M. 2001, Evolution with material symbol systems, Biosystems, 60, 95-121.
32. Rocha, L.M.; Hordijk, W. 2005, Material representations: from the genetic code to the evolution of cellular automata, Artif Life, 11, (1-2) 189-214.
33. Gabriel, C.; Wittmann, C.; Sych, D.; Dong, R.; Mauerer, W.; Andersen, U.L.; Marquardt, C.; Leuchs, G. 2010, A generator for unique quantum random numbers based on vacuum states Nature Photonics, (August 29).

34. Chaitin, G.J. 2001, Exploring randomness. Springer: London ; New York.
35. Abel, D.L. Prescriptive Information (PI) [Scirus Topic Page]. http://www.scitopics.com/Prescriptive_Information_PI.html (Last accessed May, 2011).
36. Johnson, D.E. 2010, Programming of Life. Big Mac Publishers: Sylacauga, Alabama.
37. Rocha, L.M. 1997, Evidence Sets and Contextual Genetic Algorithms: Exploring uncertainty, context, and embodiment in cognitive and biological systems. . State University of New York, Binghamton.
38. Rocha, L.M. 2000, Syntactic autonomy: or why there is no autonomy without symbols and how self-organizing systems might evolve them, Annals of the New York Academy of Sciences, 207-223.
39. Abel, D.L. The GS (Genetic Selection) Principle [Scirus Topic Page]. http://www.scitopics.com/The_GS_Principle_The_Genetic_Selection_Principle.html (Last accessed May, 2011).
40. Abel, D.L. 2011, Moving 'far from equilibrium' in a prebiotic environment: The role of Maxwell's Demon in life origin. In Genesis: Origin of Life on Earth and Planets Seckbach, J.Gordon, R., Eds. Springer: Dordrecht.
41. Barbieri, M. 2006, Semantic Biology and the Mind-Body Problem: The Theory of the Conventional Mind, Biological Theory, 1, (4) 352-356.
42. Heritch, A.J. 1992, Information theory perspective on the mind-body problem, Am J Psychiatry, 149, (3) 428-9.
43. Kim, J. 1998, Mind in a Physical World: An Essay on the Mind-Body Problem and Mental Causation. MIT Press: Cambridge, MA.
44. Schimmel, P. 2001, Mind over matter? I: philosophical aspects of the mind-brain problem, Aust N Z J Psychiatry, 35, (4) 481-7.
45. Weber, M. 2007, Indeterminism in Neurobiology, Philosophy of Science, 72, 663-674.
46. Johnson, D.E. 2010, Probability's Nature and Nature's Probability (A call to scientific integrity). Booksurge Publishing: Charleston, S.C.
47. Abel, D.L. 2006 Life origin: The role of complexity at the edge of chaos, Washington Science 2006, Headquarters of the National Science Foundation, Arlington, VA
48. Abel, D.L. 2008 The capabilities of chaos and complexity, Society for Chaos Theory: Society for Complexity in Psychology and the Life Sciences, International Conference at Virginia Commonwealth University, Richmond, VA., Aug 8-10.
49. Abel, D.L.; Trevors, J.T. 2007, More than Metaphor: Genomes are Objective Sign Systems. In BioSemiotic Research Trends, Barbieri, M., Ed. Nova Science Publishers: New York, pp 1-15
50. Yockey, H.P. 2005, Information Theory, Evolution, and the Origin of Life. Second ed., Cambridge University Press: Cambridge.
51. Joslyn, C. 2001, The semiotics of control and modeling relations in complex systems, Biosystems, 60.
52. Mayr, E. 1982, The place of biology in the sciences and its conceptional structure. In The Growth of Biological Thought: Diversity, Evolution, and Inheritance Mayr, E., Ed. Harvard University Press: Cambridge, MA, pp 21-82.
53. Mayr, E. 1988, Introduction, pp 1-7; Is biology an autonomous science? pp 8-23. In Toward a New Philosophy of Biology, Part 1, Mayr, E., Ed. Harvard University Press: Cambridge, MA.
54. Mayr, E. 2001, What Evolution Is. Basic Books: New York.
55. Prigogine, I.; Stengers, I. 1984, Order Out of Chaos. Heinemann: London, 285-287, 297-301.
56. Pattee, H.H. 1977, Dynamic and linguistic modes of complex systems, Int. J. General Systems, 3, 259-266.
57. Bohr, N. 1933, Light and life, Nature, 131, 421.
58. Mayr, E. 1997, This Is Biology: The Science of the Living World. Harvard University Press: Cambridge, MA.
59. Mayr, E. 2004, What Makes Biology Unique? : Considerations on the Autonomy of a Scientific Discipline.
60. Durston, K.K.; Chiu, D.K.; Abel, D.L.; Trevors, J.T. 2007, Measuring the functional sequence complexity of proteins, Theor Biol Med Model, 4, 47 Free on-line access at http://www.tbiomed.com/content/4/1/47.
61. Shaw, R.E.; Kadar, E.; Sim, M.; Repperger, D. 1992, The intentional spring: A strategy for modeling systems that learn to perform intentional acts - group of 3, Journal of Motor Behavior, 24, 3-28.
62. Ahl, V.; Allen, T.F.H. 1996, Hierarchy Theory: a Vision, Vocabulary, and Epistemology. Columbia University Press: New York.
63. Joslyn, C. 1998, Are meaning and life coextensive? In Evolutionary Systems, van de Vijvier, G., Ed. Kluwer: Netherlands, pp 413-422.
64. Joslyn, C. 2000, Levels of Control and Closure in Complex Semiotic Systems, Annals of the New York Academy of Sciences, 901, (Special issue on "Closure", ed. J. Chandler, G. van de Vijver) 67-74.
65. Lewis, G.N. 1930, The symmetry of time in physics, Science, 71, 569-576.
66. Leff, H.S.; Rex, A.F. 1990, Maxwell's Demon, Entropy, Information, Computing. Princeton Univer. Press: Princeton, N.J.
67. Von Baeyer, H.C. 1998, Maxwell's Demon: Why Warmth Disperses and Time Passes. Random House New York.
68. Szilard, L. 1964, On the decrease of entropy in a thermodynamic system by the intervention of intelligent beings, Behav Sci, 9, (4) 301-10.

69. Zurek, W.H. 1990, Algorithmic information content, Church-Turing thesis, and physical entropy and Maxwell's demon. In Complexity, Entropy, and the Physics of Information, Zurek, W. H., Ed. Addison-Wesley: Redwood City, pp 73-89.
70. Spinelli, G.; Mayer-Foulkes, D. 2008, New Method to Study DNA Sequences: The Languages of Evolution, Nonlinear Dynamics, Psychology, and Life Sciences, 12, (2, April) 133-151.
71. Searls, D.B. 2002, The language of genes, Nature, 420, (6912) 211-7.

The First Gene, David L. Abel, Editor 2011, pp 55-74 ISBN: 978-0-9657988-9-1

3. The Cybernetic Cut and Configurable Switch Bridge*

David L. Abel

Department of ProtoBioCybernetics/ProtoBioSemiotics
Director, The Gene Emergence Project
The Origin-of-Life Science Foundation, Inc.
113 Hedgewood Dr. Greenbelt, MD 20770-1610 USA

ABSTRACT. The Cybernetic Cut delineates perhaps the most fundamental dichotomy of reality. The Cybernetic Cut is a vast ravine. The physicodynamics of physicality ("chance and necessity") is on one side. On the other side lies the ability to choose with intent what aspects of ontological being will be preferred, pursued, selected, rearranged, integrated, organized, preserved, and used to achieve sophisticated function and utility (cybernetic formalism). The Cybernetic Cut can be traversed across the Configurable Switch (CS) Bridge. Configurable switches are especially designed and engineered physical devices that allow instantiation of nonphysical formal programming decisions into physicality. The flow of traffic across the CS Bridge is one-way-only. Physicodynamics never determines formal computational and control choices. Regulation, controls, integration, organization, computation, programming and the achievement of function or utility always emanate from the Formalism side of the Cybernetic Cut.

Correspondence/Reprint request: Dr. David L. Abel, Department of ProtoBioCybernetics/ProtoBioSemiotics
Director, The Gene Emergence Project, The Origin-of-Life Science Foundation, Inc., 113 Hedgewood Dr. Greenbelt,
MD 20770-1610 USA E-mail: life@us.net

*Sections from previously published peer-reviewed science journal papers [1-9] have been incorporated with permission into this chapter.

Introduction: Chance and Necessity cannot steer, program, compute or regulate

Neither randomness (if it is possible at all) nor the cause-and-effect determinism of nature has ever been demonstrated to generate nontrivial algorithmic utility. Physical generation of nonphysical formalisms is a logical impossibility. Cause-and-effect determinism produces highly-ordered events containing almost no uncertainty or information. These highly-ordered events can be described using a compression algorithm much shorter than the sequence of events being described. The latter ability is the very definition of sequence order, low uncertainty, and minimal information content [10-14].

Algorithmic optimization, on the other hand, requires choice contingency rather than chance contingency, and typically produces highly-informational instructions and control. Any physical matrix capable of retaining large quantities of Prescriptive Information (PI) must offer high degrees of Shannon uncertainty and high bit content [2, 8, 15]. High bit content refers only to combinatorial possibilities within the physical matrix. But it is an essential requirement of any physical medium if PI is to be instantiated into that medium.

1. What is The Cybernetic Cut?

The Cybernetic Cut is a vast ravine that runs through the center of reality. The physicodynamics of physicality ("chance and necessity") is on one side. On the other side is the ability to choose with intent what aspects of ontological being will be preferred, pursued, selected, rearranged, integrated, organized, preserved, and used to achieve sophisticated function and utility (cybernetic formalism) [4, 16]. The Cybernetic Cut [4, 16] delineates perhaps the most fundamental dichotomy of reality.

Cybernetic (control) function requires freedom of selection. All formalisms can invariably be traced back to the exercise of choice contingency and its role in decision theory. Achieving formal utility requires crossing The Cybernetic Cut [4]. The extent of this ravine is far too wide to allow any jump from physicality to formalism. Algorithmic and computational processes are necessary to traverse the chasm of The Cybernetic Cut from formalism to physicality. This is made possible only by unique devices discussed below. Appreciating the Cybernetic Cut is the key to understanding the instantiation of any type of formal creativity and engineering success into physicality.

The Cybernetic Cut extends far beyond Howard Pattee's epistemic cut [17-19] to address two major areas: 1) the gulf between formal, purposeful choices and a materialistic world limited to chance and/or necessity, and 2) crossing that great divide through the instantiation of deliberate choices into physicality to achieve algorithmic utility in the material world. Such choices constitute much more than mere constraints. Controls are needed. The difference between constraints and controls is explained in Chapter 2, Section 4. The far side of The Cybernetic Cut manifests designing and engineering-like ability to organize abstract concepts and to instantiate those concepts into a pragmatic physical reality. The far side of the Cybernetic Cut emanates instructions, prescription, and creativity. Programming choices must wisely pursue future function and be carefully integrated and managed.

Traversing The Cybernetic Cut can be clearly observed in innumerable examples of formal controls of physicality. Pattee's excellent description, measurement, and complementarity points do not fully explain this phenomenon. Table 1 shows the difference between Pattee's description-based Epistemic Cut and its extension to a much more inclusive prescription-based Cybernetic Cut. Table 2 shows the difference between physicality and those aspects of reality that traverse the Cybernetic Cut into the sphere of functional and pragmatic controls.

Single-celled organisms *seem* to make true choices (e.g., approach/ avoidance to food sources and noxious stimuli) even though they lack physical brains and formal minds. However, at a prokaryotic level, such "choices" could be pre-programmed (as with robots and AI) by their genetic instructions, molecular nanocomputers, operating systems, software, and various pre-existing epigenetic control mechanisms. We would not attribute "mind" to a robot or bacterium even though they *seem* to make choices. Preprogramming would not require true choices by the robot or bacteria. But the question is, how were bacteria pre-programmed to approach food and avoid noxious stimuli? Typically the inanimate environment gets the credit for the source of controls in abiogenesis. But environmental fluctuations do not constitute controls. The control mechanisms lie within the cell's instruction set. The programming of the cell anticipates all environmental constraints and eventualities, and wisely responds to them. To give the inanimate environment credit for preferring improved formal function is ludicrous.

We are hard-pressed to provide empirical evidence, rational justification, or references showing *how* programming can be accomplished without intentional choices of mind (crossing The Cybernetic Cut). It is only one's materialistic metaphysical commitments that make this fact difficult to acknowledge, not anything scientific. What we repeatedly observe is that cybernetics is accomplished only across bona fide decision nodes, highly specific logic gate configurations and intentional configurable switch settings that integrate circuits and achieve formal computational halting.

Table 1.   The difference between Pattee's description-based Epistemic Cut and its extension to a much more inclusive prescription-based Cybernetic Cut.

The Epistemic Cut	The Cybernetic Cut
Knowledge based	Decision-node based
Constraint based	Control based
Description based	Prescription based
Measurements taken of existing constraints	Constraints are deliberately chosen
Uses laws	Uses rules
Learns	Instructs
End-user based	Programmer based
Non-creative	Creative
Cause and effect	Choice with intent steers the path
Observational	"Makes things happen"
Self-ordering events	Organizational
Describes causal chains of "necessity"	Optimization of genetic algorithms
No choices required	Requires choice with intent
Uses existing laws of motion	Programs configurable switches
Reads semantic information	Writes prescriptive information
Follows orders	Managerial

2. What is the Configurable Switch (CS) Bridge?

Through "configurable" switch settings, formal choice contingency can become a source of physical causation. The setting of these configurable switches and logic gates constitutes the building of the Configurable Switch (CS) Bridge [4, 16] across the vast ravine of materialistically untraversable Cybernetic Cut.

Nonphysical formalism itself can never be physical. As we have seen in previous chapters, the chance and necessity of physicality cannot steer objects and events towards formal utility. Chance and necessity cannot compute or make programming choices. Mere constraints cannot control or regulate. The inanimate environment does not desire or pursue function over nonfunction.

So how does physicality ever get organized into usefulness of any kind? How does stone and mortar ever become a building? The answer lies in our ability to build a CS Bridge from the far side of The Cybernetic Cut—the formal side of reality—to the near side—the physicodynamic (physical) side of the ravine. The scaffolding needed to build this bridge consists of devices that allow instantiation of formal choices into physical recordations of those choices.

Table 2. The difference between physicality and those aspects of reality that traverse the Cybernetic Cut into the sphere of functional and pragmatic controls.

Physicodynamics	Traversing the Cybernetic Cut
Physical	Nonphysical & Formal
Incapable of making decisions	Decision-node based
Constraint based	Control based
Natural-process based	Formal prescription based
Constraints just "happen"	Constraints are deliberately chosen
Forced by laws & Brownian movement	Writes and voluntarily uses formal rules
Incapable of learning	Learns and instructs
Product of cause-and-effect chain	Programmer produced
Determined by inflexible law	Directed by choice with intent
Blind to practical function	Makes functional things happen
Self-ordering physicodynamics	Formally organizational
Chance and necessity	Choice
No autonomy	Autonomy
Inanimacy cannot program algorithms	Programs configurable switches
Oblivious to prescriptive information	Writes prescriptive information
Blind to efficiency	Managerially efficient
Non-creative	Creative
Values and pursues nothing	Values and pursues utility
Cannot pursue algorithmic optimization	Optimizes genetic algorithms

This is accomplished through the construction of physical logic gates—the equivalent of Maxwell's demon's trap door. The gate can be opened or closed by agent choice at different times and in difference contextual circumstances. The open or shut gate corresponds to "yes" vs. "no," "1" vs. "0." Because the gate can be opened or closed by the operator at will, we call it "configurable." It's the equivalent of an "On" or "Off" configurable switch. We saw such a switch in Chapter 2, Figure 1a. No physical force determines how the configurable switch is set. On a horizontal circuit board with old-fashioned binary switches, the forces of gravity and electromagnetism work equally on either possible setting of these switches. The only other forces of physics, the strong and weak nuclear forces, are also irrelevant to how configurable switches are set. Only one thing determines how they are set—choice contingency. The deliberate, purposeful setting of a single binary configurable switch constitutes crossing The Cybernetic Cut across the CS Bridge.

Another means of crossing the CS Bridge across The Cybernetic Cut is to select physical symbol vehicles (tokens) from an alphabet of tokens available in a material symbol system. Like configurable switches, the tokens are unique physical devices. Each token is specially marked with a particular formal symbol. Scrabble tokens, for example, theoretically could be "randomly selected" (technically a self-contradictory nonsense phrase), just as configurable switches theoretically could be "randomly set." But universal empirical experience has long since taught humanity, including the scientific community, that "random selections" never produce or improve sophisticated programming function. "Garbage in, Garbage out!" Mutations cannot be distinguished from "garbage." The one and only factor that produces or improves sophisticated function is purposeful and wise choice contingency. The specifically symbolized tokens have to be deliberately chosen from an alphabet of "physical symbol vehicles" [20-23] to spell a meaningful message. Similarly, configurable switches have to be deliberately set to integrate a circuit or to successfully program computational success. The essence of crossing the CS Bridge across the vast ravine of The Cybernetic Cut is *purposeful choice contingency.*

3. The one-way-only nature of traffic across the CS Bridge

The need for "semantic closure" between natural physicodynamics and the seemingly very unnatural (abstract, conceptual, formal) control functions employed by life has been widely known for some time [18, 19, 24-35]. The hope for a naturalistic semantic closure, complementarity and "code duality" [36-39] is usually pursued along the lines of blurring the clear distinctions between categories of constraints vs. controls. Despite decades of trying to bridge the gap, The Cybernetic Cut [4, 16] remains untraversed except across the unidirectional CS (Configurable Switch) Bridge [4, 16]. Traffic flow across this bridge has thus far been observed to be one-way-only. Said Howard H. Pattee,

> The amazing property of symbols is their ability to control the lawful behavior of matter, while the laws, on the other hand, do not exert control over the symbols or their coded references. [40]

Formalism can be instantiated into physicality. But physicality cannot reverse the traffic flow across the CS Bridge to invade the world of formal controls. The reason is that physicality offers nothing but constraints and chance contingency with which to attempt programming controls, computation, circuit integration, complex machine generation, algorithmic optimization, organization, and sophisticated utility of any kind. Neither chance nor necessity can steer toward "usefulness," pragmatism, or generate nontrivial formal function.

Physicality cannot choose which way to throw a horizontal binary switch knob to produce desired function. The physical environment might be able to constrain the switch knob to be thrown in a certain direction if the switch comes near a magnet, for example. But if the switch just happened to be near a magnet, no formal determinism would be in play that would program that switch setting for potential formal function. And if multiple switches just happened to be near the magnet, all of the metal switches would be set the same way. A program consisting of all 1's, or of all 0's would result that could not integrate a circuit or compute anything functional. All the switches would be set to "open," or all the switches would be set to the "closed" position *by law*. Programming would be impossible. Freedom from law is necessary to program. Yet chance contingency cannot program switch settings either. Thus physicality (chance and necessity) on the near side of The Cybernetic Cut cannot generate nonphysical formal controls and regulation. Physicodynamics cannot generate programming choices so as to generate sophisticated (nontrivial) formal function. Thus to generate any kind of formal, cybernetic, computational, utilitarian function requires choice contingency, not chance contingency or law. The introduction of choice contingency into physicality requires traveling across the CS Bridge. All traffic across this Configurable Switch Bridge flows in one direction only—from the nonphysical formal world of abstract conceptuality, organizational specification, and engineering into the physical world.

4. Evidence that The Cybernetic Cut has been traversed

Nonphysical formal programming can *use* physicodynamics to accomplish its ends. But the programming decisions themselves are intangible. A cybernetic switch is physical. The flipping of that switch is also a dynamic process through time. The selection of a certain option from among multiple options that the switch offers, however, is as formal as mathematics itself. The consideration and choice of switch positions precedes the physicodynamic action of actually flipping that switch. Choice contingency has the ability to determine future dynamic effects. But the intent of which choice commitment will be made (using the switch to accomplish some utilitarian purpose) is entirely nonphysical—non-physicodynamic. And it is not merely descriptive. It is *prescriptive*. Whatever switch position is chosen will *determine the degree of utility* of the physical integrated circuit. Function is determined by the formal computational success of the system. But computational success is accomplished by passing through a series of individual decision nodes. In addition, overall integration of those individual decisions must be made with purpose and intent to bring about holistic success (e.g., metabolism). Coordination of solitary configurable switch settings into holistic function constitutes even more abstract meta-control. But this control is instantiated into physicality at the point of each purposeful decision-node selection.

Over the last ten years, this author has published in numerous peer-reviewed papers many versions of the following null hypothesis: "If these decision-node programming choices are made randomly or by law rather than with purposeful intent, no nontrivial (sophisticated) utility will result." [41]. It would take only one exception (without behind-the-scenes steering) to falsify this null hypothesis. At the time of the writing of this book, so far this repeatedly published null hypothesis has never been falsified. The hypothesis now can and should be extended into a formal testable scientific prediction: *"No nontrivial algorithmic/computational utility will ever arise from chance and/or necessity."* How can such a bold, dogmatic prediction be made? The answer is that it arises from logical necessity, not from empirical observation alone or inductive reasoning. The prediction is a logically sound inference based on prior deductive absoluteness within its own axiomatic system. The only possibility of error on the deductive side would be an axiomatic one where a presupposition is "out of touch with reality" (as theoretical physics is sometimes accused of being). Since no axiom is ever proven, we are forced to consider the deduction best-thus-far, and the prediction that flows from it tentatively valid. After another decade or two with no worldwide success at falsification, this formal scientific prediction should become a mature generalized theory, if not a tentative law of science, which I shall name in advance, ***"The Law of Physicodynamic Insufficiency."*** This proposed tentative law states that physicodynamics is completely inadequate to generate, or even explain, formal processes and procedures leading to sophisticated function. Chance and necessity alone, in other words, cannot steer, program or optimize algorithmic/computational success to provide desired nontrivial utility. When we see sophisticated function of any kind, we have strong evidence suggesting that the Cybernetic Cut has been traversed across the one-way-only CS Bridge. Nonphysical formalism (purposeful choice contingency) has been instantiated into physicality via logic gates, configurable switch settings, the purposeful selection of tokens from an alphabet of tokens, or cooperative integration of physical components into formal systems and conceptually complex machines.

Whenever we observe nontrivial conceptually-complex function, programming that leads to computational success, design, engineering, integrated circuits, or sophisticated organization of any kind of physical components, we know that The Cybernetic Cut has been traversed across the one-way-only CS Bridge from formalism to physicality. Physicality can self-order. But it cannot organize itself into formal algorithmic systems. Physicodynamics cannot integrate parts into holistic, cooperative, functional metasystems. Physicodynamics does include spontaneous non-linear phenomena, but it cannot produce the formal *applied*-science known as "non-linear dynamics." The latter is produced only by agents, not by inanimate nature.

5. Life traverse's the Cybernetic Cut

Base-pairing of existing positive nucleotide single strands to form double strands is a purely physicodynamic phenomenon. Base pairing is mediated by simple hydrogen bonds which themselves are not directly related to informational syntax. Montmorillonite adsorption of ribonucleosides and other forms of templating in primordial models of life-origin are also purely physicodynamic [42]. What physicalism cannot explain, however, is how each template or original positive strand acquired its own prescriptive informational *sequencing*. Physicodynamics such as base-pairing appears to play no role in the determination of which particular monomer is added next to a forming positive single-stranded instructional biopolymer. Neither the individual nucleotide selections in these positive single strands nor optimization of life's literal genetic algorithms proceeds according to the laws of physics and chemistry. Life provides the very basis for the notion of artificial genetic algorithms [43-45]. Sequencing (primary structure) instructs the folding of ribozymes, the prescription of structural proteins, catalysts, ribosomes, and regulatory ncRNAs. Life uses these strings of dynamically-inert configurable switch settings to record formal programming selections. Nothing is more highly-informational than the instructional organization of the components of life. Even supposedly-epigenetic regulatory proteins and ncRNAs are genetically prescribed by a vast syntax of sequential nucleotide selections. Such programming cannot be an effect of physical "necessity." Any law-based selection (e.g., clay surface adsorption) would produce only low-informational redundancy (e.g., a polyadenosine with near zero Shannon uncertainty [46]). For highly-prescriptive information content to be instantiated into any physical matrix, high Shannon combinatorial uncertainty is required. This in turn requires freedom from law and necessity. Yet in the absence of physicochemical causation and spontaneous self-ordering, equally nonfunctional "noise" would occur in the form of stochastic ensembles. Noise produces no more formal function than redundant low-informational laws. Spontaneous self-ordering of Progogine's dissipative structures (in chaos theory) is very different from formal organization. Genetic prescription requires uncoerced—arbitrary, yet non-random—selection of monomers.

The sequencing of initially non-templated positive strands is thus "dynamically incoherent" or "dynamically decoupled" [19, 21, 28]. Turing and von Neumann were inspired by, and modeled early computer technology after, the dynamic inertness of genetic cybernetics [47, 48]. Each single-stranded nucleotide selection represents a new "dynamically inert" configurable switch setting. Any of the four nucleotides is polymerized with relatively equal physicodynamic difficulty. Genes are sequences of specifically set quaternary (four-way) decision-node logic gates. While many selections *seem* inconsequential, others are essential to achieving computational function. Because of recently discovered overlapping of linear digital prescription and reads in the opposite direction, sections previously thought to be inconsequential for one protein are now known to be highly prescriptive of regulation and other overlapping prescriptions.

Genetic instruction requires freedom to make efficacious biological programming selections at the genetic level. Open-ended evolution (OEE) [18, 49, 50] is impossible without such freedom of selection of physical symbol vehicles. Each logic gate must be freely configurable. Nucleotide selection and sequencing cannot be determined by chance or necessity. Nucleotides are physical symbol vehicles in a material symbol system (MSS) [21, 29, 51, 52]. The sequencing of these physical symbol vehicles is critical to how the DNA positive strand instructs protein translation. In addition, most DNA is transcribed into regulatory RNA in which the sequencing of ribonucleosides is also critical. Functional Sequence Complexity (FSC) [8, 53] rather than Ordered Sequence Complexity (OSC) or Random Sequence Complexity (RSC) is instantiated into the physical linear, digital, resortable, physical-symbol-vehicle syntaxes known as genes [13, 14, 54, 55] (See chapter 5 by Kirk Durston and David Chiu). A great deal of sophisticated editing of DNA is required to piece together genes from remote sites. This only adds to the layers of Prescriptive Information employed by living organisms. But only the instantiation of formal Prescriptive Information (PI) [6, 56] into physicality makes genetic control possible. The nucleic acid of living organisms contains extraordinarily sophisticated linear digital programming. We are only just beginning to understand the many superimposed dimensions of Prescriptive Information found in this programming [6, 57, 58]. Particular monomeric sequencing is crucial to life. More than any other characteristic, computational linear digital pre-scribed algorithms distinguish life from non-life [54] [59]. Says Yockey,

> The existence of a genome and the genetic code divides living organ-isms from non-living matter. In living matter chemical reactions are directed by sequences of nucleotides in mRNA. . . . There is nothing in the physico-chemical world that remotely resembles reactions being determined by a sequence and codes between sequences. [55, pg. 54]

Küppers [60, pg 166] makes the same point as Jacques Monod [61], Ernst Mayr [62, 63], and Hubert Yockey [13, 64], that physics and chemistry do not explain life. Niels Bohr argued that "Life is consistent with, but undecidable from physics and chemistry"[65]. What exactly is the missing ingredient that sets life apart from inanimate physics and chemistry? The answer lies in the fact that *life, unlike inanimacy, emanates from the far side of The Cybernetic Cut.*

These specific switch settings also determine how RNA strands fold back onto themselves, forming helices, bulges, loops, junctions, coaxial stacking, etc. Not even the hypothesized pre-RNA World and RNA World escape the formal linear digital algorithmic governance of computational function. The generic chemical properties alone of nucleic acid and protein are insufficient to generate life.

In molecular biology, "The 'meaning' (significance) of prescriptive in-formation is the function that information instructs or produces at its metabolic destination" [8]. Szostak has used the term "functional information" [66]. Prescriptive information includes instruction and algorithmic/ computational programming, not just description. Genes provide instructions and algorithmic prescription of computational function. The oft used term "complexity" in life-origin literature is grossly inadequate to define the nature of genetic con-trol [2, 8, 9, 53, 67]. As Hoffmeyer and Emmeche point out [39, pg. 39], "Bio-logical information is not a substance." Later they repeat, "But biological in-formation is not identical to genes or to DNA (any more than the words on this page are identical to the printers ink visible to the eye of the reader). Infor-mation, whether biological or cultural, is not a part of the world of substance." [39, pg. 40]. As stated earlier, the formal, nonphysical, prescriptive selections instantiated into configurable switch settings (nucleotide selections in this case) must never be confused with the physicality of those configurable switches themselves.

Most information theorists are trained to define information from the per-spective of an observer. The problem with this perspective is that in the absence of an observer, no information can exist. Yet clearly information was at work in the organization of early life. No observers existed >3.5 billion years ago [68]. Real prescriptive information, therefore, has to have predated animal observation. Certain types of prescriptive information must *objectively* exist [2]. Early prokaryotic genetic programming cannot be reduced to the subjective mental constructs or observation of any animal knower/observer [2]. *A purely epistemological definition of biological Prescriptive Information (PI) is grossly inadequate.*

The previous maximum length of oligoribonucleotides in aqueous solution was only 8-10 mers [69]. Recently the number has been increased through heating [70, 71]. But these oligomers are homopolymers, not potential informational messenger molecules. And they are cyclical, not linear.

The genetic programming of longer strands is certainly not "blind." Stochastic ensembles of single-stranded small RNAs or of polyamino acids do not fold into functional shapes. Yet both single nucleotide and dipeptide overall frequencies are close to random in living organisms [72, 73]. Biomessages are unique in nature in that they are formally and functionally sequenced. They are not randomly sequenced, and they are not ordered by physical laws. They are sequenced so as to encrypt programmed instructions for the undeniable goal of achieving homeostatic metabolism. The realization of this goal requires algorithmic processing with nanocomputers, operating systems and software [57]. Transcription is required, along with transcriptional editing, decryption (translation), folding, and sometimes even post-translational editing [74]. These processes are fundamentally formal—as formal as their underlying mathematics. The genome and its editing algorithmic processes not only prescribe, but directly and indirectly compute the end product.

In a Peptide or Protein World model of life origin, efficacious selection of each amino acid must be explained at the level of covalent peptide bond formation. Polyamino acid primary structure (sequence) is formed prior to folding. Primary structure is the main determinant of how the strand will fold. Thus functional shapes must be prescribed by linear digital sequencing. The covalent bonds of these highly-informational strings are "written in stone" prior to when weak hydrogen-bond folding secondarily occurs. Instructive sequencing must be completed before tertiary shape and function can occur. The Genetic Selection (GS) Principle obtains [2, 8]. This Principle states that selection must operate at the genetic level, not just at the phenotypic organismal level, to explain the origin of genetic prescription of structural and regulatory biological function. This is the level of configurable switch settings (nucleotide selection). Selection must first occur at each decision node in the syntactical string. Initial programming function cannot be achieved by chance plus after-the-fact selection of the already-existing fittest programs (phenotypes). Evolution is nothing more than differential survival and reproduction of already-existing fittest phenotypes. *The computational programming proficiency* that produced each and every phenotype must first be explained. Programming takes place at the genetic level. Even epigenetic prescription, development, and regulation ultimately trace back to the genetic programming of ncRNAs and regulatory proteins. Thus far, no "natural-process" explanation has been published for selection at the decision-node, configurable-switch, nucleotide-selection level [5].

Even the translated polyamino acid language is physically nonfunctional while forming until after it dynamically folds according to the instructions contained within its linear digital programming (its primary structure). Only later does this syntax of covalently (rigidly) bound monomeric sequencing determine minimum-Gibbs-free-energy folding. Even then, not even three-dimensional shape, or tertiary structure, is selectable by the environment. A far more holistic context of differential organismic survival and reproduction are required for natural selection to kick in.

In molecular biology, recipe code is translated from nucleotide sequence language into a completely different conceptual amino acid language via code bijection. Bijection is a correspondence of representational meaning between arbitrary alphanumeric symbols in different symbol systems. Codons are not "words"! Each triplet codon is a Hamming "block code" for a single letter (amino acid) of a very long protein word [13], the longest known to be around 30,000 amino acids but most in the few hundred range. A prescriptive codon prescribes a certain amino acid letter at the receiver upon decoding. It is often argued that the symbol system and code bijection (translation) of molecular biology are only heuristic. Yet the correspondence between the codon-block-code sequencing and amino-acid sequencing is clearly both real and nonphysicalistic. Nucleotide sequencing is physicodynamically arbitrary and re-sortable. Bijection is formal, not physicodynamic. In addition, whatever the initial alphabet was, the Shannon Channel Capacity Theorem [75] guarantees that it was at least as symbolically complex as today's codon alphabet [57, pg. 112, 76, pg. 104]. No binding or physicochemical reaction occurs between nucleotide symbols and the amino acid symbols they represent. Anticodon and amino acid are on opposite ends of each tRNA. Amino acyl synthetases are also independent enzyme molecules that have no direct binding affinity to codons. Neither fixed laws nor chance contingency can explain the integration of 20 different kinds of each formally-linked entity: amino acyl synthetase, the specific amino-acid end of each tRNA molecule, the specific anticodon opposite end of each tRNA, and the Hamming "block code" of each triplet codon.

The number of permutations is staggering. The spontaneous integration of all these individual entities into a *formal association* capable of promoting even a protometabolism is statistically prohibitive. In addition, the hypothesis of self-organization of all these integrated biochemical pathways and cycles into a holistic cooperative metabolic scheme can be definitively falsified by the Universal Plausibility Principle [77].

The key to life is controls, not constraints. Because life is so dependent upon true controls rather than mere constraints, life clearly traverses the Cybernetic Cut. It cannot be reduced to physicodynamics.

6. Even the laws of inanimate physics and chemistry traverse the Cybernetic Cut

Mathematical formalism gave birth to the formulaic relationships known as natural laws. The unreasonable effectiveness of mathematics [78-81] can be explained only by formalism's organizational governance of physicality. Albert Einstein in his Sidelights on Relativity mused, "How is it possible that mathematics, a product of human thought that is independent of experience, fits so excellently the objects of physical reality?" How did Einstein determine that mathematical formalism could only be the product of human thought? Was not Einstein's $e = mc^2$ objectively in effect long before the *Homo sapiens* species came along to discovery it? Was not $F = ma$ in effect prior to human existence? Before life formed, did physicality not obey Boyle's "Law"? How could Einstein have been so blind as to think that humans are the source of such mathematical and formal controls of physical relationships?

The singularity's differentiation into gravity and the other three fundamental forces of physicality (electromagnetism, weak and strong nuclear forces) presumably occurred with the realization of the first units of Planck length and Planck time following the Big Bang at 10^{-43} seconds [82, 83]. Our current understanding of space and time breaks down prior to that. But the mathematically predictable physical relationships that appeared at 10^{-43} seconds can hardly be attributable to a human consciousness that did not yet exist. Formal mathematical logic would have needed not only to pre-date, but to organize the instantiation of formalism into physicality by 10^{-43} seconds.

To count, measure and weigh in physics traverses The Cybernetic Cut, as does the scientific method itself. The modeling and successful predicting of physical interactions using mathematics points to an underlying rational and cybernetic structure to physicality. Physics (the study of physicality) functionally consists of 80% mathematics. Mathematics is a formal concept, not a physical entity. The percentage of physics that does not employ mathematics certainly employs language and logic—both of which are formalisms—not mere physicodynamics. It is therefore not just biology that has to explain the controlling formal components of reality. All of these realizations support the validity of the Formalism > Physicality (F > P) Principle, which is covered in Chapter 12.

If general relativity and quantum mechanics are ever successfully unified into a formal theory of quantum gravity, the model will be mathematical, though likely with some form of "new math." But physicodynamics itself cannot explain the Big Bang or any other cosmogony. Physicality cannot explain its own origin, or the nonphysical formal relationships that govern and predict that physicality.

We have no reason to doubt, and every reason to believe, that the mathematical structure of physical relationships predates human consciousness. Numerology seems to objectively exist in the periodic table. All sorts of formalisms in objective reality (e.g., geometric formulae like the volume of spheres) cannot be reduced to human mental construction. Geometric relationships predated their discovery. Such formalisms lie on the far side of The Cybernetic Cut. Traversing The Cybernetic Cut cannot be limited to human consciousness or even to biology. The structure of the known universe is mathematical. It is only our solipsistic metaphysical tendencies, not anything scientific, that make us think that this mathematical structure of physicality was of *our* making. It would be arrogant and anthropocentric of us to think for one second that the mathematical nature of physicality is nothing more than a purely subjective human mental construction.

The contention that Pattee's "semantic closure" can be accomplished in the absence of intentionality has never been fully explained. Examples of spontaneous semiotic closure in the inanimate "real world" are also sorely lacking. While Pattee and Rosen never denied the existence of intentionality, prescription, control and creativity, neither investigator has succeeded in explaining *the derivation* of these phenomena from physicality itself. The same is true of Rocha, Barbieri and many others who believe in infogenesis from within physicality alone. A big part of the problem comes from confusion over the definition of information. Mere combinatorial uncertainty most certainly can arise within physicodynamics alone, but not Prescriptive Information (PI) [1-9, 53, 56, 67]. The major challenge to naturalistic science is to elucidate how cause- and-effect physicodynamics (including heat agitation and quantum uncertainty) could have generated the intentionality of biotic messages.

Figure 1. Null hypotheses that need falsifying.

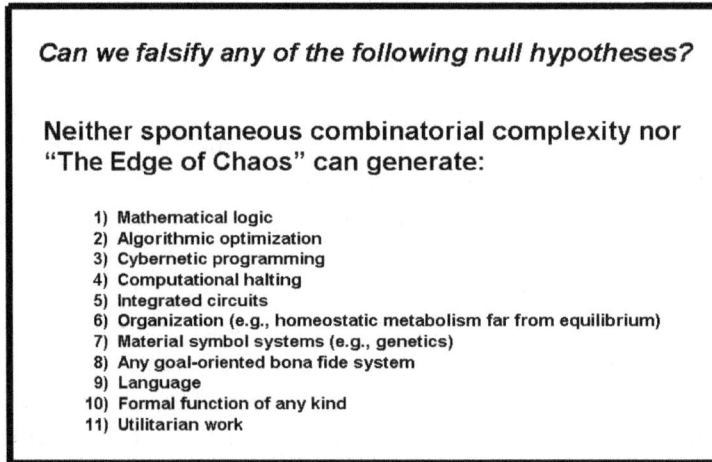

Can we falsify any of the following null hypotheses?

Neither spontaneous combinatorial complexity nor "The Edge of Chaos" can generate:

1) Mathematical logic
2) Algorithmic optimization
3) Cybernetic programming
4) Computational halting
5) Integrated circuits
6) Organization (e.g., homeostatic metabolism far from equilibrium)
7) Material symbol systems (e.g., genetics)
8) Any goal-oriented bona fide system
9) Language
10) Formal function of any kind
11) Utilitarian work

Used with Reprinted with permission from: Abel DL: The capabilities of chaos and complexity. *Int J Mol Sci* 2009, 10:247-291

Inanimate physicodynamics cannot generate the phenomenon of choice with intent. The particular setting of configurable switches to achieve formal function is beyond physics to explain. The latter statement is the essence of the meaning of "metaphysical." The answer to the riddle of *cybernetic causal determinism* lies only in the arena of *formal choice contingency*—of *control*, not the arena of physicodynamic constraints, fixed forces and highly-ordered relationships. Until naturalistic science is willing to acknowledge this fact of reality, progress will be thwarted in many investigative specialties. A Kuhnian paradigm rut prevails: "Physicality (e.g., the cosmos) is all there is, ever was, or ever will be." The scientific method itself cannot be practiced with such a naïve and misguided metaphysical pontification governing science.   In opposition to this religious materialistic belief system is the supervening role of formal mathematics, logic theory, language, and cybernetics so universally employed and required by science. "Information is information, not matter or energy," said Norbert Wiener. "No materialism which does not admit this can survive at the present day." [84, pg. 132]   "Biological information is not a substance," say Hoffmeyer and Emmeche [39, pg. 39].   "Biological information is not identical to genes or to DNA (any more than the words on this page are identical to the printers ink visible to the eye of the reader). Information, whether biological or cultural, is not a part of the world of substance" [39, pg. 40].

7. Conclusion

The Cybernetic Cut is the great divide between physicality and formalism. On the one side of this canyon lies everything that can be explained by the chance and necessity of physicodynamics. On the other side lies those phenomena than can only be explained by formal choice contingency and decision theory—the ability to choose with intent what aspects of ontological being will be preferred, pursued, selected, rearranged, integrated, organized, preserved and used.   Nonphysical formalisms can be instantiated into physicality by traversing The Cybernetic Cut across the one-way-only CS bridge (Configurable Switch Bridge) from formalism into physicality. Not only life, but even inanimate physicality traverses The Cybernetic Cut by virtue of the mathematical organization of physical relationships.   The F > P Principle (Formalism > Physicality Principle) which will be covered in Chapter 12 states that Formalism precedes, instructs, prescribes, organizes, controls, regulates, governs and predicts the unfoldings of Physicodynamics. Formalism, not physicality, is the primary source of and the most fundamental aspect of reality.   Any attempt to falsify this most fundamental Principle of physics only employs the very formalism that falsification quest seeks to deny.

References

1. Abel, D.L. 2009, The capabilities of chaos and complexity, Int. J. Mol. Sci., 10, (Special Issue on Life Origin) 247-291 Open access at http://mdpi.com/1422-0067/10/1/247

2. Abel, D.L.; Trevors, J.T. 2006, More than metaphor: Genomes are objective sign systems, Journal of BioSemiotics, 1, (2) 253-267.

3. Abel, D.L. 2007, Complexity, self-organization, and emergence at the edge of chaos in life-origin models, Journal of the Washington Academy of Sciences, 93, (4) 1-20.

4. Abel, D.L. 2008, 'The Cybernetic Cut': Progressing from description to prescription in systems theory, The Open Cybernetics and Systemics Journal, 2, 234-244 Open access at www.bentham.org/open/tocsj/articles/V002/252TOCSJ.pdf

5. Abel, D.L. 2009, The GS (Genetic Selection) Principle, Frontiers in Bioscience, 14, (January 1) 2959-2969 Open access at http://www.bioscience.org/2009/v14/af/3426/fulltext.htm.

6. Abel, D.L. 2009, The biosemiosis of prescriptive information, Semiotica, 2009, (174) 1-19.

7. Abel, D.L. 2010, Constraints vs. Controls, Open Cybernetics and Systemics Journal, 4, 14-27 Open Access at http://www.bentham.org/open/tocsj/articles/V004/14TOCSJ.pdf.

8. Abel, D.L.; Trevors, J.T. 2005, Three subsets of sequence complexity and their relevance to biopolymeric information., Theoretical Biology and Medical Modeling, 2, 29 Open access at http://www.tbiomed.com/content/2/1/29.

9. Abel, D.L.; Trevors, J.T. 2006, Self-Organization vs. Self-Ordering events in life-origin models, Physics of Life Reviews, 3, 211-228.

10. Kolmogorov, A.N. 1965, Three approaches to the quantitative definition of the concept "quantity of information", Problems Inform. Transmission, 1, 1-7.

11. Chaitin, G.J. 1988, Algorithmic Information Theory. Revised Second Printing ed., Cambridge University Press: Cambridge.

12. Li, M.; Vitanyi, P. 1997, An Introduction to Kolmogorov Complexity and Its Applications. 2 ed., Springer-Verlag: New York.

13. Yockey, H.P. 1992, Information Theory and Molecular Biology. Cambridge University Press: Cambridge.

14. Yockey, H.P. 2002, Information theory, evolution and the origin of life, Information Sciences, 141, 219-225.

15. Chaitin, G.J. 2001, Exploring randomness. Springer: London ; New York.

16. Abel, D.L. The Cybernetic Cut [Scirus Topic Page]. http://www.scitopics.com/The_Cybernetic_Cut.html (Last accessed May, 2011).

17. Pattee, H.H. 1972, The Evolution of Self-Simplifying Systems. In The Relevance of General Systems Theory, Laszlo, E., Ed. George Braziller: New York, pp 32-41; 193-195.

18. Pattee, H.H. 1982, Cell psychology: an evolutionary approach to the symbol-matter problem, Cognition and Brain Theory, 5, 325-341.

19. Pattee, H.H. 1995, Evolving Self-Reference: Matter, Symbols, and Semantic Closure, Communication and Cognition-Artificial Intelligence, 12, 9-28.

20. Rocha, L.M. 2001, The physics and evolution of symbols and codes: reflections on the work of Howard Pattee, Biosystems, 60, 1-4.

21. Rocha, L.M. 2001, Evolution with material symbol systems, Biosystems, 60, 95-121.

22. Rocha, L.M.; Hordijk, W. 2005, Material representations: from the genetic code to the evolution of cellular automata, Artif Life, 11, (1-2) 189-214.

23. Rocha, L.M.; Joslyn, C., 1998, Simulations of embodied evolving semiosis: Emergent semantics in artificial environments In Simulations of embodied evolving semiosis: Emergent semantics in artificial environments, Proceedings of the 1998 Conference on Virtual Worlds and Simulation, 1998; Landauer, C.Bellman, K. L., Eds. The Society for Computer Simulation International: pp 233-238.

24. Pattee, H.H. 1979, The complementarity principle and the origin of macromolecular information, Biosystems, 11, (2-3) 217-26.

25. Pattee, H.H. 1979, Complementarity vs. reduction as explanation of biological complexity, Am J Physiol, 236, (5) R241-6.

26. Emmeche, C. 2000, Closure, function, emergence, semiosis, and life: the same idea? Reflections on the concrete and the abstract in theoretical biology, Ann N Y Acad Sci, 901, 187-97.

27. Pattee, H.H. 1978, The complementarity principle in biological and social structures, Journal of Social and Biological Structure, 1, 191-200.

28. Umerez, J. 1995, Semantic Closure: a guiding notion to ground Artificial Life. In Advances in Artificial Life, Moran, F.; Moreno, J. J.Chacon, P., Eds. Springer-Verlag: Berlin, pp 77-94.

29. Rocha, L.M. 2000, Syntactic autonomy: or why there is no autonomy without symbols and how self-organizing systems might evolve them, Annals of the New York Academy of Sciences, 207-223.

30. Sarkar, S. 2003, Genes encode information for phenotypic traits. In Comtemporary debates in Philosophy of Science, Hitchcock, C., Ed. Blackwell: London, pp 259-274.

31. Lemke, J.L. 2000, Opening up closure. Semiotics across scales, Ann N Y Acad Sci, 901, 100-11.

32. Joslyn, C. 2000, Levels of Control and Closure in Complex Semiotic Systems, Annals of the New York Academy of Sciences, 901, (Special issue on "Closure", ed. J. Chandler, G. van de Vijver) 67-74.

33. Hoffmeyer, J. 1995, The semiotic body-mind. In Essays in Honor of Thbomas A. Sebeok, Tasca, N., Ed. Porto: pp 367-383.

34. Hoffmeyer, J. 1996, Signs of Meaning in the Universe Nature: The natural history of signification. Indiana U. Press: Bloomington.

35. Hoffmeyer, J. 1997, Biosemiotics: Towards a new synthesis in biology, European Journal for Semiotic Studies, 9, 355-376.

36. Hoffmeyer, J. 2000, Code-duality and the epistemic cut, Ann N Y Acad Sci, 901, 175-86.

37. Hoffmeyer, J. 2002, Code duality revisited, SEED, 2, 1-19.

38. Hoffmeyer, J.; Emmeche, C. 1991, Code duality and the semiotics of nature. In On Semiotic Modeling, Anderson, M.Merrell, F., Eds. Mouton de Gruyter: New York, pp 117-166.

39. Hoffmeyer, J.; Emmeche, C. 2005, Code-Duality and the Semiotics of Nature, Journal of Biosemiotics, 1, 37-91.

40. Pattee, H.H.; Kull, K. 2009, A biosemiotic conversation: Between physics and semiotics, Sign Systems Studies 37, (1/2).

41. Abel, D.L.; Trevors, J.T. 2007, More than Metaphor: Genomes are Objective Sign Systems. In BioSemiotic Research Trends, Barbieri, M., Ed. Nova Science Publishers: New York, pp 1-15

42. Ferris, J.P. 2002, Montmorillonite catalysis of 30-50 mer oligonucleotides: laboratory demonstration of potential steps in the origin of the RNA world, Origins of Life and Evolution of the Biosphere, 32, (4) 311-32.

43. Mitchell, M. 1998, An Introduction to Genetic Algorithms. Bradford Books: Cambridge, MA.

44. Davis, L. 1991, Handbook of genetic algorithms. Van Nostrand Reinhold: New York.

45. Goldberg, D.E. 1989, Genetic Algorithms in Search, Optimization and Machine Learning. Addison-Wesley: Reading, MA.

46. Ertem, G.; Ferris, J.P. 2000, Sequence- and regio-selectivity in the montmorillonite-catalyzed synthesis of RNA, Orig Life Evol Biosph, 30, (5) 411-422.

47. Quastler, H., 1958, A Primer on Information Theory in Symposium on Information Theory in Biology In A Primer on Information Theory in Symposium on Information Theory in Biology, The Gatlinburg Symposium, Gatlinburg, 1958; Yockey, H. P.; Platzman, R. P.Quastler, H., Eds. Pergamon Press: Gatlinburg, p pages 37 and 360.

48. von Neumann, J.; Burks, A.W. 1966, Theory of Self-Reproducing Automata. University of Illinois Press: Urbana,.

49. Pattee, H.H. 1969, How does a molecule become a message? In Communication in Development; Twenty-eighth Symposium of the Society of Developmental Biology., Lang, A., Ed. Academic Press: New York, pp 1-16.

50. Pattee, H.H. 1972, Laws and constraints, symbols and languages. In Towards a Theoretical Biology, Waddington, C. H., Ed. University of Edinburgh Press: Edinburgh, Vol. 4, pp 248-258.

51. Rocha, L.M. 1998, Selected self-organization and the semiotics of evolutionary systems. In Evolutionary Systems: Biological and Epistemological Perspectives on Selection and Self-Organization, Salthe, S.; van de Vijver, G.Delpos, M., Eds. Kluwer: The Netherlands, pp 341-358.

52. Rocha, L.M. 1997, Evidence Sets and Contextual Genetic Algorithms: Exploring uncertainty, context, and embodiment in cognitive and biological systems. . State University of New York, Binghamton.

53. Abel, D.L. 2002, Is Life Reducible to Complexity? In Fundamentals of Life, Palyi, G.; Zucchi, C.Caglioti, L., Eds. Elsevier: Paris, pp 57-72.

54. Yockey, H.P. 2000, Origin of life on earth and Shannon's theory of communication, Comput Chem, 24, (1) 105-123.

55. Yockey, H.P. 2002, Information theory, evolution, and the origin of life. In Fundamentals of Life, Palyi, G.; Zucchi, C.Caglioti, L., Eds. Elsevier: Paris, pp 335-348.

56. Abel, D.L. Prescriptive Information (PI) [Scirus Topic Page]. http://www.scitopics.com/Prescriptive_Information_PI.html (Last accessed May, 2011).

57. Johnson, D.E. 2010, Programming of Life. Big Mac Publishers: Sylacauga, Alabama.

58. Wells, J. 2011, The Myth of Junk DNA. Discovery Institute Press: Seattle.

59. Turvey, M.T.; Kugler, P.N. 1984, A comment on equating information with symbol strings, Am J Physiol, 246, (6 Pt 2) R925-7.

60. Küppers, B.-O. 1990, Information and the Origin of Life. MIT Press: Cambridge, MA.

61. Monod, J. 1972, Chance and Necessity. Knopf: New York.

62. Mayr, E. 1988, Introduction, pp 1-7; Is biology an autonomous science? pp 8-23. In Toward a New Philosophy of Biology, Part 1, Mayr, E., Ed. Harvard University Press: Cambridge, MA.

63. Mayr, E. 1982, The place of biology in the sciences and its conceptional structure. In The Growth of Biological Thought: Diversity, Evolution, and Inheritance Mayr, E., Ed. Harvard University Press: Cambridge, MA, pp 21-82.

64. Yockey, H.P. 2005, Information Theory, Evolution, and the Origin of Life. Second ed., Cambridge University Press: Cambridge.

65. Bohr, N. 1933, Light and life, Nature, 131, 421.

66. Szostak, J.W. 2003, Functional information: Molecular messages, Nature, 423, (6941) 689.

67. Abel, D.L. 2006 Life origin: The role of complexity at the edge of chaos, Washington Science 2006, Headquarters of the National Science Foundation, Arlington, VA

68. Mojzsis, S.J.; Arrhenius, G.; McKeegan, K.D.; Harrison, T.M.; Nutman, A.P.; Friend, G.R.L. 1996, Evidence for life on Earth before 3,800 million years ago., Nature, 384, 55-59.

69. Joyce, G.F.; Orgel, L.E. 1999, Prospects for understanding the origin of the RNA World. In The RNA World, Second ed.; Gesteland, R. F.; Cech, T. R.Atkins, J. F., Eds. Cold Spring Harbor Laboratory Press: Cold Spring Harbor, NY, pp 49-78.

70. Grant, R.P. 2010, Alphabet soup, The Scientist, 24, (7, July 1) 23

71. Costanzo, G.; Pino, S.; Ciciriello, F.; Di Mauro, E. 2009, RNA: Processing and Catalysis: Generation of Long RNA Chains in Water, J. Biol. Chem., 284, 33206-33216.

72. Kok, R.A.; Taylor, J.A.; Bradley, W.L. 1988, A statistical examination of self-ordering of amino acids in proteins, Origins of life and evolution of the biosphere, 18, (1-2) 135-42.

73. Weiss, O.; Jimenez-Montano, M.A.; Herzel, H. 2000, Information content of protein sequences, J Theor Biol, 206, (3) 379-86.

74. Roth, M.J.; Forbes, A.J.; Boyne, M.T., 2nd; Kim, Y.B.; Robinson, D.E.; Kelleher, N.L. 2005, Precise and parallel characterization of coding polymorphisms, alternative splicing, and modifications in human proteins by mass spectrometry, Molecular & cellular proteomics, 4, (7) 1002-8.

75. Shannon, C. 1948, Part I and II: A mathematical theory of communication, The Bell System Technical Journal, XXVII, (3 July) 379-423.

76. Johnson, D.E. 2010, Probability's Nature and Nature's Probability (A call to scientific integrity). Booksurge Publishing: Charleston, S.C.

77. Abel, D.L. 2009, The Universal Plausibility Metric (UPM) & Principle (UPP), Theor Biol Med Model, 6, (1) 27 Open access at http://www.tbiomed.com/content/6/1/27.

78. Einstein, A. 1920, Sidelights on Relativity. Dover: Mineola, N.Y.

79. Wigner, E.P. 1960, The unreasonable effectiveness of mathematics in the natural sciences, Comm. Pure Appl., 13 (Feb).

80. Hamming, R.W. 1980, The unreasonable effectiveness of mathematics, The American Mathematical Monthly, 87, (2 February) 81-90.

81. Steiner, M. 1998, The Applicability of Mathematics as a Philosophical Problem. Harvard University Press: Cambridge, MA.

82. Planck, M. 1899, Über irreversible Strahlungsvorgänge, Sitzungsberichte der Preußischen Akademie der Wissenschaften, 5, 479

83. Planck, M. 1960, A Survey of Physical Theory. Dover: New York.

84. Wiener, N. 1961, Cybernetics, its Control and Communication in the Animal and the Machine. 2 ed., MIT Press: Cambridge.

The First Gene, David L. Abel, Editor 2011, pp 75-116 ISBN: 978-0-9657988-9-1

4. What Utility Does Order, Pattern or Complexity Prescribe?*

David L. Abel

Department of ProtoBioCybernetics/ProtoBioSemiotics
Director, The Gene Emergence Project
The Origin-of-Life Science Foundation, Inc.
113 Hedgewood Dr. Greenbelt, MD 20770-1610 USA

Abstract: "Order," "pattern," "complexity," "self-organization," and "emergence" are all terms used extensively in life-origin literature. Sorely lacking are precise and quantitative definitions of these terms. Vivid imagination of spontaneous creativity ensues from mystical phrases like "the adjacent other" and "emergence at the edge of chaos." More wish-fulfillment than healthy scientific skepticism prevails when we become enamored with such phrases. Nowhere in peer-reviewed literature is a plausible hypothetical mechanism provided, let alone any repeated empirical observations or prediction fulfillments, of bona fide spontaneous "natural process self-organization." Supposed examples show only one of two things: 1) spontaneous physicodynamic self-ordering rather than formal organization, or 2) behind-the-scenes investigator involvement in steering experimental results toward the goal of desired results. The very experiments that were supposed to prove spontaneous self-organization only provide more evidence of the need for artificial selection. Patterns are a form of order. Neither order nor combinatorial uncertainty (complexity) demonstrate an ability to compute or produce formal utility. Physical laws describe low-informational physicodynamic self-ordering, not high-informational cybernetic and computational utility.

Correspondence/Reprint request: Dr. David L. Abel, Department of ProtoBioCybernetics/ProtoBioSemiotics, The Origin-of-Life Science Foundation, Inc., 113 Hedgewood Dr. Greenbelt, MD 20770-1610 USA E-mail: life@us.net

*Sections from previously published peer-reviewed science journal papers [1-9] have been incorporated with permission into this chapter.

Introduction: Order vs. Complexity

A great deal of confusion exists about the roles of order and complexity in explaining sophisticated function. We presume, incorrectly, that anything organized must be highly ordered and complex. Worse yet, we equate the two. It is true that the more organized something becomes, the more complex it tends to become. It is also true that we usually find some compressible order within Functional Sequence Complexity (FSC) [8] and Prescriptive Information (PI) [6].

But, a serious problem arises when trying to conflate order with complexity. Order and complexity are antithetical. The more ordered the conglomerate, the less complex it is; the more complex the conglomerate, the less ordered it is.   This is easier to understand in one dimension (See Figure 1). When we progress from linear complexity into two- and three- dimensional complexity, quantifying the degree of complexity can quickly become intractable [10]. Thus, let us begin by precisely defining linear sequence complexity. Later we will expand this understanding into additional dimensions.

1. What is complexity?

As extensively as "Complexity" is used in life origin literature, the term is almost never defined. "Complexity" tends to be highly elastic term we use to explain everything we don't understand and cannot reduce.

An unequivocal, pristine, mathematical definition of linear "complexity" already exists in scientific literature [8, 11-13]: maximum complexity in a linear string, oddly enough, is randomness if one were to attempt to algorithmically decipher patterns or function. A sequence is maximally complex when it cannot be algorithmically compressed.   Random sequences are maximally complex because they lack pattern and order [14, 15]. Thus complexity can be quantified by measuring algorithmic compressibility. The more ordered a linear digital sequence, the less complex it becomes, and the fewer bits of potential information that string can retain.

Randomness is antithetical to order. Randomness represents maximum uncertainty. Uncertainty can be measured in Shannon bits. Bits of uncertainty increase as we move from high order on the left of Figure 1 towards randomness on the right. As we move towards the left, away from complexity towards increasing order and pattern, bits of uncertainty decrease. A random string (Random Sequence Complexity, RSC) [8] is the most complex because its sequence cannot be enumerated using any algorithmically compressive string shorter than itself.

Notice that this precise definition of linear complexity has nothing to do with meaning or function. Complexity in linear digital strings is fully measurable by the degree to which each string can be algorithmically compressed. This is true whether the string *does* anything useful or not. Figure 2 helps to show the relationship between compressibility and the bidirectional vector of order vs. complexity.

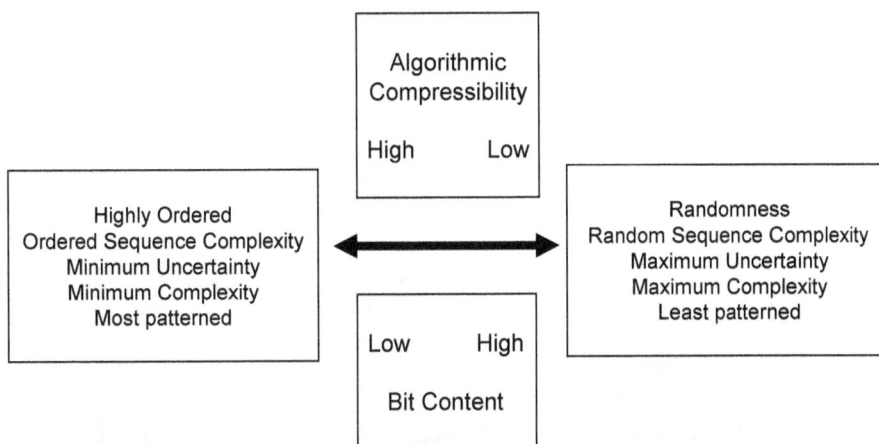

Figure 1. An antithetical relationship exists between linear sequence order and complexity. Randomness affords the greatest measure of complexity. The more ordered and patterned a sequence, the less uncertain are its components, and the less complex the sequence. Neither order nor complexity generates formal meaning or utility, both of which lie in a completely different dimension from order/complexity measures.

Used with permission from: Abel, D.L.; Trevors, J.T. 2005, Three subsets of sequence complexity and their relevance to biopolymeric information., Theoretical Biology and Medical Modeling, 2, 29.

Cellulose is a highly ordered molecule consisting of a string of d-glucose (dextrose) molecules. It can be algorithmically compressed by saying, "Give me a dextrose molecule, repeat X times." A theoretical stochastic ensemble of multiple types of polymerized simple sugars, however, might have no compression algorithm shorter than enumerating the random sequence of sugars itself [14-16]. It possesses no order or pattern, and therefore no compressibility. Thus a reliable definition of complexity is provided by its degree of algorithmic compressibility. The less ordered and the more random is a sequence, the more complex it is.

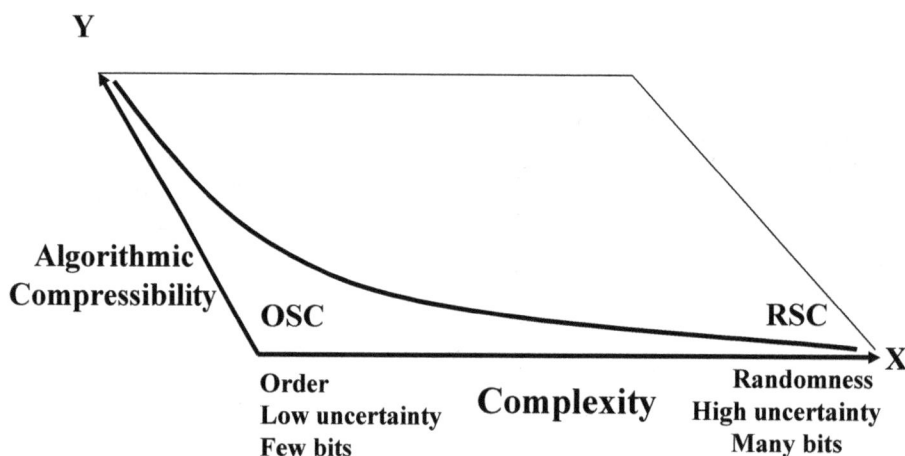

Figure 2. The adding of a second dimension to Figure 1 allows visualization of the relationship of Kolmogorov algorithmic compressibility to order and complexity. The more highly ordered (patterned) a sequence, the more highly compressible that sequence becomes. The less compressible a sequence, the more complex it is. OSC = Ordered Sequence Complexity. RSC = Random Sequence Complexity. A random sequence manifests no Kolmogorov compressibility. This reality serves as the very definition of a random, highly complex string.

Table 1: The difference between "order" and "complexity"

Order	Complexity
Regular	Irregular
Repeating	Non-repeating
Redundant	Non-redundant
Predictable	Non-predictable
Symmetrical	Asymmetrical
Periodic	Aperiodic
Monotonous	Variable
Crystal-like patterning	Little patterning; none if random
Reducible	Largely irreducible
Compressible	Largely non-compressible

High order, such as we see in homopolymers of nucleic acid (with all the nucleotides in the string being the same), possesses little or no uncertainty. High order, therefore, possesses little ability to retain information. Ribozymes and genomes are not homopolymers of DNA for good reason. Sequencing is highly variable depending upon the prescribed function. Highly specific sequencing matters. The reason life uses carbon chemistry to instantiate Prescriptive Information (PI) is the ability of carbon-based polymers to provide unlimited Shannon uncertainty in biopolymer combinatorialism. Consequently, high complexity comes closer to explaining sophistication than does high order. We can see that to attribute sophisticated organization, function and work to high degrees of order doesn't make sense. And to appeal to high degrees of order and complexity simultaneously is logically impossible. Is it complexity that explains organization and sophisticated function, then? If we *had* to pick between the two as the source of organization, we would have to go with complexity rather than order.

But now we have a new problem. Maximum complexity is randomness. Maximum complexity is noise. Since when did randomness and noise ever produce anything with sophisticated function? Would the results of an algorithm that produces supposedly random numbers ever produce a meaningful computational program? Neither order nor complexity is the answer when it comes to explaining the derivation of organization, sophisticated function, computation, algorithmic optimization, and useful work. As mentioned above, it is certainly true that sophisticated formal systems are almost always complex. But complexity is just a secondary feature of highly organized systems, not the cause of that organization. Randomness and noise (maximum complexity) cannot organize anything. Something major is missing from the equation.

We have invested so much confidence and anticipation in "complexity" as a potential source of spontaneous prescriptive information and organization that our senses should be jolted by the pristine mathematical definition of sequence complexity:

$$H = \sum_{i=1}^{M} p_i(-\log_2 p_i) \tag{1}$$

This, of course, is Shannon's basic measurement of uncertainty in linear sequence complexity. This metric of uncertainty also measures linear digital complexity in bits. The minus sign is necessary to invert the log of a fraction (a probability) so as to render the measurement of negative uncertainty in positive bits. Normally, the minus sign is just placed in front of the entire expression.

This mathematical measurement should remind us that maximum complexity is nothing more than maximum uncertainty and randomness. Complexity, therefore, has nothing to do with generating formal function. Complexity possesses no creative or computational talents. No justification exists for attributing exquisite formal organization to mere complexity.

The relation of order and complexity remains the same as we begin to add dimensions. The secondary structure of ribozymes, for example, is represented by a linear digital string of ribonucleotides folded back onto itself into two dimensions forming helices, bulges, hairpin loops, internal loops, and junctions [17, pg. 683]. The more ordered (redundant the sequence of nucleotides) in the initial polymer, the less sophisticated secondary structures tend to form. Of course stochastic ensembles of ribonucleotides (random strings) don't tend to form functional folds either. The initial primary structure of linear digital sequence of ribonucleosides in catalytic ribozymes is highly varied; they are unpredictable with a non-deterministic, non-ordered sequence of nucleotides. But, this complexity is only secondary. The real issue is that of having a particular needed sequence (the *programming*) that allows a string to fold back onto itself to form the requisite secondary and tertiary structures. High order (unimaginative redundancy of nucleotide or amino acid sequencing) only limits sophistication of three-dimensional molecular machine folding.   The lack of order (randomness) does no better at prescribing catalytic secondary structures.

Shannon equations only quantify uncertainty and reduced uncertainty (the measured uncertainty of the "before" state minus the measured uncertainty of the after state = acquired knowledge).   Randomness (maximum complexity) contains the maximum number of bits of non-compressible Shannon uncertainty. The objective of Shannon theory is to compare two sequences: the one sent by the transmitter with the other received at the receiver.   Shannon quantifications have nothing to do with meaning or function [18].   Shannon uncertainty measurements should never be used to refer to information.   Shannon himself objected to calling his theory of communication engineering, "information theory" [19].   Shannon stated very clearly from the beginning of his work that his measurements of uncertainty would have nothing to do with meaning, function and intuitive information [13, 18].   The von Weizsäckers [20] pointed out that Weaver understood very well that the negative logarithm of an event's probability could not differentiate a meaningful message from nonsense: "Two messages, one heavily loaded with meaning, and the other pure nonsense, can be equivalent as regards information." [21].   Yet Weaver still unfortunately allowed himself to refer to quantifiable uncertainty as "information." Reduced uncertainty (mutual entropy) does not provide what Abel has termed, "Prescriptive Information (PI)" [2, 8, 22].   PI either instructs or produces (with formal algorithmic processing) nontrivial optimized function.

As we have pointed out in previous chapters, the missing ingredient needs to explain the phenomenon of "organization" and the achievement of sophisticated function/work.   This missing ingredient is *formal choice contingency* at bona fide decision nodes, logic gates, and configurable switch settings.   Without choice contingency, nothing will get formally organized from randomness/uselessness into nontrivial usefulness.

Complexity can arise from chance contingency or choice contingency. Table 1 fails to differentiate between the two kinds of contingency.   Chance contingency has minimal patterning, reducibility, or compressibility.   Choice contingency tends to reuse programming modules or linguistic constructions, giving rise to more patterning, reducibility and compressibility than randomness.   The highly functional programs resulting from choice contingency are much more complex than ordered.   They are more proximate to the high complexity, high uncertainty end of the bidirectional vector in Figure 1 than to the ordered end.   But the complexity of the sequence is not the cause of its functionality.   It is only a secondary result.

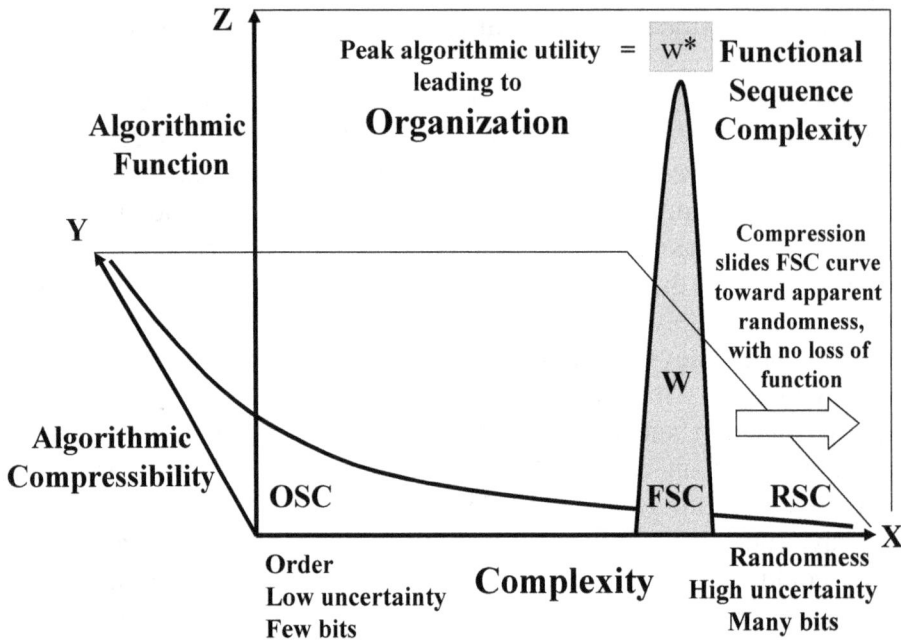

Figure 3.   Superimposition of Functional Sequence Complexity (FSC) onto Figure 2.   The Y axis plane plots the decreasing degree of algorithmic compressibility as complexity increases from order towards randomness. The Z axis plane shows where along the same complexity gradient (X-axis) that highly instructional sequences are generally found.   The Functional Sequence Complexity (FSC) curve includes all algorithmic sequences that work at all (W, inside the shaded steep curve). The peak of this curve (w*) represents the particular algorithmic sequence that "works *best*." The FSC curve is usually quite narrow and is located closer to the random end than to the ordered end of the complexity scale. Compression of an instructive sequence slides the FSC curve towards the right (away from order, towards maximum complexity, maximum Shannon uncertainty, and seeming randomness) *with no loss of function* (assuming decompression).

When we see sophisticated function of any kind, we have strong evidence suggesting that the Cybernetic Cut has been traversed across the one-way CS Bridge [4, 7] [See chapter 3]. Nonphysical formalisms are the product of purposeful choice contingency [4, 7, 23]. Choice contingency is instantiated into physicality via logic gates and configurable switch settings. The purposeful selection of tokens from an alphabet of "physical symbol vehicles" (tokens) is a second means of instantiating choice contingency into physicality. A third way is cooperative integration of physical components into formal systems or conceptually complex machines [1, 3, 4, 7, 8, 24]. Mere physicodynamic constraints can accomplish none of the above examples of formal organization. Organization and sophisticated function in the physical world are all the products of formalisms instantiated into physicality. Physicality cannot generate nonphysical formalisms. Figure 3 shows the relation of order, complexity, and compressibility with functionality.

The third dimension of utility and organization is when each alphabetical token in the linear string is selected for meaning or potential function. The string becomes either language or a cybernetic program capable of computation only when signs/symbols/tokens are chosen to represent utilitarian configurable switch settings. What is the common denominator to all aspects of design and engineering function? Choice contingency: not chance contingency, not law, not physicodynamics, but formal choice contingency—traversing The Cybernetic Cut across the one-way CS Bridge from formalism to physicality (Chapter 3). The FSC curve is usually quite narrow and is located closer to the random end than to the ordered end of the complexity scale. Compression of an instructive sequence slides the FSC curve towards the right (away from order, towards maximum complexity, maximum Shannon uncertainty, and seeming randomness) with no loss of function. This further demonstrates that neither order nor complexity is the determinant of algorithmic function. Functionality arises in a third dimension of selection that is unknown to the second dimension of compressibility. This is one of most poorly understood realities in information theory and life-origin science. Selection alone produces functionality. Without selection, evolution would be impossible.

Suppose stochastic ensembles of oligoribonucleotides were forming out of sequence space in an hypothesized "primordial soup." Since only 4 different nucleosides could be added next to a forming single positive strand, then in Equation 1 above would $M = 4$. Suppose next that the prebiotic availability p_i for adenine was 0.46, and the p_i's for uracil, guanine, and cytosine were 0.40, 0.12, and 0.02 respectively. This is being presumptious for cytosine, given that cytosine would have been extremely difficult to make in any prebiotic environment [25]. Using these hypothetical base-availability probabilities, the Shannon uncertainty would have been equal to:

Adenine	0.46 (- log$_2$ 0.46)	= 0.515
Uracil	0.40 (- log$_2$ 0.40)	= 0.529
Guanine	0.12 (- log$_2$ 0.12)	= 0.367
Cytosine	0.02 (- log$_2$ 0.02)	= 0.113
	1.00	1.524 bits

Notice how unequal availability of the four nucleotides (*a form of ordering*) greatly reduces Shannon uncertainty at each locus, and in the entire sequence, of any biopolymeric stochastic ensemble (Figure 1). Maximum uncertainty would occur if all four base availability probabilities were 0.25. Under these equally available base conditions, Shannon uncertainty would have equaled 2 bits per independent nucleotide addition to the strand. A stochastic ensemble formed under aqueous conditions of mostly adenine availability, however, would have had little information-retaining ability because of its high order [8].

As pointed out in the above reference, even less information-retaining ability would be found in an oligoribonucleotide adsorbed onto montmorillonite [26-31]. Clay surfaces would have been required to align ribonucleotides with 3' 5' linkages. The problem is that only polyadenosines or polyuridines tend to form. Using clay adsorption to solve one biochemical problem creates an immense informational problem (e.g., high order, low complexity, low uncertainty, and low information retaining ability. See Figure 1). High order means considerable compressibility. The Kolmogorov [11] algorithmic compression program for clay-adsorbed biopolymers (Figure 2) would read: "Choose adenosine; repeat the same choice fifty times." Such a redundant, highly-ordered sequence could not begin to prescribe even the simplest protometabolic contributor. Such "self-ordering" phenomena would not be the key to life's early organization or algorithmic programming.

The RNA Word and pre-RNA World models [17, 32] still prevail despite daunting biochemical problems. Life-origin models also include clay life [33-36]; early three-dimensional "genomes" [37, 38]; "Metabolism/Peptide First" [39-42]; "Co-evolution" [43-46]; "Simultaneous nucleic acid and protein" [47-49]; and "Two-Step" models of life-origin [50-52]. In virtually all of these life origin models, "self-ordering" is confused with "self-organizing." No mechanism is provided for the development of a linear digital prescription and oversight system to integrate metabolism. No known life form exists that does not depend upon such genetic instruction.

Well, what about a *combination* of order *and* complexity? Doesn't that explain how Prescriptive Information (PI) or true organization comes into being? Combinations of stochastic elements with ordered structures do nothing more to generate utility than either separately. No basis exists for steering events toward computational success or algorithmic optimization in a combination of order and mere combinatorial complexity. No goal of pragmatism exists. Function is not even perceived or valued in such an inanimate system of mixed order and random combinatorial elements.

2. Pattern vs. Noise

If order is not the key to formal function, what about pattern? SETI looks for patterns as evidence of extraterrestrial intelligence. Couldn't patterning prescribe function? The answer is no! To understand why, we must also define pattern. Starting with a single dimension, pattern in a sequence is defined by an increasing probability of occurrence of a single symbol or symbol sequence. As the probability of an event increases towards 1.0, its Shannon uncertainty decreases towards 0 bits [18]. So a recurring pattern is found to be a form of order. The more patterned a sequence, the more ordered. The more ordered, the fewer bits of uncertainty, and therefore the less information retaining potential the signal would have. Highly patterned sequences contain minimal complexity. Recurring pattern is therefore antithetical to complexity. They lie at opposite extremes of the same bidirectional vector found in Figure 1. [8, 13, 53]. The literature is filled with misunderstanding of the relationship between pattern and complexity, and how they both relate to formal function.

What is noise? Noise is pollution of a meaningful/functional message by random combinatorial influences. Both randomness and noise may be defined by some as extremely complex, poorly understood interactive physicodynamic necessity. But randomness and noise still spell a non-choice-contingent, nonsensical degradation of the transmission of recorded purposeful choices. We go to great lengths to protect our meaningful language transmissions and programming decisions from noise pollution. Chance contingency is the enemy of meaningful communication and instruction. To whatever degree chance contingency replaces purposeful choice contingency, function, usefulness and biologic metabolism will deteriorate concomitantly.

It also true that most messages manifest relative degrees of both combinatorial complexity and patterns in varying segments. But, like complexity, patterning in and of itself does not account for the functionality of the message. They are just secondary results of re-use of linguistic letter associations (e.g., "qu" in English), words, phrases, and programming modules. Any language has frequencies of letter reuse. But the reason a sequence is able to impart Prescriptive Information (PI) and programming function is because that sequence instantiates *cybernetic determinism*, not physicodynamic determinism, into its formal material symbol system (See chapter 6). Combinatorial complexity is just the secondary effect of unconstrained freedom of choice. Theoretically, choice contingency can be just as complex as chance contingency. A string of programming choices can appear patternless and measure the same number of Shannon bits as a random string. But clearly complexity alone does not provide the answer of why choice contingency is able to generate such high degrees of utility.

Highly patterned strings can be greatly compressed algorithmically. In nature, ordered strings frequently contain repeating patterns. The most patterned string is exampled by a string of identical letters, or, in nature, by a sugar polymer or a DNA homopolymer consisting of all adenosines. A polyadenosine has maximum order, no uncertainty, and therefore no complexity. A polymer of 200 adenosines can be fully enumerated by the very short compression algorithm, "Give me an adenosine; repeat 200 times." This compression algorithm for a polyadenosine contains almost no uncertainty, and therefore almost no information potential. It is an example of Ordered Sequence Complexity (OSC)[8]. Note that this polyadenosine can base pair its full length with thymine. What makes DNA important to life is not just base-pairing or the double-helix structure of DNA. What is most important is the programming of the particular sequence of nucleosides in the single positive prescriptive strand of DNA, before ordinary base-pairing ever occurs.

Of course, we now know that the supposed "anti-sense" complementary strand sequence that is base-paired by hydrogen bonding to the "sense" strand also contains additional layers and dimensions of Prescriptive Information (PI). Both micro- and coding m- RNAs exist in this complementary strand, for example. The complementary microRNA can regulate many other metabolic activities, including in some cases the regulation of protein production prescribed by its own complementary "sense" strand sequence [54, 55]. He et al. [56] found that individual transcripts are derived from both the plus and minus strands of chromosomes. These authors found evidence for antisense transcripts in 2900 to 6400 human genes. Human cells are a long way from theoretical protocells. But, such formal organization and multi-layered, multi-dimensional PI could not have arisen out of chance and necessity. It also could not have arisen out of natural selection which operates only at the post facto phenotypic level of favoring the fittest pre-programmed, already-living organisms (see *The GS Principle* reviewed in Chapter 7).

Formal organization is not limited to linear digital prescription. Genetic structure exposes genes to regulation and chromosomal cross-communication [57]. Say Duan et al.:

Layered on top of information conveyed by DNA sequence and chromatin are higher order structures that encompass portions of chromosomes, entire chromosomes, and even whole genomes. Interphase chromosomes are not positioned randomly within the nucleus, but instead adopt preferred conformations. Disparate DNA elements co-localize into functionally defined aggregates or 'factories' for transcription and DNA replication. [58]

Synergistic regulation between the multi-units has long since become apparent in molecular biology [59]. "Bi-enzyme nanomachines exist where the binding partner is crucial for ligand-binding processes" [60]. Wang [61] goes into great detail in pointing out just how much the cell is like a computer, discussing the "multi-step information flow from storage level to the execution level." He goes on to say,

Functional similarities can be found in almost every facet of the retrieval process. Firstly, common architecture is shared, as the ribosome (RNA space) and the proteome (protein space) are functionally similar to the computer primary memory and the computer cache memory, respectively. Secondly, the retrieval process functions, in both systems, to support the operation of dynamic networks—biochemical regulatory networks in cells and, in computers, the virtual networks (of CPU instructions) that the CPU travels through while executing computer programs. Moreover, many regulatory techniques are implemented in computers at each step of the information retrieval process, with a goal of optimizing system performance. Cellular counterparts can be easily identified for these regulatory techniques. [61]

Wang utilized theoretical insight from computer system design principles to sketch "an integrative view of the gene expression process, that is, how it functions to ensure efficient operation of the overall cellular regulatory network." Wang found the computer analogy to be a credible source of information with which to decipher regulatory logics underneath biochemical network operation. The credibility of those who deny the role and necessity of Prescriptive Information (PI) to organize cellular life is becoming increasing strained with each new month of journal publications.

Returning to the far simpler discussion of contrasting mere order and pattern from PI, a pulsar signal has abundant order and pattern. But it doesn't DO anything useful. It contains no meaningful or functional message. It knows nothing of decision nodes, choice contingency programming, or PI. The signal generates no formal utility at the receiver.

The probability of encountering the next element of a repeating pattern like polyadenosine is high; the probability of coming across any uniqueness is low. Note that statistical order and pattern have no more to do with function and formal utility than does maximum complexity (randomness). Neither order nor complexity can program, compute, optimize algorithms, or organize.

Three subsets of linear complexity have been defined in any environment [8]. These subsets are very helpful in understanding potential sources of Functional Sequence Complexity (FSC) as opposed to mere Random Sequence Complexity (RSC) and Ordered Sequence Complexity (OSC) [8]. FSC requires a third dimension not only to detect, but to produce formal utility. Neither chance nor necessity (nor any combination of the two) has ever been observed or demonstrated to produce nontrivial FSC [9]. Nontrivial useful function and formal work arise only within the narrow FSC curve seen in Figure 3.

Durston and Chiu, at the University of Guelph, developed a method for measuring what they call *functional uncertainty* (H_f) [62]. They extended Shannon uncertainty to measure a *joint variable* (X, F), where X represents the variability of data, and F the variable of functionality. This explicitly incorporated the empirical knowledge of embedded function into the measure of sequence complexity:

$$H(X_f(t)) = - \sum P(X_f(t)) \log P(X_f(t)) \qquad (2)$$

where X_f denotes the conditional variable of the given sequence data (X) on the described biological function f which is an outcome of the variable (F). The state variable t, representing time or a sequence of ordered events, can be fixed, discrete, or continuous. Discrete changes may be represented as discrete time states. Mathematically, the above measure is defined precisely as an outcome of a discrete-valued variable, denoted as $F=\{f\}$. The set of outcomes can be thought of as specified biological states.

Using this method allowed Durston and Chiu to compare quantifications of 2,442 aligned sequences of proteins belonging to the Ubiquitin protein family, among many other protein families evaluated. All of these sequences satisfied the same specified function f, which might represent the known 3-D structure of the Ubiquitin protein family, or some other function common to Ubiquitin. The definition of functionality used by Durston and Chiu relates to the whole protein family. Thus this data can be inputted from readily available databases. Even subsets (e.g., the active sites) of the aligned sequences all having the same function can be quantified and compared. The tremendous advantage of using $H(X_f(t))$ is that *slight changes* in the functionality characteristics of biosequences can be incorporated and analyzed.

Subsequently, Durston and Chiu have developed a theoretically sound method of actually quantifying Functional Sequence Complexity (FSC) [63]. This method holds great promise in being able to measure the increase or decrease of FSC through evolutionary transitions of both nucleic acid and proteins. This FSC measure, denoted as ζ, is defined as the change in functional uncertainty from the ground state $H(X_g(t_i))$ to the functional state $H(X_f(t_i))$, or

$$\zeta = \Delta H \left(X_g(t_i), X_f(t_j) \right) \qquad\qquad 3)$$

The *ground state g* of a system is the state of presumed highest uncertainty permitted by the constraints of the physical system, when no specified biological function is required or present. Durston and Chiu wisely differentiate the ground state *g* from the *null state $H_\emptyset$*. The null state represents the absence of *any* physicodynamic constraints on sequencing. The null state produces bona fide stochastic ensembles, the sequencing of which is *dynamically inert* (physicodynamically decoupled or incoherent [64, 65]). The FSC variation in various protein families, measured in Fits (Functional bits), showed that the highest value sites correlate with the primary binding domain [63].

As we add dimensions, highly patterned waveforms, signal structures, short sequences of events, or crystalline structures might form [12].   But all of the multidimensional high redundancy structures preclude information retaining ability in any object. Repeating patterns generate high order, low complexity, few bits of uncertainty, and little information retaining possibility.

A law of physics also contains very little information because the data it compresses is so highly ordered. The best way to view a parsimonious physical law is as a compression algorithm for reams of data. This is an aspect of valuing Ockham's razor so highly in science. Phenomena should be explained with as few assumptions as possible. The more parsimonious a statement that reduces (algorithmically compresses) all of the data, the better [66, 67]. A sequence can contain much order with frequently recurring patterns, yet manifest no utility. Neither order nor recurring pattern is synonymous with meaning or function.

Those trained in information theory will be quick to point out at this point that "information is always defined in terms of an observer or knower." They argue that information is not in the law's parsimonious statement or equation, but in the difference (R) between all of the uncertainty of the raw data, and the lesser amount of uncertainty generated by knowing the law. But the problem with this concept of information is that for most of life's history, linear digital genetic instructions have been prescribing exquisite metabolic organization long before any observers or knowers existed on earth. Observers and knowers themselves would not exist except for the extraordinary amount of cellular programming and organization that produced human brains. Prescriptive Information (PI) [6] cannot be reduced to an exlcusive endeavor of human creation or epistemology. To attempt to define information solely in terms of human generation, observation and knowledge is grossly inadequate. Such anthropocentrism blinds us to the reality of life's *objective* genetic programming, regulatory mechanisms, and biosemiosis using symbol systems such as codon translation [2, 68-76].

3. Structure vs. chaos

Well what about "structure"? Surely structure is the answer as to what makes functional physical objects. Look at proteins. Is not folded protein structure the key to what enables molecular machines to work?

The answer is, "Yes and No." Some structures are functional and others are not. What makes the difference? In the everyday practical world, we might want to ask, "What specific structure?" "How was the structure assembled, and for what purpose?" "What was the structure designed and engineered to do?" Structures performing sophisticated functions don't just spontaneously self-assemble. Returning to the ordinary paper clip analogy from Chapter 1, we have never observed a single simple paper clip spontaneously spring from the iron ore in the ground in *any* environment. A paper clip is nothing more than a long solid cylinder of relatively constant diameter that is folded back onto itself in a certain way so as to make it useful for temporarily binding sheets of paper. Even if a paper clip spontaneously formed out of inanimate nature, an agent would still have to choose to *use* the paper clip for its optimized purpose. Like proteins, the ore had to be processed into an alloy (the correct mix and sequence of 20 left-handed, biological-only amino acids, all with peptide-only bonds) and elongated into exceedingly long cylinders (wires) of uniform diameter. The wire than had to be folded in 3 dimensions (with one dimension kept deliberately constant). A chaperone-like bending machine then had to made that could make 3 bends at the best places using 3 wheels, all according to the specifications that would engineer an optimized clasp for paper.. The wire then had to be cut appropriately. Why have we never observed or demonstrated a single spontaneous occurrence of a paper clip from inanimate nature? How much more conceptually organized and efficacious is a bacterium than a paper clip? If paper clips do not spontaneously form, why would we entertain the ridiculous notion that sophisticated protein molecular machines, along with their highly specific contribution to cooperative metabolic schemes, would spontaneously spring from happenstantial molecular interactions? How does this blind belief differ from superstition?

Inorganic crystals provide a great deal of highly ordered/patterned three-dimensional structure. Even these well-ordered structures contain noise pollution in the form of occasional crystal irregularities. The Cairns-Smith clay-life model fell by the wayside mostly because clay crystals are so regular, so ordered and patterned, that they cannot have any significant amount of PI formally instantiated into their physicality. Crystal irregularities provide the only freedom with which to program, but if those irregularities are only generated by chance contingency rather than choice contingency, no meaningful/functional PI will be contained in those crystal irregularities.

Thus the mere presence of structure as opposed to heat agitation-like molecular chaos tells us little about function and utility. Many rigid, sustained structures exhibit no function. In chaos theory, candle flames and tornadoes manifest seemingly sustained structure from a continual string of momentary self-ordered dissipative states. Neither kind of structure computes or optimizes any algorithmic function. None of Prigogine's "dissipative structures" generates a Sustained Functional System (SFS) [23]. It is for good reason that Prigogine named them "dissipative." But, regardless of how long dissipative structures last, they certainly produce no sophisticated functions. SFS's do.

Neither chaos or the edge of chaos is a
1) Calculus
2) Algorithm
3) Program that achieves computational halting
4) Organizer of formal function
5) Bona fide system

Chaos is a bounded state of *disorganization* that is extremely sensitive to the effects of initial conditions. Note that chaos is a disorganized state of matter, *not a disordered* state of matter. A considerable amount of order can arise spontaneously out of chaos. This is what chaos theory is about. All we have to do to observe spontaneous self-ordering is to pull the stopper out of our bathtub drain. Water molecules quickly self-order into a swirl—a vortex—from purely physicodynamic complex causation. We mistakenly call this self-organization, but the vortex is not in the least bit organized. It is only self-ordered [9]. What is the difference? No decision nodes are required for a bathtub swirl to self-order out of seemingly random Brownian motion. Proficient programming choices are not required for heat agitation of water molecules to self-order into a vortex. No configurable switches have to be purpose-fully set, each in a certain way, to achieve self-ordering. No pursuit of a goal is involved. No algorithmic optimization is required. In addition, Prigogine's dissipative structures do not DO anything formally productive. They possess no ability to achieve computational halting. They do not construct sophisticat-ed Sustained Functional Systems (SFS) [23]. Dissipative structures are mo-mentary. They only appear sustained (e.g., a candle flame) because of we ob-serve through time a long string of momentary dissipative events or structures. This is where their name comes from. They cannot generate a sustained func-tional machine with optimized functionality.

Chaos is capable of producing incredibly complex physicodynamic behav-ior. But we must never confuse this complexity with formal function. Order spontaneously appears out of disorder in the complete absence of any formal creative input or cybernetic management. But no algorithmic organization is produced by a candle flame. What seems to be a totally random environment is in fact a caldron of complex interaction of multiple force fields. The complex-ity of interactive causation can create the illusion of randomness, or of very real self-ordering. There may also be as-of-yet undiscovered physical causes. But dissipative structures self-order; they do NOT self-organize. The dissipa-tive structures of chaos theory are unimaginative. Highly ordered structures contain very little information. Information retention in any physical medium requires freedom of selection of configurable switch settings. Switches must be "dynamically inert" with respect to their function to serve as logic gates.

Dissipative structures are
1) highly ordered
2) monotonous
3) predictable
4) regular (vortices, sand piles)
5) low informational
6) strings of momentary states

Dissipative structures are usually destructive, not cybernetically construc-tive (e.g., tornadoes, hurricanes). Trying to use "chaos" and "complexity" to provide mechanism for "self-organization" is like trying to use the Shannon transmission engineering to explain intuitive information, meaning and func-tion. Shannon's equations define negative "uncertainty," not positive "surpris-al." Functional "surprisal" requires the acquisition of positive specific seman-tic information. Just as we cannot explain and measure "intuitive information" using Shannon combinatorial uncertainty, we cannot explain a truly organized system appealing to nothing but a mystical "edge of chaos." Reduced uncertainty ("mutual entropy") in Shannon theory comes closer to semantic information. To achieve this, however, we have to mix in the formal elements of human knowledge gained by mathematical subtraction of "after uncertainty" from "before uncertainty." We measure the reduced uncertainty of *our knowledge*. Prior background knowledge and agent processing of that knowledge is already at play. At that point, we are no longer talking about objective information in nature. We are only talking about human epistemology. Human consciousness is highly subjective. The second we insist on defining information solely in terms of a human observer and knower, we have destroyed all hope of elucidating the derivation of objective information in evolutionary history, especially at the intra-cellular level.

The disorganization of chaos is characterized by conceptual uncertainty and confusion. Disorganization lacks sophisticated steering and control. Disorganization pursues no purpose. Even if chaos had purpose, it would lack all means of accomplishing purpose. If chaos by definition is a bounded state of disorganization, how could we possibly attribute self-organization to chaos? No scientific basis exists for granting formal capabilities to chaos, complexity or catastrophe. None of these three has ever been observed to produce formal integration and algorithmic organization of any kind.

Scientists accomplish impressive feats using the applied science of "nonlinear dynamics." But the capabilities of this applied science all-too-easily get confused with the capabilities of chaos itself. Chaos generates nothing close to formal function. We overlook the considerable degree of "investigator involvement" and artificial steering that goes into nonlinear dynamic experiments. Formal mathematics is invariably employed by agents to achieve some goal.

Sophisticated algorithmic optimization has never been achieved by duplication plus mere random variation. Function must be "selected for" at the logic-gate programming level prior to the realization of improved function. Selection *for potential* fittest function is necessary to achieve computational success. This is called The GS Principle (Genetic Selection Principle) [5]. Natural selection favors only the fittest already-computed, already-living phenotypes. Configurable switches are "set in stone" with rigid covalent bonds before folding begins. The selection of each nucleoside constitutes a rigid programming choice that, in turn, determines eventual folding and functional structure.

Three-dimensional conformation of molecular machines is largely determined by the minimum-free-energy sinks of primary structure folding. The primary structure of any protein or sRNA is the already-covalently-bound sequence of particular monomers that serve as configurable switch settings.

4. Self-ordering vs. the illusion of "Self-organization"

The term "self-organization" is, unfortunately, in widespread use in the literature. The terms "organization" and "self-ordering" should not be confused [2, 8]. No empirical evidence exists of unaided algorithmic self-optimization or spontaneous bona fide self-organization [9].

Organization ≠ order. Disorganization ≠ disorder. Organization is abstract, conceptual, nontrivial and algorithmic. Organization is formal, not merely physicodynamic. Organization requires choice contingency at bona fide decision nodes. Organization integrates and correlates cybernetic choices into holistic functional systems. Organization typically contains high ratios of Prescriptive Information (PI) to noise. PI either instructs or indirectly produces (through algorithmic processing) nontrivial optimized function at its destination. Prescription requires choice contingency rather than chance contingency or necessity. Organized phenomena are typically informationally and cybernetically complex, not just combinatorially complex (high uncertainty as to how components might come together). They are prescriptively complex and programmatically highly optimized. Prescriptive complexity typically requires intentional choices at bona fide decision nodes. The null hypothesis we seek to falsify is this: "Any form of nontrivial organization traverses the Cybernetic Cut, requiring formal choices with intent to explain." Inanimacy cannot "organize" itself. Inanimacy can only self-order. "Self-organization" is without empirical and prediction-fulfilling support.

Organization often utilizes a sign/symbol/token system to represent multiple configurable switch settings. Physical switch settings allow instantiation of nonphysical selections for function into physicality. Switch settings represent choices at successive decision nodes that integrate circuits and instantiate cooperative management into conceptual physical systems. Switch positions must be freely selectable to function as logic gates. Switches must be set according to rules, not laws.

Self-ordering phenomena are not examples of self-organization. Self-ordering phenomena are simple and redundant [7, 8]. Self-ordered structures, whether sustained (e.g. crystals) or dissipative (e.g., the chaos theory first investigated by Prigogine) *contain no organization at all.* Self-ordering events occur spontaneously every day. But they do not involve decision nodes or dynamically-inert, purposeful, configurable switch settings. No logic gates need to be programmed with self-ordering phenomena. Self-ordering events involve no steering toward algorithmic success or "computational halting." Self-ordering phenomena are purely physicodynamic and incapable of organizational attempts. Laws and fractals are both compression algorithms containing minimal complexity and information. Inanimate physicodynamics cannot exercise purposeful choices or pursue potential function. No model of undirected evolution pursues the goal of future utility.

Order cannot compute. Much life-origin literature appeals to "yet-to-be discovered laws of self-organization." Laws, however, describe highly ordered/patterned behavior. Because they are parsimonious compression algorithms of data, they contain very little information. Given the high information content of life, expectation of a new law to explain sophisticated genetic algorithmic programming is ill-founded. Considerable peer-reviewed published literature is erroneous because of failure to appreciate that the "complexity of life" could never arise from such highly "ordered," low informational physicodynamic patterning. Tremendous combinatorial uncertainty is required. The complexity of life will never be explained by the highly-ordered behavior that is reducible to the low-informational laws of physics and chemistry.

A crystal is highly-ordered. Its description can be easily algorithmically compressed. A crystal is about as far from being "alive" as any physical state we could suggest. Every member of a 300-monomer string of adenosines (a homopolymer) can be specifically enumerated by stating: "Give me a set of adenosine molecules; repeatly connect one to another 300 times." This is called a compression algorithm. The simplicity and shortness of this compression algorithm is a measure of the extremely low complexity and uncertainty of this polymer. Such a parsimonious statement of the full sequence is only possible because that sequence is so highly patterned. Such a highly ordered sequence lacks uncertainty, complexity, and the ability to instantiate prescriptive information. Such a parsimonious compression algorithm can enumerate each and every member of the 200-mer string with only seven words. This reality defines high order or pattern along with low information retaining potential.

The spontaneous self-organization of ever-improving hypercylces [77-81], stoichiometric self-assemblies [82], and Ganti's chemotons [83] have never been observed, let alone repeatedly observed. No prediction fulfillments have ever been realized. "Self-organization" provides no mechanism and offers no detailed verifiable explanatory power. The hypotheses of chemotons ever-growing capabilities are not even falsifiable. No lack of evidence or the repeated observation of hypercycle's failure to arise is capable of providing falsification. So the notion is conveniently and indefinitely protected from any scientific challenge. It must just be accepted by blind faith. Any scientist who raises an eyebrow of healthy scientific skepticism is immediately labeled a heretic from the hierarchy of scientism's presupposed imperative of metaphysical naturalism.

We value Ockham's Razor in science because we wish to reduce physical reality down to concise reductive statements. We consider a law to be elegant and beautiful because of its ability to compress our interpretation of reams of empirical data down to a single parsimonious equation. When we look for new laws of physics, we look for new algorithms as a proxy for compressing reams of data.

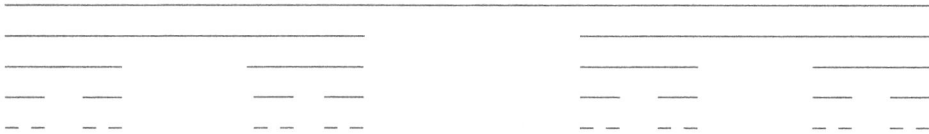

Figure 4. The Cantor Dust Fractal representative of the high order content of all fractals. Fractals create the illusion of high complexity. Their low complexity is demonstrated by the simple Kolmogorov compression algorithm: "Take a line segment, remove the center third. Repeat N times."

Used with permission from: Abel DL, Trevors JT: Self-Organization vs. Self-Ordering events in life-origin models. *Physics of Life Reviews* 2006, 3:211-228

For biology, however, we encounter not only the highest degree of complexity known, we encounter linear, digital, cybernetic encoding along with Prescriptive Information (PI) of the most sophisticated, abstract, and conceptual nature. Complexity gets all the press. But, complexity is not the remarkable issue. The phenomenon of cybernetics and how it could have arisen in a prebiotic chance-and-necessity environment is the issue. The world's fastest, parallel-architecture main-frame computer systems (e.g., the K computer) cannot hold a candle to the central nervous system of any mammal. The "processing" units and the interconnect fabric of that computer has nowhere near the connections of, for example, the human brain (10^{15} neuraltransmitter/receivers. No yet-to-be-discovered parsimonious law will ever be able to explain the programming found in a single cell [84]. Neither order nor mere combinatorial complexity can generate algorithmic organization. Bona fide organization results from algorithmic optimization. The best solutions to any problem must be selected from a formal "possible solution space" to achieve optimization. Apart from such purposeful selection, noise will increase within any system. A tendency toward randomization and loss of function unfolds from noise. Complexity increases while algorithmic optimization decreases. Any attempt to exclude choice-with-intent from the mix results in the deterioration of programming function, computational halting, integration, and organization.

Table 2: The difference between spontaneous "self-ordering phenomena" and "organized systems" in living organisms.

SELF-ORDERING PHENOMENA	ORGANIZED SYSTEMS
Increases redundancy	Decreases redundancy
Increases predictability	Decreases predictability
Increases symmetry	Decreases symmetry
Increases periodicity	Decreases periodicity
Increases monotony	Decreases monotony
Produces crystal-like patterns	Produces linguistic-like patterns
Decreases complexity	Increases complexity
Short-lived (highly dissipative)	Long-lasting (minimal dissipation)
Produced by cause-and-effect	Still lacking natural process mechanism
Observed	Bona fide self-organization unobserved
Consistent with 2nd Law	Seems inconsistent with the 2nd Law
Non-integrative	Integrative
Non-conceptual	Conceptual
Not particularly functional	Produces extraordinary function

Fractals are often cited as evidence of self-organized complexity arising out of simple order. But fractals are examples of neither complexity nor organization. They only create the illusion of complexity. And, fractals certainly have nothing to do with organization. Organization requires steering cybernetic choices at bona fide decision nodes and logic gates. Representative symbols can be used to denote antithetical binary choices (e.g., On vs. Off, Yes vs. No, 0 vs. 1). Figure 4 helps visualize the fact that fractals are really nothing more than highly ordered, highly compressible, low-informational, redundancies.

Self-ordering of many kinds occurs spontaneously every day in nature in the absence of any organization. Spontaneous bona fide self-organization, on the other hand, has never been observed. Certainly no *prediction* of bona fide *self*-organization from unaided physicodynamics has ever been fulfilled. Of course, if we fail through sloppy definitions to discern between self-ordering phenomena and organization, we will *think* that evidence of self-organization is abundant. We will point to hundreds of peer-reviewed papers with "self-organization" in their titles. But when all of these papers are carefully critiqued with proper scientific skepticism, embarrassment only grows with each exposure of the blatant artificial selection that was incorporated into each paper's experimental design. Such investigator involvement is usually readily apparent right within Materials and Methods of the paper. Organization depends upon PI. The self-ordering phenomena of physical nature provide no PI, and therefore no bona fide organization.

5. Can spontaneous combinatorial complexity generate organization?

Order, pattern, noise and complexity have little to do with prescription of function. Attempts to demonstrate self-organization via mere combinatorial complexity are too numerous to cite [9, 85-88]. Under careful scrutiny, however, these papers seem to universally incorporate investigator agency into their experimental designs. To demonstrate the viability of any molecularly evolved hypothetical scenario, it is imperative that we provide *stand-alone natural process* evidence of nontrivial self-organization at the edge of chaos. We must demonstrate on sound scientific grounds the *formal capabilities* of naturally-occurring physicodynamic complexity. So-called evolutionary algorithms, for example, must be stripped of all artificial selection and the purposeful steering of iterations toward desired products. The latter intrusions into natural process clearly violate sound evolution theory [89, 90]. Undirected evolution has no goal [91, 92]. Evolution provides no steering toward *potential* computational and cybernetic function [3-5, 9, 24, 93, 94].

The theme of naturalistic ProtoBioCybernetics is the active pursuit of falsification of the following null hypothesis: *"Physicodynamics alone cannot self-organize itself into formal, functional systems that would require algorithmic optimization, computational halting, and circuit integration."* At first glance the falsification of this hypothesis might seem like a daunting task. But a single exception of nontrivial, unaided, spontaneous optimization of formal function by truly natural process would quickly falsify this null hypothesis. Such falsification would once and for all silence Intelligent Design intrusions into naturalistic science.

Science celebrates positive and parsimonious descriptions of presumed objectivity. But we must never forget that our knowledge is only "best thus far." Even the most fundamental laws of physics technically must be viewed as "tentative." We rightly eschew diatribes of metaphysical pontifications. Science proceeds through open-mindedness and the falsification of null hypotheses, not through the rhetorical pronouncement of dogmas. Popper— and many since—have exposed the problems associated with trying to prove any positive hypothesis [95, 96]. Neither induction nor deduction is foolproof. Theses that cannot be proven ought not to be proclaimed as absolute statements of fact.

At the same time, naturalistic science has spent most of the last century, and especially the first decade of the new millennium, arguing to the lay community that science *has* proved the current biological paradigm. Unfortunately, very few in the scientific community seem critical of this indiscretion. One would think that if all this evidence is so abundant, it would be quick and easy to falsify the null hypothesis put forward above. If, on the other hand, no falsification is forthcoming, a more positive thesis might become rather obvious by default. Any suggestion that programming is required would only be labeled metaphysical by true-believers in spontaneous self-organization. Those same true-believers would disingenuously fail to acknowledge the purely metaphysical nature of the current Kuhnian paradigm rut [97]. A better tact is to thoroughly review the evidence. Let the reader provide the supposedly easy falsification of the above null hypothesis. Inability to do so should cause pangs of conscience in any scientist who equates metaphysical materialism with science. On the other hand, providing the requested falsification of this null hypothesis would once-and-for-all end a lot of unwanted intrusions into science from philosophies competing with metaphysical materialism.

While proof may be evasive, science has an obligation to be honest about what the entire body of evidence clearly *suggests*. We cannot just keep endlessly labeling abundant evidence of formal prescription in nature "apparent." The fact of purposeful programming at multiple layers gets more "apparent" with each new issue of virtually every molecular biology journal [98-100]. Says de Silva and Uchiyama:

> Molecular substrates can be viewed as computational devices that process physical or chemical 'inputs' to generate 'outputs' based on a set of logical operators. By recognizing this conceptual crossover between chemistry and computation, it can be argued that the success of life itself is founded on a much longer-term revolution in information handling when compared with the modern semiconductor computing industry. Many of the simpler logic operations can be identified within chemical reactions and phenomena, as well as being produced in specifically designed systems. Some degree of integration can also be arranged, leading, in some instances, to arithmetic processing. These molecular logic systems can also lend themselves to convenient reconfiguring. Their clearest application area is in the life sciences, where

their small size is a distinct advantage over conventional semi-conductor counterparts. Molecular logic designs aid chemical (especially intracellular) sensing, small object recognition and intelligent diagnostics. [100]

What scientific evidence exists of spontaneous physicodynamics ever having programmed a single purposeful configurable switch-setting? If we cannot present any such evidence, we should be self-honest enough to ask ourselves, "How long are we going to try to maintain this ruse that the cybernetic programming we repeatedly observe is only 'apparent' rather than real?"

Has "natural process" ever been observed to write conceptual instructions? Neither reason nor empiricism has justified believing in spontaneous algorithm-writing and optimization by inanimate nature. The inanimate environment does not generate meaning, or program and optimize sophisticated formal function. Physics and chemistry do not symbolize meaning or pursue and prescribe ideal utility. Physicodynamics does not translate linear digital PI from one language into another. All of these functions are as nonphysical and as formal as mathematics itself.

The association of complexity *or* patterns with most forms of bona fide organization should never be confused with causation [101]. Neither order nor complexity is a cause of organization or any other form of formal algorithmic optimization. We sling the words "chaos," "complexity," "order" and "pattern" around with vivid imagination and a great deal of blind faith in their capabilities. None of the latter states has ever been observed to produce the slightest amount of algorithmic organization. Stand-alone chaos and complexity have absolutely nothing to do with generating formal function. Neither do order and pattern. Self-ordering phenomena produce boring, unimaginative redundancy. Self-ordering phenomena, just like chaos and complexity, have never been observed to achieve

1) programming,
2) computational halting,
3) creative engineering,
4) symbol systems,
5) language
6) bona fide organization [9].

The latter are all formal processes, not physicodynamic processes.

"Self-organization" is logically a nonsense term. Inanimate objects cannot organize themselves into integrated, cooperative, holistic schemes. Schemes are formal, not physical. To organize requires choice contingency, not just chance contingency and law-like necessity. Sloppy definitions lead to fallacious inferences, especially to category errors. Organization requires

1) decision nodes
2) steering toward a goal of formal function
3) algorithmic optimization
4) selective switch-setting to achieve potential integration of a circuit
5) choice with intent

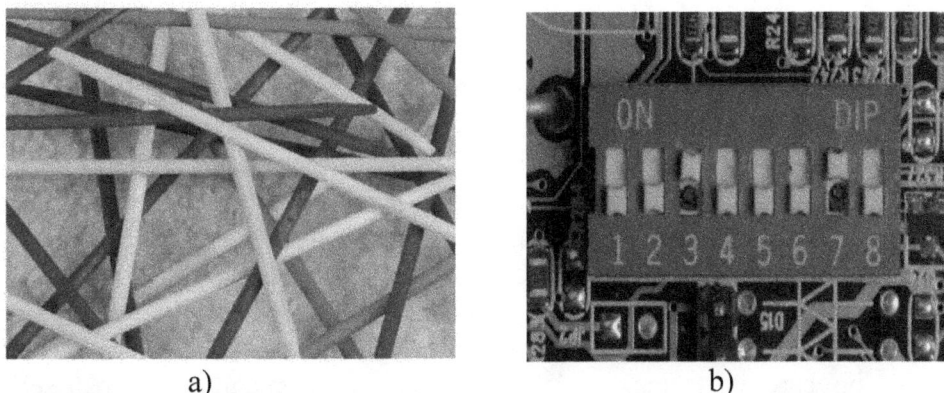

a)   b)

Figure 5 a) Complexity is often confused with programming controls and formal organization. The degree of three-dimensional structural complexity within a pile of pick-up sticks is staggering. But what exactly does this enormous degree of structural complexity DO? If we poured glue on this pile to freeze its structure, what sophisticated formal function would this complex pile of objects generate? Mere combinatorial complexity must never be confused with organization or formal utility. **b)** A row of dip switch settings depicts a different category of complexity—algorithmic, cybernetic programming complexity. Choice contingency is incorporated into purposeful configurable switch settings that collectively prescribe and integrate formal function.

Used with permission from: Abel DL: The capabilities of chaos and complexity. *Int J Mol Sci* 2009, 10:247-291

The only entity that might be able to organize itself is an agent. But not even an agent self-organizes. Agents organize things and events in their lives. They do not organize their own molecular biology, cellular structure, organs and organ systems. Agents do not organize their own being. Agents do not create themselves. They merely make purposeful choices with the brains and minds with which they find themselves. Artificial intelligence does not organize itself either. It is invariably programmed by agents to respond in certain ways to various environmental challenges found in the artificial life data base.

Thus the reality of self-organization is highly suspect on logical and analytic grounds even before facing the absence of empirical evidence of any spontaneous formal self-organization.

If formal self-organization phenomena are so objectively real and common, one would think that we could make abundant reliable predictions of very rudimentary future instances of self-organization. If all of the incredibly integrative cybernetic complexities of tens of millions of different species have resulted from spontaneous physicodynamic self-organization, we should find *exhaustive* empirical evidence on a daily basis of fulfillment of minor self-organization predictions. How many such prediction fulfillments has the scientific community observed?

Prediction fulfillment is a cardinal parameter of scientific investigation. The complete absence of prediction fulfillment is strong evidence that spontaneous self-organization of formal utility in nature is the product of vivid imagination rather than repeated observation of a presumed objective reality. No prediction fulfillments have been realized of progressing hypercycles, spontaneously computational neural nets, genetic and evolutionary algorithms, or cellular automata self-organizing out of inanimate physicodynamic interactions.

To evaluate whether any such predictions have been fulfilled, we would have to be careful to make sure that what we are calling formal self-organization isn't in reality mere physicodynamic self-ordering events. The latter have no integrative ability. Self-ordering phenomena cannot organize anything into potential nontrivial formal utility. No basis exists in cause-and-effect physicodynamic interactions for organizing events into needed or desired or expedient formal function. The laws of physics do not consider utilitarian expediency.

6. The mystical "Edge of Chaos" and the magical "Adjacent Other"

If chaos is inadequate to explain self-organization, what about "the Edge of Chaos" or "the Adjacent Other." [3, 88, 93, 102-124] The edge of chaos is somehow much more appealing to us than just plain chaos. The edge of chaos is more poetic. And "the Adjacent Other," now there's some scientific content we can really sink our teeth into! Both phrases are wonderfully inviting. They offer incredible mystical allure. The question is, does either phrase actually exist as a physical state? If these states are objectively real, what exactly are they? Where in time/space can we find them, what are their initial conditions, what are their physical characteristics, and what exactly can they independently DO? Are "the Edge of Chaos" or "the Adjacent Other" scientifically addressable? How would we go about falsifying such nebulous metaphysical notions? How do they differ from superstition?

Let us first examine the potential interface of the "edge" of chaos with natural order—with the regularities of nature described by the physical laws. Can "order" program configurable switches? If "order" programmed configurable switches, they would all be programmed the same way. They would all be set to "On's," OR. . . they would all be set to "Off's." Either way, the configurable switches would not be formally programmable into any algorithmic function. No more creativity would exist at the interface of chaos with forced order than in either single entity. No reason exists to expect any increased cybernetic potential at the edge of chaos than squarely in the middle of chaos (bounded disorganization). The fact that chaos is extremely sensitive to the effects of initial conditions adds no formal attributes. The latter certainly increases its changeability and the number of bits of uncertainty in the bounded state. But mere changeability and combinatorial uncertainty provide no optimization of formal function.

"The edge of chaos" [88, 102-107] affords mesmerizing visions of potential accomplishment. While poetic and wonderfully inviting, the concept is sorely lacking in scientific content. The functional reality of "the edge of chaos" has been challenged [3, 93, 110, 117]. Have we had any prediction fulfillments since it was first described in 1992 by Waldrop[102]? Is the notion of vast formal capabilities arising from the edge of chaos falsifiable? One has to wonder if the notion is worthy of serious discussion in a peer-reviewed science journal paper. It would not be were it not for the fact that so many peer-reviewed papers already cite this nebulous dream as an objective source of self-organization.

What about the interface of the bounded state of disorganization with heat agitation and Brownian motion? Maximum complexity would set all configurable switches randomly. What synergistic capabilities could emerge from the interface of disorganization with randomness? The two are not synonymous. But neither contributes anything to programming proficiency.

Switches must be set a certain way to achieve integrated circuits. If chaos sets configurable switches, the result will predictably "blue screen," as is known in Microsoft's "crash" terminology. Without steering towards sophisticated function at each decision node, sophisticated function has never been observed to arise spontaneously. Only destructive self-ordering (e.g., tornadoes) and disorganization accumulates. No prediction fulfillments have been realized of cooperative integration of biofunction arising spontaneously in nature. Dreaming of an "edge of chaos" doesn't help.

What scientific substance does "the Adjacent Other" provide? What is this magical "other"? Is this "other" physical? Is it observational reality? Can it compute and organize systems? What is the logic behind such hoped-for capabilities of this "Adjacent Other"? What empirical support do we have of formal function arising spontaneously from the interface of "otherness" with physicodynamic chance and necessity? How can an otherwise intelligent and skeptical scientific community possibly buy into such an unadulterated fairy tale?

Unfortunately, neither "the edge of chaos" nor "the adjacent other" mysticisms has provided detailed scientific mechanism to explain the efficacious selection of pragmatic configurable switch settings. Organization requires algorithmic optimization. The latter requires expedient decision-node commitments that are instantiated into specific physical configurable switch settings. To explain life origin requires elucidating how these particular logic gates were selected at the genetic level. Phenotypes must first be computed before the fittest living organisms can exist to be preferred by any environment.

In every case that provides the illusion of spontaneous emergence, investigator involvement can be demonstrated in the Materials and Methods section of so-called "evolutionary algorithm" papers. The experimenter's goal and steering are apparent in faulty experimental designs. This is usually evident in the choice of each successive iteration to pursue. Undirected evolution has no goal. Iterations cannot be steered toward experimenters' goals (e.g., a desired ribozyme using SELEX [125-127]). Quality science requires brutal self-honesty. We must be open-minded enough to consider the possibility that emergence and self-organization are closer to metaphysical presuppositions than observed scientific facts."

7. What about neural nets?

Bipartite graphs of neural nets showing vertices (nodes) and edges (lines) are frequently featured in attempts to explain the derivation of self-organization. "Buttons and strings" are supposed to provide the answer for how circuitry and selective switching arises. But bipartite graphs show only the Aristotelian "final" edges and connections that allow computational success and optimized function. Such graphs correspond to mere Descriptive Information (DI), not Prescriptions Information (PI). No explanation is provided by these graphs as to *how* the elements got connected in their unique functional relationships. Graphing the state of functional affairs does not provide the ability to generate such a state. A description of a Lamborgini automobile does not provide the ability to manufacture Lamborginis.

How do the signals get selectively steered through the circuitry of a spontaneously self-assembled neural net? How did the nodes get *functionally* associated? How is successful computation accomplished? Neural net graphs fail to explain efficacious programming choices at a single decision node, let alone all of them working in concert to achieve a potential formal goal.

How did a prebiotic environment make so many wise programming decisions? All of the programming errors and "wild goose chases" are never shown; only the final product. Even then the picture of the final product does not tell us how it steers signals and computes. "Why did this edge (string connecting the buttons) [104] form rather than the 293 other possible edges that this node could have formed?" Or why did the signal traverse only certain edges from the node, and not all the other edges that emanate from that node? The model of a bunch of interconnected buttons and strings does not address the question of selective Boolean logic gating. Yet selective gate openings and closings is the essence of programming, circuitry, control and regulation. To prescribe all of the integrated controls through the connections (edges) between elements (nodes) in bipartite graphs requires programming choices that neural network graphs do not provide.

If all nodes in a neural net fire with any impulse introduction into the net, as with an all-or-none muscle depolarization, no selectivity, steering or integration of circuitry is possible. Pathways of conduction must be specific and uniquely selectable. A "buttons and strings" model [104] never explains the phenomenon of circuit integration or computational success.

8. Systems theory

Evolutionary (undirected) biological systems theory regularly presupposes the metaphysical belief of physicodynamic self-organization into formal function. One would think that systems theorists could readily offer a crystal-clear definition of "system." Sadly, this is not the case. It is not surprising, therefore, that chaos and such phenomena as weather fronts are also referred to as "systems" with no eyebrows raised. Bona fide systems require organizational controls. True systems are cybernetic. The definition of "system" is "an organized assembly of parts and/or controlled procedures designed and engineered to produce utility." Any attempt to eliminate parts of this definition results in a breakdown of the system and a compromise of optimized functionality. A term like, "Chaotic system," therefore, is an oxymoron—a self-contradiction. If it is chaotic, it cannot be a bona fide system. If it is a system, it cannot be chaotic.

Weather fronts are at best self-ordered by complex degrees of interactive physicodynamic causation. They are not formally controlled or organized to achieve sophisticated utility of any kind. A weather front is a physicodynamic interface complete with criticality and phase changes. A weather front may become a highly self-ordered tornado or a hurricane. But it's not a true system because it is not formally organized or cybernetically programmed. No representational symbol system is used. No abstract conceptualizations are employed by weather fronts. They are simply physicodynamic interfaces totally lacking in algorithmic organization. We simply "murder the King's English" by referring to a weather front as a system. Such sloppy word usage may not cause problems in everyday usage. But it leads to a great deal of confusion in understanding fundamental physics and the relation of physics to biology.

The temporary and local circumvention of the 2nd Law is made possible by the formal algorithmic processes that comprise true systems. Chaos is neither organized nor a bona fide system, let alone "self-organized." As pointed out above, organization is not the same as order. A bona fide *system* requires *organization.* Chaos by definition lacks organization. That's why we call it "chaos" even though it manifests extensive self-ordering tendencies. What could possibly be more self-ordered than a massive hurricane? But what formal functions does a hurricane perform? A hurricane possesses no PI [6]. It has no programming talents or creative instincts. A hurricane is not a participant in Decision Theory. A hurricane does not set logic gates according to rules of inference or deduction from axioms. A hurricane has no specifically designed physicodynamically-decoupled configurable switches. No means exists to instantiate formal choices or function into physicality. A highly self-ordered hurricane destroys organization. To call a hurricane "self-organized" constitutes one of the most egregious errors in science stemming from sloppy definitions, category errors, and non-sequiturs.

Complexity is not a system, either, as we saw in the highly complex pile of pick-up sticks (Figure 5). No programming is involved. No algorithms are optimized. No steering toward formal function occurs. A true PI-based system requires organization.

Systems theory is literally taking over in biology today. "Systems biology" is considered to be the in-vogue descriptor by research institutes and academics. A growing number of molecular and cell biologists are arguing not only for the insufficiency of genetic control, but even of genomic regulation, to explain the holistic integration of metabolic pathways and cycles. The organization of the cell is attributable to many epigenetic factors and subsystems in addition to genomics. All of these subsystems contribute to an overarching, conceptual, cooperative system. Development, error repair, control and regulation all contribute to a formal *metasystem* within each prokaryotic cell. These metasystems only grow with eukaryotes, and grow even more astoundingly with multi-cellular organisms.

Epigenetic factors do not negate the continuing reality of extensive genetic and genomic controls. No rational or empirical justification exists for attributing linear, digital, encrypted, genetic recipes to stochastic ensembles OR to physical laws in *any* amount of time. Yet thousands of peer-reviewed papers exist in the literature appealing to materialistic "self-organization." The latter cannot generate formally organized bona fide systems.

A phenomenal amount of *objective* Prescriptive Information (PI) instructs and organizes each cell. A great deal more objective PI is required to integrate cell systems, organs, organ systems, and holistic organisms. From an evolutionary history perspective, no observers or knowers were around when bacteria were being prescribed and their metabolisms organized into replicating formal metasystems. Human observers are Johnny-come-lately *discoverers* of biological Prescriptive Information (PI) [6]. Human epistemology is not an essential component of what objective genetic prescriptive information *is* in nature. Nor is human mentation a factor in a "systems biology" that predates the very existence of *Homo sapiens*.

Many scientists across a wide array of disciplines exercise a surprisingly blind faith in the amazing formal capabilities of spontaneous molecular chaos and combinatorial complexity. Phenomenal systems are just blindly believed to self-organize. Achieving sophisticated formal function consistently requires regulation and control. Control always emanates from choice contingency and intentionality, not from spontaneous molecular chaos.

9. Cells are exquisitely organized systems that accomplish formal work

Formal work is not just heat transfer. Formal work achieves functionality. If ever there were an example of an object achieving formal functionality, it is a living cell. Staying alive has ultimate value to a living organism. Organisms pursue the goal of remaining alive and reproducing. Metabolism is the most highly integrated, holistic, conglomerate of organized formal functions known to science. How did life get so organized and goal-oriented? What forces integrated life's formal systems?

The answer is that life is programmed. The scientific community has been invited in many peer-reviewed science journal papers over the last decade to falsify the following null hypothesis: "All known life is cybernetic." [5-7, 84, 128] Cybernetic simply means "formally controlled" rather than merely "physicochemically constrained." All that is needed to falsify this null hypothesis is a single living cell that is free of metabolic and reproductive controls. No falsification has been provided.

Biological controls are accomplished through the use of material symbol systems, logic gates, configurable switches, and the functional organization of cellular components. Say Ramakrishnan and Bhalla,

Just as complex electronic circuits are built from simple Boolean gates, diverse biological functions, including signal transduction, differentiation, and stress response, frequently use biochemical switches as a functional module. [129]

Genetic cybernetics inspired Turing's, von Neumann's, and Wiener's development of computer science [130-136]. Genomic and epigenomic cybernetics cannot be explained by models that metaphysically pre-assume the all-sufficiency of mass-energy interactions and the chance and necessity of physicodynamics alone. Genetic and genomic algorithmic controls are fundamentally formal, not physical. But like other formalisms, they can be instantiated into a physical medium of retention and channel transmission using a material symbol system or dynamically-inert configurable switches. Neither parsimonious law nor mere combinatorial complexity can program the efficacious decision-node logic-gate settings of algorithmic organization observed in all known living organisms.

Any life-origin chemist, whether a nucleic acid RNA-World advocate, or Peptide/Polypeptide-First or Lipid Metabolism-First advocate, can readily relate innumerable nightmares of cross-reactions and catastrophes experienced as bench scientists have tried to model theoretical abiogenesis models. The extreme difficulty of making cytosine, even with the best minds in the world steering biochemical events, is a classic example [25]. Difficulties in making common components like ribose sugar, and its instability once made, are constantly throwing a wrench into any theoretical mechanism of spontaneous generation of any protosystem, let alone life. Says Shapiro with regarding the formation of D-ribose on prebiotic earth,

> Polymerization of formaldehyde (the formose reaction) has been the single reaction cited for prebiotic ribose synthesis. . . .The complex sugar mixture produced in the formose reaction is rapidly destroyed under the reaction conditions. Nitrogenous substances (needed for prebiotic base synthesis) would interfere with the formose reaction by reacting with formaldehyde, the intermediates, and sugar products in undesirable ways. The evidence that is currently available does not support the availability of ribose on the prebiotic earth, except perhaps for brief periods of time, in low concentration as part of a complex mixture, and under conditions unsuitable for nucleoside synthesis. [137]

Homochirality issues arise in trying to generate pure populations of right-handed sugars and left-handed amino acids. Activation of monomers necessary for polymerization of long chains is no small issue. Polymerization of more than ten residues in aqueous solution is almost impossible when dehydration synthesis is needed for polymerization. When heat is applied, cyclical cAMPs and cGMPs can form with up to 100 mers, but these homopolymers are informationless [138]. Adsorption of nucleosides onto montmorillonite clay surfaces allows polymerization of chains of 30-50 monomers [139]. But these too are informationless homopolymers, usually polyadenosines or polyuridines. These spontaneous reactions are so physicodynamically ordered that they cannot have any significant PI instantiated into them. Information instantiation into any physical matrix requires Shannon *uncertainty*. This, is turn, requires freedom from self-ordering physicodynamic determinism. Sequencing must be arbitrary (freely selectable) and inert (physicodynamically indeterminate; decoupled from and incoherent with physical causation.) Homopolymers, therefore, could not possibly be the source of highly informational genetic instructions.

Multiple chicken-and-egg dilemmas arise such as the need for protein to supplement the extremely sophisticated ribozyme component of the ribosome—a veritable molecular computer. Yet no protein can be made for the ribosome's construction without the ribosome itself already being there to make those protein components. Multiple problems arise in trying to develop the genetic code piecemeal over a long period of time. Francis Crick's Central Dogma will not be overturned by any amount of empirical evidence. Formally absolute mathematical prohibitions exist for trying to build the genetic code table from bottom up (with fewer codons) rather than top down [13, 84].

We could go on ad infinitum with the train wrecks that occur in any abiogenesis model given real-world biochemical and formal realities. The bottom line is that any Composomal or Metabolism-First model of life-origin must have control mechanisms in place almost from the first instance of any protometabolism for any hint of progress to be made towards an imagined protolife. There must be manifold directionality to system composed of cooperative functional processes. These processes must be steered toward energy utilization and other utilitarian formal goals if any hope of organization of protometabolism is possible. Highly selective active transport across bilipid pseudomembranes is needed very early on in any miceller model of a developing protocell. Osmotic pressure alone is enough to kill one's vivid imagination of the first protocell. Nothing is more crucial to life or any envisioned protolife, than control and regulation mechanisms. As we saw in Chapter 3, such controls cannot be generated or explained by mere physicodynamic constraints

[4, 7]. The only exception is when certain constraints are deliberately chosen by experimenters in their experimental design so as to steer outcomes toward the experimenters' desired results. But this does not model "natural process" in inanimate prebiotic nature. This is artificial selection. The latter is nothing less than human engineering. It hardly qualifies as a naturalistic life-origin model.

10. Conclusion.

Chance and necessity produce no useful nontrivial organization or work. Will some yet-to-be discovered new law be able to explain or produce sophisticated utility? Nontrivial formal functions require high levels of Prescriptive Information (PI) to steer, control and regulate. High levels of PI require high levels of physical combinatorial uncertainty into which to record purposeful choices. Law-like physicodynamic behavior, on the other hand, manifests minimal uncertainty. It is highly ordered, patterned and redundant. The high degree of order found in "necessity" (the regularities of nature described by the "laws" of physics) only restricts PI instantiation into any physical medium. No yet-to-be-discovered law, therefore, will ever be able to explain the high information content of even short prescriptive programs or algorithms. In a deductively absolute sense, no new law will be able to generate nontrivial pragmatic work. The latter requires *formal* control of physicality.

The mind/body problem has only become more enigmatic with the latest and best neurophysiological research findings [140-149]. Mind cannot be adequately explained with the physical brain alone. Even more perplexing are the many phenomena in pre-vertebrate biological nature with extraordinary programming and integrational attributes. Life is undeniably cybernetic at almost every stage and level. The organization of even *Mycoplasmal* life is as choice-contingent as the intentional operation of Maxwell's Demon's trap door.

References:

1. Abel, D.L. 2009, The capabilities of chaos and complexity, Int. J. Mol. Sci., 10, (Special Issue on Life Origin) 247-291 Open access at http://mdpi.com/1422-0067/10/1/247
2. Abel, D.L.; Trevors, J.T. 2006, More than metaphor: Genomes are objective sign systems, Journal of BioSemiotics, 1, (2) 253-267.
3. Abel, D.L. 2007, Complexity, self-organization, and emergence at the edge of chaos in life-origin models, Journal of the Washington Academy of Sciences, 93, (4) 1-20.
4. Abel, D.L. 2008, 'The Cybernetic Cut': Progressing from description to prescription in systems theory, The Open Cybernetics and Systemics Journal, 2, 234-244 Open access at www.bentham.org/open/tocsj/articles/V002/252TOCSJ.pdf
5. Abel, D.L. 2009, The GS (Genetic Selection) Principle, Frontiers in Bioscience, 14, (January 1) 2959-2969 Open access at http://www.bioscience.org/2009/v14/af/3426/fulltext.htm.
6. Abel, D.L. 2009, The biosemiosis of prescriptive information, Semiotica, 2009, (174) 1-19.
7. Abel, D.L. 2010, Constraints vs. Controls, Open Cybernetics and Systemics Journal, 4, 14-27 Open Access at http://www.bentham.org/open/tocsj/articles/V004/14TOCSJ.pdf.
8. Abel, D.L.; Trevors, J.T. 2005, Three subsets of sequence complexity and their relevance to biopolymeric information., Theoretical Biology and Medical Modeling, 2, 29 Open access at http://www.tbiomed.com/content/2/1/29.
9. Abel, D.L.; Trevors, J.T. 2006, Self-Organization vs. Self-Ordering events in life-origin models, Physics of Life Reviews, 3, 211-228.
10. Adami, C. 2002, What is complexity?, Bioessays, 24, (12) 1085-94.
11. Kolmogorov, A.N. 1965, Three approaches to the quantitative definition of the concept "quantity of information", Problems Inform. Transmission, 1, 1-7.
12. Li, M.; Vitanyi, P. 1997, An Introduction to Kolmogorov Complexity and Its Applications. 2 ed., Springer-Verlag: New York.
13. Yockey, H.P. 1992, Information Theory and Molecular Biology. Cambridge University Press: Cambridge.
14. Chaitin, G.J. 2001, Exploring randomness. Springer: London ; New York.
15. Chaitin, G.J. 1990, Information, randomness and incompleteness : papers on algorithmic information theory. 2nd ed., World Scientific: Singapore ; New Jersey.
16. Chaitin, G.J. 2002, Conversations with a mathematician : math, art, science, and the limits of reason : a collection of his most wide-ranging and non-technical lectures and interviews. Springer: London ; New York.
17. Gesteland, R.F.; Cech, T.R.; Atkins, J.F. 2006, The RNA World. 3 ed., Cold Spring Harbor Laboratory Press: Cold Spring Harbor.
18. Shannon, C. 1948, Part I and II: A mathematical theory of communication, The Bell System Technical Journal, XXVII, (3 July) 379-423.
19. Shannon, E.C. 1951, Intersymbol influence reduces information content, Bell. Sys. Tech. J., 30, 50.
20. Weizsäcker_von, E.; Weizsäcker_von, C. 1998, Information, evolution and 'error-friendliness', Biological Cybernetics, 79, 501-506.
21. Weaver, W. 1949, The mathematics of communication., Sci Am, (July).
22. Trevors, J.T.; Abel, D.L. 2004, Chance and necessity do not explain the origin of life, Cell Biology International, 28, 729-739.
23. Abel, D.L. 2011, Moving 'far from equilibrium' in a prebiotic environment: The role of Maxwell's Demon in life origin. In Genesis - In the Beginning: Precursors of Life, Chemical Models and Early Biological Evolution Seckbach, J.Gordon, R., Eds. Springer: Dordrecht.
24. Abel, D.L. 2002, Is Life Reducible to Complexity? In Fundamentals of Life, Palyi, G.; Zucchi, C.Caglioti, L., Eds. Elsevier: Paris, pp 57-72.
25. Shapiro, R. 1999, Prebiotic cytosine synthesis: a critical analysis and implications for the origin of life, Proc Natl Acad Sci U S A, 96, (8) 4396-4401.
26. Ferris, J.P.; Huang, C.H.; Hagan, W.J., Jr. 1988, Montmorillonite: a multifunctional mineral catalyst for the prebiological formation of phosphate esters, Orig Life Evol Biosph, 18, (1-2) 121-133.
27. Ferris, J.P.; Ertem, G. 1992, Oligomerization of ribonucleotides on montmorillonite: reaction of the 5'-phosphorimidazolide of adenosine, Science, 257, (5075) 1387-1389.
28. Ferris, J.P. 1993, Catalysis and prebiotic RNA synthesis, Orig Life Evol Biosph, 23, (5-6) 307-15.
29. Ferris, J.P.; Hill, A.R., Jr.; Liu, R.; Orgel, L.E. 1996, Synthesis of long prebiotic oligomers on mineral surfaces, Nature, 381, (6577) 59-61.
30. Miyakawa, S.; Ferris, J.P. 2003, Sequence- and regioselectivity in the montmorillonite-catalyzed synthesis of RNA, J Am Chem Soc, 125, (27) 8202-8208.

31. Huang, W.; Ferris, J.P. 2003, Synthesis of 35-40 mers of RNA oligomers from unblocked monomers. A simple approach to the RNA world, Chem Commun (Camb), 12, 1458-1459.

32. Gilbert, W. 1986, Origin of life -- the RNA World, Nature, 319, 618.

33. Cairns-Smith, A.G. 1990, Seven Clues to the Origin of Life. Canto ed., Cambridge University Press: Cambridge.

34. Cairns-Smith, A.G. 1966, The origin of life and the nature of the primitive gene, J Theor Biol, 10, (1) 53-88.

35. Cairns-Smith, A.G. 1977, Takeover mechanisms and early biochemical evolution, Biosystems, 9, (2-3) 105-109.

36. Cairns-Smith, A.G.; Walker, G.L. 1974, Primitive metabolism, Curr Mod Biol, 5, (4) 173-186.

37. Segre, D.; Ben-Eli, D.; Lancet, D. 2000, Compositional genomes: prebiotic information transfer in mutually catalytic noncovalent assemblies, Proc Natl Acad Sci U S A, 97, (8) 4112-7.

38. Segre, D.; Lancet, D.; Kedem, O.; Pilpel, Y. 1998, Graded autocatalysis replication domain (GARD): kinetic analysis of self-replication in mutually catalytic sets, Orig Life Evol Biosph, 28, (4-6) 501-14.

39. Guimaraes, R.C. 1994, Linguistics of biomolecules and the protein-first hypothesis for the origins of cells, Journal of Biological Physics, 20, 193-199.

40. Shapiro, R. 2000, A replicator was not involved in the origin of life, IUBMB Life, 49, (3) 173-176.

41. Freeland, S.J.; Knight, R.D.; Landweber, L.F. 1999, Do proteins predate DNA?, Science, 286, (5440) 690-2.

42. Rode, B.M. 1999, Peptides and the origin of life, Peptides, 20, (6) 773-86.

43. Wong, J.T. 1975, A co-evolution theory of the genetic code, Proc Natl Acad Sci U S A, 72, (5) 1909-1912.

44. Wong, J.T. 1976, The evolution of a universal genetic code, Proc Natl Acad Sci U S A, 73, (7) 2336-40.

45. Wong, J.T. 2005, Coevolution theory of the genetic code at age thirty, Bioessays, 27, (4) 416-25.

46. Wong, J.T. 2007, Question 6: coevolution theory of the genetic code: a proven theory, Orig Life Evol Biosph, 37, (4-5) 403-8.

47. Zhao, Y.F.; Cao, P.-s. 1994, Phosphoryl amino acids: Common origin for nucleic acids and protein, Journal of Biological Physics, 20, 283-287.

48. Zhou, W.; Ju, Y.; Zhao, Y.; Wang, Q.; Luo, G. 1996, Simultaneous formation of peptides and nucleotides from N-phosphothreonine, Orig Life Evol Biosph, 26, (6) 547-60.

49. Nashimoto, M. 2001, The rna/protein symmetry hypothesis: experimental support for reverse translation of primitive proteins, J Theor Biol., 209, (2, Mar 21) 181-7.

50. Dyson, F.J. 1998, Origins of Life. 2nd ed., Cambridge University Press: Cambridge.

51. Dyson, F., 1999, Life in the Universe: Is Life Digital or Analog? In Life in the Universe: Is Life Digital or Analog?, NASA Goddard Space Flight Center Colloquiem, Greenbelt, MD, 1999; Greenbelt, MD,

52. Dyson, F.J. 1982, A model for the origin of life, J Mol Evol, 18, (5) 344-50.

53. Chaitin, G.J. 1988, Algorithmic Information Theory. Revised Second Printing ed., Cambridge University Press: Cambridge.

54. Beiter, T.; Reich, E.; Williams, R.; Simon, P. 2009, Antisense transcription: A critical look in both directions, Cellular and Molecular Life Sciences (CMLS).

55. Seila, A.C.; Calabrese, J.M.; Levine, S.S.; Yeo, G.W.; Rahl, P.B.; Flynn, R.A.; Young, R.A.; Sharp, P.A. 2008, Divergent transcription from active promoters, Science, 322, (5909) 1849-51.

56. He, Y.; Vogelstein, B.; Velculescu, V.E.; Papadopoulos, N.; Kinzler, K.W. 2008, The Antisense Transcriptomes of Human Cells, Science, 322, (5909) 1855-1857.

57. Tanizawa, H.; Et_al 2010, Mapping of long-range associations throughout the fission yeast genome reveals global genome organization linked to transcriptional regulation. , Nucleic Acids Research. Published online before print October 28, 2010.

58. Duan, Z.; Andronescu, M.; Schutz, K.; McIlwain, S.; Kim, Y.J.; Lee, C.; Shendure, J.; Fields, S.; Blau, C.A.; Noble, W.S. 2010, A three-dimensional model of the yeast genome, Nature, 465, (7296) 363-7.

59. Missiuro, P.V.; Liu, K.; Zou, L.; Ross, B.C.; Zhao, G.; Liu, J.S.; Ge, H. 2009, Information Flow Analysis of Interactome Networks, PLoS Comput Biol, 5, (4) e1000350.

60. Fatmi, M.Q.; Chang, C.-e.A. 2010, The Role of Oligomerization and Cooperative Regulation in Protein Function: The Case of Tryptophan Synthase, PLoS computational biology, 6, (11) e1000994.

61. Wang, D. 2008, Discrepancy between mRNA and protein abundance: Insight from information retrieval process in computers. , Computational Biology and Chemistry 32, 462-468.

62. Durston, K.K.; Chiu, D.K.Y. 2005, A functional entropy model for biological sequences, Dynamics of Continuous, Discrete & Impulsive Systems, Series B.

63. Durston, K.K.; Chiu, D.K.; Abel, D.L.; Trevors, J.T. 2007, Measuring the functional sequence complexity of proteins, Theor Biol Med Model, 4, 47 Free on-line access at http://www.tbiomed.com/content/4/1/47.

64. Rocha, L.M. 2001, Evolution with material symbol systems, Biosystems, 60, 95-121.

65. Rocha, L.M.; Hordijk, W. 2005, Material representations: from the genetic code to the evolution of cellular automata, Artif Life, 11, (1-2) 189-214.

66. Vitányi, P.M.B.; Li, M. 2000, Minimum Description Length Induction, Bayesianism and Kolmogorov Complexity, IEEE Transactions on Information Theory, 46, (2) 446 - 464.

67. Swinburne, R. 1997, Simplicity as Evidence for Truth Marquette University Press: Milwaukee, Wisconsin.

68. Barbieri, M. 2003, The Organic Codes: An Introduction to Semantic Biology. Cambridge University Press: Cambridge.
69. Barbieri, M. 2004, Biology with information and meaning, History & Philosophy of the Life Sciences, 25, (2 (June)) 243-254.
70. Barbieri, M. 2006, Is the Cell a Semiotic System? In Introduction to Biosemiotics: The New Biological Synthesis, Barbieri, M., Ed. Springer-Verlag New York, Inc.   : Secaucus, NJ, USA
71. Barbieri, M. 2008, Biosemiotics: a new understanding of life, Naturwissenschaften, 95, 577-599.
72. Pattee, H.H. 1968, The physical basis of coding and reliabiity in biological evolution. In Prolegomena to Theoretical Biology, Waddington, C. H., Ed. University of Edinburgh: Edinburgh.
73. Pattee, H.H. 1969, How does a molecule become a message? In Communication in Development; Twenty-eighth Symposium of the Society of Developmental Biology., Lang, A., Ed. Academic Press: New York, pp 1-16.
74. Pattee, H.H. 1972, Physical problems of decision-making constraints, Int J Neurosci, 3, (3) 99-106.
75. Pattee, H.H. 2001, The physics of symbols: bridging the epistemic cut, Biosystems, 60, (1-3) 5-21.
76. Pattee, H.H. 2005, The physics and metaphysics of Biosemiotics, Journal of Biosemiotics, 1, 303-324.
77. Eigen, M.; Gardiner, W.C., Jr.; Schuster, P. 1980, Hypercycles and compartments. Compartments assists--but do not replace--hypercyclic organization of early genetic information, J Theor Biol, 85, (3) 407-411.
78. Eigen, M.; Schuster, P. 1981, Comments on "growth of a hypercycle" by King (1981), Biosystems, 13, (4) 235.
79. Eigen, M.; Schuster, P.; Sigmund, K.; Wolff, R. 1980, Elementary step dynamics of catalytic hypercycles, Biosystems, 13, (1-2) 1-22.
80. Smith, J.M. 1979, Hypercycles and the origin of life, Nature, 280, (5722) 445-446.
81. Ycas, M. 1999, Codons and hypercycles, Orig Life Evol Biosph, 29, (1) 95-108.
82. Melendez-Hevia, E.; Montero-Gomez, N.; Montero, F. 2008, From prebiotic chemistry to cellular metabolism--the chemical evolution of metabolism before Darwinian natural selection, J Theor Biol, 252, (3) 505-19.
83. Munteanu, A.; Sole, R.V. 2006, Phenotypic diversity and chaos in a minimal cell model, J Theor Biol, 240, (3) 434-42.
84. Johnson, D.E. 2010, Programming of Life. Big Mac Publishers: Sylacauga, Alabama.
85. Grandpierre, A. 2005, Complexity, information and biological organization, Interdiscinplinary Description of Complex systems, 3, 59-71.
86. Chandler, J.L. 2000, Complexity IX. Closure over the organization of a scientific truth, Ann N Y Acad Sci, 901, 75-90.
87. Wimsatt, W.C. 1974, Complexity and organization. In SA-1972 (Boston Studies in the Philosophy of Science), Reidel: Dordrecht, Vol. 20, pp 67-86.
88. Kauffman, S. 1995, At Home in the Universe: The Search for the Laws of Self-Organization and Complexity. Oxford University Press: New York.
89. Mayr, E. 1988, Introduction, pp 1-7; Is biology an autonomous science? pp 8-23. In Toward a New Philosophy of Biology, Part 1, Mayr, E., Ed. Harvard University Press: Cambridge, MA.
90. Mayr, E. 1982, The place of biology in the sciences and its conceptional structure. In The Growth of Biological Thought: Diversity, Evolution, and Inheritance Mayr, E., Ed. Harvard University Press: Cambridge, MA, pp 21-82.
91. Monod, J. 1972, Chance and Necessity. Knopf: New York.
92. Mayr, E. 2001, What Evolution Is. Basic Books: New York.
93. Abel, D.L. 2006 Life origin: The role of complexity at the edge of chaos, Washington Science 2006, Headquarters of the National Science Foundation, Arlington, VA
94. Abel, D.L. 2008 The capabilities of chaos and complexity, Society for Chaos Theory: Society for Complexity in Psychology and the Life Sciences, International Conference at Virginia Commonwealth University, Richmond, VA., Aug 8-10.
95. Popper, K. 1963, Conjectures and Refutations. Harper: New York.
96. Popper, K.R. 1972, The logic of scientific discovery. 6th impression revised. ed., Hutchinson: London.
97. Kuhn, T.S. 1970, The Structure of Scientific Revolutions. 2nd 1970 ed., The University of Chicago Press: Chicago.
98. Dinger, M.E.; Pang, K.C.; Mercer, T.R.; Mattick, J.S. 2008, Differentiating Protein-Coding and Noncoding RNA: Challenges and Ambiguities, PLoS Computational Biology, 4, (11) e1000176.
99. Banks, E.; Nabieva, E.; Chazelle, B.; Singh, M. 2008, Organization of Physical Interactomes as Uncovered by Network Schemas, PLoS Computational Biology, 4, (10) e1000203.
100. de Silva, A.P.; Uchiyama, S. 2007, Molecular logic and computing, Nat Nano, 2, (7) 399-410.
101. Pearle, J. 2000, Causation. Cambridge University Press: Cambridge.
102. Waldrop, M.M. 1992, Complexity. Simon and Schuster: New York.
103. Kauffman, S.A.; Johnsen, S. 1991, Coevolution to the edge of chaos: coupled fitness landscapes, poised states, and coevolutionary avalanches, J Theor Biol, 149, (4) 467-505.
104. Kauffman, S.A. 2000, Investigations. Oxford University Press: New York.

105. Bratman, R.L. 2002, Edge of chaos, J R Soc Med, 95, (3) 165.
106. Ito, K.; Gunji, Y.P. 1994, Self-organisation of living systems towards criticality at the edge of chaos, Biosystems, 33, (1) 17-24.
107. Munday, D. 2002, Edge of chaos, J R Soc Med, 95, (3) 165.
108. Forrest, S. 1999, Creativity on the edge of chaos, Semin Nurse Manag, 7, (3) 136-40.
109. Innes, A.D.; Campion, P.D.; Griffiths, F.E. 2005, Complex consultations and the 'edge of chaos', Br J Gen Pract, 55, (510) 47-52.
110. Mitchell, M.; Hraber, P.T.; Crutchfield, J.T. 1994, Dynamics, computation, and "the edge of chaos:" a re-examination. In Complexity: Metaphors, Models, and Reality, Cowan, G. P., D. and Melzner, D. , Ed. Addison-Wesley: Reading, MA pp 1-16.
111. Albano, E.V.; Monetti, R.A. 1995, Comment on "Life at the edge of chaos", Phys Rev Lett, 75, (5) 981.
112. Baym, M.; Hubler, A.W. 2006, Conserved quantities and adaptation to the edge of chaos, Phys Rev E Stat Nonlin Soft Matter Phys, 73, (5 Pt 2) 056210.
113. Bernardes, A.T.; dos Santos, R.M. 1997, Immune network at the edge of chaos, J Theor Biol, 186, (2) 173-87.
114. Bertschinger, N.; Natschlager, T. 2004, Real-time computation at the edge of chaos in recurrent neural networks, Neural Comput, 16, (7) 1413-36.
115. Borges, E.P.; Tsallis, C.; Ananos, G.F.; de Oliveira, P.M. 2002, Nonequilibrium probabilistic dynamics of the logistic map at the edge of chaos, Phys Rev Lett, 89, (25) 254103.
116. Hiett, P.J. 1999, Characterizing critical rules at the 'edge of chaos', Biosystems, 49, (2) 127-42.
117. Mitchell, M.; Hraber, P.T.; Crutchfield, J.T. 1993, Revisiting the edge of chaos: Evolving cellular automata to perform computations, Complex Systems, 7, 89-130.
118. Legenstein, R.; Maass, W. 2007, Edge of chaos and prediction of computational performance for neural circuit models, Neural Netw, 20, (3) 323-34.
119. Melby, P.; Kaidel, J.; Weber, N.; Hubler, A. 2000, Adaptation to the edge of chaos in the self-adjusting logistic map, Phys Rev Lett, 84, (26 Pt 1) 5991-3.
120. Mycek, S. 1999, Teetering on the edge of chaos, Volunt Leader, 40, (2) 13-6.
121. Mycek, S. 1999, Teetering on the edge of chaos. Giving up control and embracing uncertainty can lead to surprising creativity, Trustee, 52, (4) 10-3.
122. Neubauer, J. 1997, Beyond hierarchy: working on the edge of chaos, J Nurs Manag, 5, (2) 65-7.
123. Schneider, T.M.; Eckhardt, B.; Yorke, J.A. 2007, Turbulence transition and the edge of chaos in pipe flow, Phys Rev Lett, 99, (3) 034502.
124. Stokic, D.; Hanel, R.; Thurner, S. 2008, Inflation of the edge of chaos in a simple model of gene interaction networks, Phys Rev E Stat Nonlin Soft Matter Phys, 77, (6 Pt 1) 061917.
125. Ellington, A.D.; Szostak, J.W. 1990, In vitro selection of RNA molecules that bind specific ligands, Nature, 346, (6287) 818-822.
126. Tuerk, C.; Gold, L. 1990, Systematic evolution of ligands by exponential enrichment -- RNA ligands to bacteriophage - T4 DNA-polymerase, Science, 249, 505-510.
127. Robertson, D.L.; Joyce, G.F. 1990, Selection in vitro of an RNA enzyme that specifically cleaves single-stranded DNA, Nature, 344, 467-468.
128. Johnson, D.E. 2010, Probability's Nature and Nature's Probability (A call to scientific integrity). Booksurge Publishing: Charleston, S.C.
129. Ramakrishnan, N.; Bhalla, U.S. 2008, Memory Switches in Chemical Reaction Space, PLoS Computational Biology, 4, (7) e1000122.
130. Turing, A.M. 1936, On computable numbers, with an application to the *entscheidungs problem*, Proc. Roy. Soc. London Mathematical Society, 42, (Ser 2) 230-265 [correction in 43, 544-546].
131. von Neumann, J. 1950, Functional operators. Princeton University Press: Princeton.
132. von Neumann, J. 1956, The general and logical theory of automata. In The World of Mathematics Vol 4, Newman, J. R., Ed. Simon and Schuster: New YHork.
133. von Neumann, J.; Aspray, W.; Burks, A.W. 1987, Papers of John von Neumann on computing and computer theory. MIT Press ; Tomash Publishers: Cambridge, Mass.
134. von Neumann, J.; Churchland, P.M.; Churchland, P.S. 2000, The computer and the brain. 2nd ed., Yale University Press: New Haven, CT.
135. Wiener, N. 1948, Cybernetics. J. Wiley: New York.
136. Wiener, N. 1961, Cybernetics, its Control and Communication in the Animal and the Machine. 2 ed., MIT Press: Cambridge.
137. Shapiro, R. 1988, Prebiotic ribose synthesis: a critical analysis, Orig Life Evol Biosph, 18, (1-2) 71-85.
138. Costanzo, G.; Pino, S.; Ciciriello, F.; Di Mauro, E. 2009, RNA: Processing and Catalysis: Generation of Long RNA Chains in Water, J. Biol. Chem., 284, 33206-33216.
139. Ferris, J.P. 2002, Montmorillonite catalysis of 30-50 mer oligonucleotides: laboratory demonstration of potential steps in the origin of the RNA world, Origins of Life and Evolution of the Biosphere, 32, (4) 311-32.

140. Barbieri, M. 2006, Semantic Biology and the Mind-Body Problem: The Theory of the Conventional Mind, Biological Theory, 1, (4) 352-356.
141. Sehon, S.R. 2005, Teleological Realism: Mind, Agency and Explanation. MIT University Press: Cambridge.
142. Green, C. 2003, The Lost Cause: Causation and the Mind-Body Problem. Oxford: Oxford Forum.
143. Blakemore, S.J.; Decety, J. 2001, From the perception of action to the understanding of intention, Nat Rev Neurosci, 2, (8) 561-7.
144. Polanyi, M. 1968, Life's irreducible structure., Science, 160, (834) 1308-12.
145. Freeman, W.J. 1997, Three Centuries of Category Errors in Studies of the Neural Basis of Consciousness and Intentionality, Neural Netw, 10, (7) 1175-1183.
146. Rosen, R. 1993, Drawing the boundary between subject and object: comments on the mind-brain problem, Theor Med, 14, (2) 89-100.
147. Rosen, R. 1995, The Mind-Brain Problem and the Physics of Reductionism. In Self-Reference in Cognitive and Biological Systems, Rocha, L., Ed. pp 29-44.
148. Schimmel, P. 2001, Mind over matter? I: philosophical aspects of the mind-brain problem, Aust N Z J Psychiatry, 35, (4) 481-7.
149. Schimmel, P. 2001, Mind over matter? II: implications for psychiatry, Aust N Z J Psychiatry, 35, (4) 488-94; discussion 495-7.

The First Gene, David L. Abel, Editor 2011, pp 117-134   ISBN: 978-0-9657988-9-1

5. Functional Sequence Complexity in Biopolymers

Kirk K. Durston & David K.Y. Chiu

Department of computer science, Bioinformatics
University of Guelph
50 Stone Road East, Guelph, ON, Canada, N1G 2W1

ABSTRACT. It is generally recognized that biopolymers such as DNA, RNA and proteins demonstrate a form of sequence complexity. Recent work has provided a more detailed insight into biopolymeric complexity by introducing three types of sequence complexity, Random Sequence Complexity (RSC), Ordered Sequence Complexity (OSC) and Functional Sequence Complexity (FSC). The primary feature of FSC that distinguishes it from RSC and OSC, is the imposition of functional controls upon the sequence. In this paper, we propose that it can be measured using an extended form of Shannon uncertainty that includes a variable of functionality. Clearly, FSC can be found in human languages and carefully designed computer code, but the measure we propose in this paper reveals that it is also found in biopolymers. In the case of proteins, the measure of FSC provides an estimate for the target size of a protein family in the amino acid sequence space, revealing that functional sequences occupy an extremely small fraction of sequence space. Due to the miniscule size of functional sequence space for a given protein family, as mutations accumulate there will be an increasing likelihood of moving the mutated sequence outside that space, with a corresponding deleterious effect on FSC.

Correspondence/Reprint request: Dr. Kirk Durston, Dept. of Computer Science, University of Guelph, 50 Stone Road East, Guelph, Ontario, Canada N1G 2W1   E-mail: kirkdurston@gmail.com

Introduction: sequence complexity in biopolymers

It has recently been pointed out that traditional notions of complexity are inadequate when applied to biosequences [1, 2]. For example, characterizing biosequence complexity in terms of algorithmic complexity fails to account for the redundancy found in numerous different sequences even when they have the same function [1]. Functional controls imposed upon a biological sequence are critical for maintaining specific functions of the sequence within the cell and, ultimately, for the existence of life. A more rigorous formulation for complexity in biosequences that incorporates functionality is therefore required. Abel and Trevors have defined three types of sequence complexity, only one of which accounts for functional controls imposed upon biosequences such as DNA, RNA and proteins. We will discuss these three types of complexity within the context of biopolymers, with a special focus on that form of sequence complexity that incorporates functionality.

1. Random sequence complexity

Abel and Trevors have defined Random Sequence Complexity (RSC) as *a linear string of stochastically linked units, the sequencing of which is dynamically inert, statistically unweighted, and is unchosen by agents; a random sequence of independent and equiprobable unit occurrence* [3]. Implicitly, four components contribute to RSC. First, the sequence is composed of sites, or loci. Second, there is the importance of the symbols that could occupy each site in the sequence. Third, there is a complete absence of constraints and controls on these symbols, statistically making all options equiprobable. Finally, the value of the symbol at each site must be independent of the values at any other site, such that no site is constrained by any other site in the sequence. An example of RSC can be found in atactic polystyrene, where the orientation of the side chains at each site appears to be completely unconstrained. In summary, if no agent or law of nature controls or constrains the outcomes of any site in a sequence, then they are presumed to be equiprobable, and the complexity of the sequence is characterized as RSC.

2. Ordered sequence complexity

Ordered Sequence Complexity (OSC) is defined as *a linear string of linked units, the sequencing of which is patterned either by the natural regularities described by physical laws (necessity) or by statistically weighted means (e.g., unequal availability of units), but which is not patterned by deliberate choice contingency (agency)* [3]. Examples of OSC are repeating patterns arising out of chaotic interactions or a string of repeating alphabet characters such as TGTGTGTGTGTG ... In nature, OSC is presumed to occur when laws of nature impose such tight constraints that there is no possibility of variation. In this case, repeatable, highly constrained sequences are produced that cannot, therefore, incorporate new functional inputs as functional information. An example of OSC is the highly ordered and repeating sequence obtained through the formation of polyadenosine absorbed onto the surface of montmorillonite clay [4].

3. Functional sequence complexity

Given the limitations discussed above, neither RSC, OSC, nor a combination of the two, are capable of producing significant levels of FSC since neither, by definition, are controlled by functionality [5]. Szostak [1] has further pointed out that, traditionally, neither algorithmic complexity [6] nor Shannon's measure of uncertainty [7] is adequate for biopolymers. Functional Sequence Complexity (FSC) is therefore defined as *a linear, digital, cybernetic string of symbols representing syntactic, semantic and pragmatic prescription; each successive symbol in the string is a representation of a decision-node configurable switch-setting---a specific selection for function* [3]. Volitional agency (control) is implicitly required to properly set each configurable-switch-position symbol to achieve functionality. Examples of FSC are said to occur in well-designed computer code and, naturally, in human languages. For biopolymers, functionality can be a result of structural requirements of protein families [17], cellular processes, or specific biochemical reactions [8]. Furthermore, biological functions can be nested in a hierarchical manner from the sub-molecular domain structure necessary for the 3D structure of an enzyme, all the way up to the global function of entire species of organisms. Comparing the differences between OSC and RSC on the one hand, and FSC on the other, it is the requirement of functionality that is the distinguishing feature between them.

Recent advances in the synthesis of RNA chains in water are encouraging so far as providing a storage medium for prescriptive information and FSC [9]. However, the much greater challenge of encoding FSC within RNA remains. If a RNA sequence is highly ordered, it will tend toward OSC. If the highly ordered sequence can mutate, it will tend toward RSC over time. To become functional, controls will be required to properly configure each switch-setting (nucleotide) to select for function.

4. Measuring FSC

The definition of FSC supplied above is essentially a definition of functional information. Shannon uncertainty is well known as a method to measure variability in data and as a measurement of information. Unfortunately, Shannon uncertainty makes no distinction between functional and nonfunctional variability and complexity, as Szostak has pointed out [1]. Hazen, Szostak *et al.* have advanced an equation for the measurement of functional information as follows:

$$I(E_x) = - \log_2[M(E_x)/W] \tag{1}$$

where E_x is the degree of function x (a measure of a sequence's functionality with regard to function x), $M(E_x)$ is the number of different sequences or configurations that meet or exceed E_x, and W is the total number of possible sequences or configurations. (Note that Hazen *et al.* use the notation N instead of W, but W is used here to be consistent with the notations and equations that follow.) Here we present an alternate method to measure functional information where an estimate of the probability distribution may be required. For example, in the case of a protein family the data may provide only the probability of each amino acid at each site, but not the total number of functional sequences. Our proposed method allows uncertainty to be managed when sequences with functionality are not exactly known. It also provides further analyses making use of the sequence distribution obtained [10].

Shannon uncertainty can be modified as a joint measure to analyze available data when the data is known to represent a particular function and is entered as input. This modified form of Shannon uncertainty we call *functional uncertainty* (H_f) [11] and is defined as:

$$H(X_f(t)) = - \sum P(X_f(t)) \log P(X_f(t)) \tag{2}$$

where X_f denotes the data specified with known functionality and t represents time. In the case of a protein family, the data X_f is in the form of a multiple sequence alignment specified by their family label when downloaded from Pfam [12]. When the dataset, composed of a multiple sequence alignment, is corrupted either by irrelevant sequences or irrelevant amino acids within an included sequence, there are methods such as 'noise cleaning' to address that problem. $P(X_f)$ is the *a posteriori* probability of the data with the given functionality F = f, or $P(X_f) = P(X \mid f)$. An explanation as to how this is calculated for proteins is given in section (5).

It may be useful to measure the change in $H(X_f)$ if certain mutations, insertions or deletions occur between time i and time j resulting in a loss, gain, or change in function. For this reason the time variable t is included in Eqn (2). For example, for a protein family that shares a common 3D structure that performs a known, specific functionality task f, X_f represents the dataset X of known sequences that satisfy the functionality f. Changes in sequence due to mutations may introduce a change in the specified functionality between time i and time j. (We are currently developing methods to address the dataset when sub-functionality is considered, such as a portion of the dataset coding structural domains within the larger structure of a protein. The sub-functionality in this case could be the function of contributing a critical structural component within the larger 3D structure that defines the larger biological function. In this view, functionality can form a nested hierarchy, composed of lower levels of different functionalities contributing to higher levels of global functionality.

There are three states of uncertainty to consider when measuring functional complexity. The *ground state g* is the state of greatest uncertainty permitted by the constraints imposed by the physical system when no biological function is required or present. Since the physical system may impose constraints on what type of sequences are permitted, it may be the case that not all sequences are equally probable. A special case of the ground state occurs when the physical system imposes no constraints on sequencing whatsoever, with the result that all possible sequences are equiprobable. This special case is classified as the *null state ø*. The third state is that which produces the function under investigation, denoted as the *functional state f*.

The measure of FSC, denoted as ζ, is the change in functional uncertainty between the ground state g and the functional state f, or

$$\zeta = \Delta H \left(X_g(t_i), X_f(t_j) \right) . \tag{3}$$

For proteins, the data suggests that actual dipeptide frequencies and single nucleotide frequencies for proteins are closer to random than ordered [13]. For this reason, the ground state g for biopolymers can be approximated by the null state $ø$. If we let the number of all possible sequences be represented by W and the length of each sequence by N and the number of options at each site in the sequence be denoted by m, then $W = m^N$. For example, for a 300 amino acid (aa) protein, if we assume 20 aa options per site, then $W = 20^{300}$. If the FSC if a single column within a multiple sequence alignment is being measured, then $N = 1$ and $W = m$. If the FSC of an interdependent cluster of sites is being measured, then N = the number of sites in the cluster. Since for the null state, all options are equally probable, $P(X_\emptyset(t_i)) = 1/W$ and

$$H(X_\emptyset(t_i)) = -\sum (1/W) \log (1/W) = \log W. \tag{4}$$

The measure of FSC, therefore, reduces to

$$\zeta = \log (W) - H(X_f(t_i)). \tag{5}$$

If one wishes to take into account the effect of the genetic code on the various *a priori* probabilities of generating the amino acids, then the probability of producing each amino acid given the genetic code can be used to compute a ground state that will be different from the null state, since all amino acids are no longer equiprobable.

With the exception discussed shortly, it is usually the case, in measuring FSC, that the variable t is constant, in which case

$$\zeta = \log (W) - H(X_f). \tag{6}$$

The value ζ is a measure of the FSC, or functional information, of any sequence, including biopolymers. As shown above, it is a measure of the change in uncertainty between the ground state and the functional state. This difference in uncertainty is closely related, as is Eqn. (2), to Shannon information and Shannon uncertainty respectively [14]. However, as previously noted, Shannon information is not concerned with function directly. FSC, on the other hand, is inseparable from function and can be regarded, therefore, as a measure of *functional information*, a necessary concept in biology [1, 2]. Since ζ is a measure of functional information, once ζ is known, it can be substituted for $I(E_x)$ in Eqn. (1) and an estimate for the total number of functional sequences $M(E_x)$ can be calculated. Also, the probability of finding a functional sequence in a single search can be estimated by solving Eqn (1) for $M(E_x)/W$.

Change in FSC can be used as a method to quantify evolutionary distance. The change can be between an existing or non-existing function f_a to a modified function f_b between time t_i and t_j described by

$$\Delta \zeta = \Delta H (X_{fa}(t_i), X_{fb}(t_j)). \tag{7}$$

The sequences corresponding to X_{fa} with initial function f_a have two components relative to that of X_{fb} (with resulting mutated function f_b). The *static component* is that portion of the sequence that must remain within the permitted sequence variation of the original biosequence with function f_a. The *mutating component* is the portion of X_{fa} that must change to achieve either the new function f_b, where the new function is to be understood as either a new level of efficiency for the existing function, or a novel function different from f_a. The mutating component can be assumed to be in the null state relative to the modified function f_b. To clarify, the mutating component, at the outset, is non-functional with respect to the novel function, so the probability of any particular amino acid at a site can be assumed to be equal to the probability of any other amino acid at that site. Since the mutating component is the only part that must change, the static component can be ignored *provided the probability of it remaining static* is included between t_i and t_j. The static probability would be assessed on the basis of the total number of mutations required for the mutating component to achieve functionality and the probability that none of those mutations occur in the static portion. There may be other factors as well in the computation of the static probability, which may also require inclusion in the calculation of the static probability.

5. Application of FSC to protein sequences.

One application of FSC is to protein families and protein structural domains. Measuring the FSC of a protein family can quantify the target size in sequence space for that family or structural domain which, itself, quantifies the degree of difficulty in locating any sequence at all that falls within that target area defined by the same 3D structure or function.

A measure of the lower bound for the FSC of a protein is to assume that each site is independent of all other sites in the sequence. This will yield an artificially low estimate as discussed shortly, and therefore gives a lower bound. First, a sequence alignment for the protein family or domain being investigated can be downloaded from a web database such as Pfam [12]. It is assumed that the data contains functional sequences, including neutral but functional mutations, with the non-functional sequences filtered out by natural selection. The next step is to compute the functional uncertainty of each site in the sequence. This is done by first calculating the probability of each amino acid occurring at each site. For example, if there are 1000 sequences in the alignment, and proline occurs a total of 235 times at a particular site, then the probability of proline occurring at that site is .235. This is done for each of the 20 commonly occurring amino acids. The functional uncertainty of that site is then computed using Eqn. (2) inputting the 20 amino acid probabilities for that site (ignoring chirality and non-biological amino acids). The functional uncertainty of the entire sequence is obtained by [15] summing all the values obtained for the functional uncertainty of each site in the sequence. The FSC of the protein is then computed using Eqn. (6).

It is much more likely to be the case, for most proteins, that certain sites within the amino acid sequence are associated with other sites in the same sequence, forming 2nd, 3rd and 4th order associations containing one or more amino acid patterns [16, 17], where a 2nd order association is an association between two sites, a 3rd order cluster is an association between three sites, and so on. These associations can be detected through various pattern discovery methods [18, 19, 20, 21]. Measure of FSC becomes more accurate when the sequence of individual sites is transformed into a sequence of individual site clusters. Within each site cluster, there may be one or more amino acid patterns. The transformation consists of replacing the sequence of sites with a series of non-overlapping site clusters. The functional uncertainty of each site cluster is obtained by observing the *a posteriori* probability from the data of each amino acid pattern within the site cluster. To clarify, Eqn. (2) is applied to each amino acid at one site with respect to what amino acids are associated with it at the other sites in the cluster. The analysis, therefore, runs horizontal across a two dimensional array containing the multiple sequence alignment, where each row represents a different functional sequence and each column represents an aligned site in the sequences representing a protein family. The functional uncertainty of that site cluster is then computed using Eqn. (2), inputting the probabilities of each of the observed amino acid patterns for each site cluster. Next, the functional complexity of each cluster must be computed, where the null state permits any possible amino acid pattern. For example, for a 4th order site cluster, there are a total of $W = 20^4$ possible patterns of amino acids. The functional uncertainty of the null state will depend upon the order of the cluster. The FSC of the site cluster is computed using Eqn. (6), but the variable W represents the total number of possible amino acid patterns, rather than the total number of possible sequences. The total FSC of the protein is then the sum of the individual FSC values for each site cluster within the sequence of sites. In summary, the primary difference between assuming site independence and site inter-dependence is that FSC is computed using probabilities of individual amino acids at individual sites in the case of site independence, and using probabilities of individual patterns of amino acids within clusters of interdependent sites in the case of site inter-dependence. In both cases, equation (2) is used but the unit of data X_f changes depending upon whether individual amino acids at individual sites are the focus (assuming site independence) or individual amino acid patterns within individual site clusters are the focus (assuming site inter-dependencies).

To illustrate the improvement in the accuracy of measuring FSC when site associations are taken into account, and to contrast the difference in measured FSC between site independence and site inter-dependence, consider a hypothetical 3rd order site cluster. Assume each site in the cluster contains all 20 amino acids and each amino acid is observed to appear an equal number of times in the 1000 sequence alignment. However, each amino acid in the first site in the cluster is uniquely associated with a specific amino acid in the other two sites. If we assume site independence, then since all 20 amino acids appear an equal number of times in all three sites, the site cluster appears to be in the null state and the FSC of the site cluster is 0, since there is no difference between the null state and the functional state in this particular case. The observed amino acid patterns, however, indicate that there are only a total of 20 aa patterns in the site cluster out of a total possible 20^3 patterns. Since each pattern occurs an equal number of times within the 1000-sequence alignment, the probability of each pattern is .05. Using Eqn. (2), the functional uncertainty of the site cluster is 1.30. The functional uncertainty of the null state is log (20^3) or 3.90. The FSC of the cluster, therefore, is 2.60, significantly higher than the lower bound of 0 in this hypothetical case. In reality, there may be fewer patterns per site cluster, some patterns may not be visible due to incomplete data, and the patterns are unlikely to occur with equal probability. Nevertheless, it should be clear as to the importance of considering interdependencies between sites when computing FSC. For the purpose of this paper, however, we shall assume the simplest case of site independence.

Using the site independent assumption and the method described above, the lower bound for the FSC of various proteins can be obtained, with results shown in Table 1. The proteins in Table 1 were chosen because all of them are universal proteins found throughout biological life. Additional results have been published by Durston et al. [22]. Results in Table (1) are slightly different from those published earlier due, primarily, to using the genetic code constraints as the ground state, rather than the null state as published earlier.

Protein Family	sites	Number of unique sequences in data	FSC (Fits)	Fits per site	Probability of locating a functional sequence in a single search for same-length sequence space	Estimate for upper limit of functional sequences $M(E_x)$
Ribosomal S12	122	1774	346	2.8	10^{-104}	10^{55}
Ribosomal S7	149	535	359	2.4	10^{-108}	10^{93}
Ribosomal S2	211	2469	465	2.2	10^{-140}	10^{135}
RecA	320	4301	976	3.0	10^{-294}	10^{122}

Table 1: FSC Results for Four Universal Protein Families using the Genetic Code frequencies as the ground state

Table (1) shows lower bound of FSC since we are ignoring any additional constraints imposed by other sites in the sequence. The probability of locating a functional sequence in a single search is derived from the FSC of the protein family. Once we have solved for ζ we can then solve for $M(E_x)/W$ using Eqn (1). This will be an upper probability limit due to the fact that we are assuming no interdependencies between sites. Site interdependencies will introduce additional constraints which will reduce the number of possible functional sequences, as illustrated earlier in this section (5). Thus, assuming site independence gives an upper limit for the number of functional sequences $M(E_x)$ and, therefore, an upper limit for the probability $M(E_x)/W$ of locating a functional sequence in a single search. It should also be noted that W is a lower limit since, as noted earlier, $W = m^n$ where, for proteins, $m = 20$ and $n = $ the length of the sequence, the total sequence space is radically reduced to just n-aa sequence space. Realistically, a search of sequence space is not limited to just the length of the sequences in the protein family being analyzed. Therefore, sequence space target size shown in Table 1 is only for the sequence space for the same length protein.

If all of amino acid sequence space is used for even just up to 300-aa sequence space, the probability of locating a functional sequence for a given protein family would be many orders of magnitude smaller, since W would be many orders of magnitude larger. In summary, if site independence is assumed, and given the artificially low value of W, the value of FSC calculated this way is artificially low and can safely be taken to be a lower bound. Similarly, the probability of locating a functional sequence within a protein family in sequence space is likely to be much smaller by numerous orders of magnitude. This, coupled with the results shown in Table 1, underscores the almost infinitesimal size of functional sequence space relative to the size of the entire sequence space for a given number of sites.

6. Relationship between RSC, OSC and FSC

Preliminary attempts have already been made to model the relationship between RSC, OSC and FSC [3, 22]. The following model improves upon those earlier attempts and is consistent with the method to measure FSC discussed earlier. It may not be the only way to model this relationship, but may provide a helpful model for the comparison of RSC, OSC and FSC.

Figure 1 models one approach to describing the relationship between the three types of sequence complexity, portrayed as a three dimensional coordinate system, with the X coordinate representing RSC, the Y coordinate representing OSC, and the Z coordinate representing FSC.

A very short repeating sequence would be an example of OSC and would be placed closer to the axis than a longer repeating sequence. Similarly, a short random sequence could be an example of RSC and would be placed closer to the axis than a longer random sequence. A sequence consisting of repeating random sequences would have components of both RSC and OSC and could be placed somewhere on the XY plane, as would any non-functional sequence that contained a mix of RSC and OSC.

For both OSC and RSC, the magnitude of their values is contingent upon the sequence length. This is not the case for FSC. From Eqn. (2), the magnitude of FSC is a function of the probability of finding a functional sequence in a single blind search, also a function of target size in sequence space, as has already been discussed. That probability is determined, in the simplest case, by the ratio of the number of sequences that will produce the function, $M(E_x)$, over the total number of sequence options W, both functional and non-functional, as used in Eqn. (1). This ratio also represents the target size, in sequence space, of the region that produces the function.

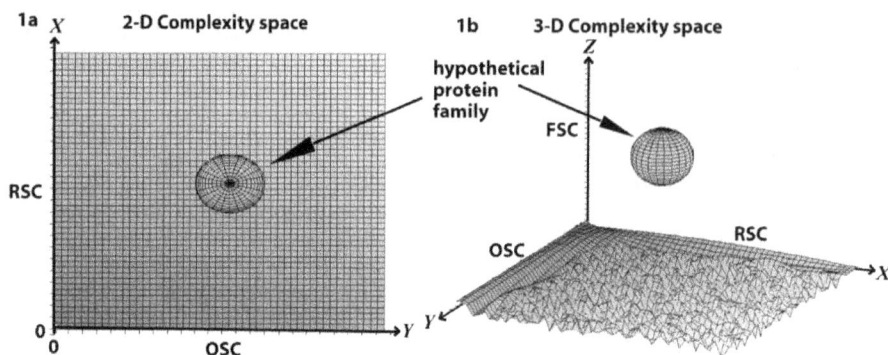

Figure 1: Relationship between RSC, OSC and FSC. In 1a, 2-D complexity space, composed of RSC and OSC, is inadequate to distinguish FSC from RSC and OSC. A third coordinate is necessary, representing the information required to achieve the function, which is a function of probability from Eqn. (1). In 1b, the uneven surface of the XY plane represents low-level, statistically insignificant FSC that can stochastically occur without any controls imposed on the generated sequences. The FSC of a hypothetical protein family can clearly be distinguished from RSC and OSC in this 3-D coordinate model of complexity space.

The location of FSC relative to the horizontal XY plane is plotted according to the combination of RSC and OSC within the sequences when the sequence, or set of sequences in the case of a protein family, is assumed to be non-functional. The size of the FSC of a protein family, along the Z-coordinate, allows for some variation of efficiency about the optimum value. From the examples in Table 1 and from Eqn. (2) and (6) it can be seen that the greater the FSC, the less probable a functional sequence becomes which, therefore, results in a greater quantum jump from the horizontal X-Y plane. This leads to the conundrum of how functional biopolymeric sequences such as protein families can be discovered in the overall sequence space when, necessarily, the higher the FSC, the less probable it becomes and, from Table 1, those probabilities are quite miniscule.

7. Biological FSC is unique in nature

Although FSC is found within human languages and computer code, the only known occurrence in nature occurs within biopolymers such as DNA, RNA and proteins [3]. To appreciate the oddity of FSC in biopolymers, it is essential to understand the distinction between constraints and controls. *Constraints* are imposed upon any physical system found in nature by the deterministic laws of nature, relevant initial conditions, and probabilistic limitations to possible outcomes [23]. The deterministic outcome of constraints is often referred to as *necessity*. Deterministic necessity steers a process toward repeatability and order. Thus, sequences determined by the constraints thus described will be highly ordered, demonstrating high OSC. The relaxation of constraints, or necessity, permits greater degrees of freedom in the outcome and is often referred to as *chance*. Chance, itself, is not causal but is permitted by the relaxation of the constraints imposed by the laws of nature, initial conditions, and probabilistic boundaries [24]. Chance, or the relaxation of necessity, in sequence formation, permits RSC. Thus, natural processes can be summarized by chance and necessity, the interplay of which produces OSC and RSC. Natural processes, therefore, are known to be limited to producing effects that lay in the XY/OSC-RSC plane in Figure 1. These effects determine the ground state of any sequence, as described earlier. FSC, however, requires a deviation from the ground state to achieve a particular function, as shown in Eqn. (3). To clarify, FSC requires a deviation from the normal results of natural processes operating in the XY plane (Figure 1). Small deviations from the ground state are statistically possible, such that low-level FSC can be achieved for very low-level functions, but the greater the FSC value ζ required by a particular function, the greater the deviation from the ground state. FSC, therefore, represents an *anomaly* within nature, a deviation from mere chance and necessity. To clarify, FSC represents something that one would not predict given the natural processes of OSC and RSC; it is improbable and the higher the FSC, the more improbable it becomes as shown experimentally (column 6, Table 1). Biological proteins require very high levels of FSC, as illustrated in Table 1. For this reason, chance and necessity define the ground state of a sequence, but are inadequate to produce the kind of cybernetic algorithms found within the genomes of life [25].

The cybernetic requirements of biological life, encoded within biopolymers such as DNA, require the addition of *controls* which can steer the physical system away from the ground state to produce the desired function [23]. Each step, or site, in a sequence represents a decision node that determines the course of physicodynamic events in such a way that the physical system can be steered in the direction of the desired function. Computer code, for example, is an illustration of FSC in the encoding of prescriptive information, using a physical medium, to steer the physical system (a computer) such that a desired function is achieved. FSC is only known to be achieved by encoding formal choices into a physical information storage medium, such as DNA through a series of volitional decisions at each configurable switch symbol. The large deviation from the ground state determined by constraints is an indicator that it has been achieved through the application of controls that arise out of choice contingency or volition. It is easy to understand why human languages and de- signed computer code exhibit such a high degree of FSC, since human intelli- gence provides the necessary volitional agency required to select the proper switch configurations. The high level FSC observed in biopolymers, however, successfully locates miniscule areas of sequence space, as shown in column 6 of Table 1, requiring very tight controls. The results are high-FSC sequences representing massive deviations from the ground state, deviations from the normal OSC-RSC of natural processes. These examples of FSC, therefore, rep- resent something that is unique in nature. In other words, FSC breaks out of the XY plane in Figure 1 and into the Z dimension, deviating from natural pro- cesses of chance and necessity that produce OSC and RSC.

8. The effect of mutations on FSC

The set of all sequences $M(E_x)$ that can perform a given function forms the functional sequence space within the larger sequence space. For example, Table 1 shows that the universal protein RecA has an estimated number of 10^{122} functional sequences within the larger 320-site sequence space, assuming site independence. Site interdependency will substantially reduce this number, as discussed earlier, but site independence will be assumed here for the sake of simplicity. These 10^{122} RecA sequences form the functional sequence space for RecA. Mutations that merely move the sequence from one area of functional sequence space to another area of functional space will have little effect on the FSC of the sequence. As discussed in Section 5, FSC is not measured on the basis of a single sequence, but on a large set of functional sequences. Some sequences, however, may be more efficient than others, as pointed out by Ha- zen et al. [2]. If that is the case, then there can be some variation in the value ζ for the FSC of a protein family. This is represented in Figure 1 as a range of Z- values in the FSC of the hypothetical protein family shown in the model.

Therefore, mutations within functional sequence space may reduce FSC as they become functionally less efficient until the mutations reduce the functionality of the sequence below the threshold required by biological life [2]. At that point, any further change can be measured using Eqn. (7) relative to the original function.

With regard to a protein with function f_a evolving into a completely novel structure with novel function f_b, once a sequence has mutated out of f_a sequence space, the evolutionary path may have to traverse a region of non-folding, non-functional sequence space. As pointed out by Blanco, natural selection is of no use in navigating non-functional sequence space, so further mutations will be unguided and take the form of a random walk [26]. Given the results shown in Table 1, and given experimental evidence [27], the area of functional sequence space for stable, folding functional proteins may be so miniscule, that attempting to locate them without the aid of controls may exceed an objective universal plausibility cut-off [28]. Mutations that occur within functional sequence space can move a functional sequence to an area within functional sequence space that may enhance fitness. However, given the miniscule size of functional sequence space, as suggested by Table 1, an obvious prediction is that an accumulation of mutations will tend to be harmful.

Since natural constraints tend to produce repeatable results, non-random mutations imposed by natural constraints will tend to move the sequence in the direction of order, or OSC. As shown in Figure 1, this constitutes movement in the XY plane which can move the sequence outside the area of functionality shown in Figure 1b. Thus, as can be seen in Figure 1b, both a limited amount of random and non-random mutations can be permitted provided the sequence remains within the functional area. But both random and non-random mutations can render the sequence non-functional, collapsing the Z component to zero, if the mutations move the sequences outside of the small functional sphere representing a hypothetical protein family.

9. Conclusion

FSC can be measured by extending Shannon uncertainty to include the joint variables of data and function. This measure can provide an estimate of the variability and hence the size of the functional sequence space for a specific functional protein. It also can measure change in a sequence due to mutation relative to the required functionality. The information calculated from an observed sequence ensemble constrained by the specified functionality then reflects the underlying sub-molecular information structure that could be used to reconstruct the structural or functional properties of the molecule. FSC thus provides a foundational measure that can form the basis for more detailed analysis.

References

1. Szostak, J.W. 2003, Functional information: Molecular messages, Nature, 423, (6941) 689.
2. Hazen, R.M.; Griffin, P.L.; Carothers, J.M.; Szostak, J.W. 2007, Functional information and the emergence of biocomplexity, Proc Natl Acad Sci U S A, 104 Suppl 1, 8574-81.
3. Abel, D.L.; Trevors, J.T. 2005, Three subsets of sequence complexity and their relevance to biopolymeric information, Theor Biol Med Model, 2, 29.
4. Ferris, J.P. 2002, Montmorillonite catalysis of 30-50 mer oligonucleotides: laboratory demonstration of potential steps in the origin of the RNA world, Orig Life Evol Biosph, 32, (4) 311-32.
5. Abel, D.L.; Trevors, J.T. 2006, Self-Organization vs. Self-Ordering events in life-origin models., Physics of Life Reviews, 3, 211-228.
6. Gammerman, A.; Vovk, V. 1999, Kolmogorov complextiy: sources, theory and applications, The Computer Journal, 42, 252-255.
7. Shannon, C. 1948, Part I and II: A mathematical theory of communication, The Bell System Technical Journal, XXVII, 379-423.
8. Karp, P. 2000, An ontology for biological function based on molecular interactions, Bioinformatics Ontology, 16, (3) 269-285.
9. Costanzo, G.; Pino, S.; Ciciriello, F.; Di Mauro, E. 2009, Generation of long RNA chains in water, J Biol Chem, 284, (48) 33206-16.
10. Durston, K.K. 2010, Statistical analyses of site variability and site inter-dependencies in sub-molecular hierarchical protein structuring. University of Guelph, Guelph.
11. Durston, K.K.; Chiu, D.K.Y. 2005, A functional entropy model for biological sequences, Dynamics of Continuous, Discrete & Impulsive Systems, Series B.
12. Finn, R.D.; Tate, J.; Mistry, J.; Coggill, P.C.; Sammut, S.J.; Hotz, H.R.; Ceric, G.; Forslund, K.; Eddy, S.R.; Sonnhammer, E.L.; Bateman, A. 2008, The Pfam protein families database, Nucleic Acids Res, 36, (Database issue) D281-8.
13. Weiss, O.; Jimenez-Montano, M.A.; Herzel, H. 2000, Information content of protein sequences, J Theor Biol, 206, (3) 379-86.
14. Schneider, T.D. 2006, Claude Shannon: biologist. The founder of information theory used biology to formulate the channel capacity, IEEE Eng Med Biol Mag, 25, (1) 30-3.
15. Wong, A.K.; Liu, T.S.; Wang, C.C. 1976, Statistical analysis of residue variability in cytochrome c, J Mol Biol, 102, (2) 287-95.
16. Lui, T.W.H.; Chiu, D.K.Y. 2009, Multi-value association patterns and data mining. In Foundations of Compuational Intelligence, Abraham, A.; Hassanien, A. E.; de Carvalho, A. P.Snael, V., Eds. Springer-Verlag: Vol. 6: Data Mining.
17. Lui, T.W.H.; Chiu, D.K.Y. 2010, Associative classification using patterns from nested granules, International Journal of Granular Computing, Rough Sets and Intelligent Systems, 1, (4) 393-406.
18. Wong, A.K.C.; Chiu, D.K.Y.; Huang, W. 2001, A discrete-valued clustering algorithm with applications to biomolecular data, Information Sciences, 139, 97-112.
19. Au, W.H.; Chan, K.C.; Wong, A.K.; Wang, Y. 2005, Attribute clustering for grouping, selection, and classification of gene expression data, IEEE/ACM Trans Comput Biol Bioinform, 2, (2) 83-101.
20. Chiu, D.K.; Wang, Y. 2006, Multipattern consensus regions in multiple aligned protein sequences and their segmentation, EURASIP J Bioinform Syst Biol, 35809.
21. Chiu, D.K.Y.; Lui, T.W.H. 2002, Integrated use of multiple interdependent patterns for biomolecular sequence analysis, International Journal of Fuzzy Systems, 4, (3) 766-775.
22. Durston, K.K.; Chiu, D.K.; Abel, D.L.; Trevors, J.T. 2007, Measuring the functional sequence complexity of proteins, Theor Biol Med Model, 4, 47.
23. Abel, D.L. 2010, Constraints vs Controls, The Open Cybernetics & Systemics Journal, 4, 14-17.
24. Pearle, J. 2000, Causation. Cambridge University Press: Cambridge.
25. Trevors, J.T.; Abel, D.L. 2004, Chance and necessity do not explain the origin of life, Cell Biol Int, 28, (11) 729-39.
26. Blanco, F.J.; Angrand, I.; Serrano, L. 1999, Exploring the conformational properties of the sequence space between two proteins with different folds: an experimental study, J Mol Biol, 285, (2) 741-53.

27.   Axe, D.D. 2004, Estimating the prevalence of protein sequences adopting functional enzyme folds, J Mol Biol, 341, (5) 1295-315.

28.   Abel, D.L. 2009, The Universal Plausibility Metric (UPM) & Principle (UPP), Theor Biol Med Model, 6, 27.

The First Gene, David L. Abel, Editor 2011, pp 135-160 ISBN: 978-0-9657988-9-1

6. Linear Digital Material Symbol Systems (MSS)[*]

David L. Abel

Department of ProtoBioCybernetics/ProtoBioSemiotics
Director, The Gene Emergence Project
The Origin-of-Life Science Foundation, Inc.
113 Hedgewood Dr. Greenbelt, MD 20770-1610 USA

Abstract. Nonphysical, formal, linear digital symbol systems can be instantiated into physicality using physical symbol vehicles (tokens) in Material Symbol Systems (MSS). Genetics and genomics employ a MSS, not a two-dimensional pictorial "blueprint." Highly functional molecular biological MSS's existed prior to human consciousness in tens of millions of species. Genetic code is conceptually ideal. Not all signals are messages. Encoding employs a conversionary algorithm to represent choices using a symbol system. Encoding/decoding is formal, not physicodynamic. Symbols must be purposefully chosen from alphabets of symbols to generate meaning, instructions, and control. Formal rules must first be generated, and then both sender and receiver must voluntarily adhere to those arbitrary rules. Neither law nor random variation of duplications can generate a meaningful/functional MSS. All known life is cybernetic (controlled, not just constrained) and semiotic (message dependent). Even protocells would require controls, biosemiosis, regulation, and an extraordinary degree of organization that mere mass/energy interactions, or chance and necessity, cannot produce.

Correspondence/Reprint request: Dr. David L. Abel, Department of ProtoBioCybernetics/ProtoBioSemiotics, The Origin-of-Life Science Foundation, Inc., 113 Hedgewood Dr. Greenbelt, MD 20770-1610 USA E-mail: life@us.net

*Sections from previously published peer-reviewed science journal papers [1-9] have been incorporated with permission into this chapter:

Introduction: Linear digital sign/symbol/token systems

Linear refers to a uni-dimensional, sequential string of representational command characters. The simplest computer programming, for example, is directed by such a linear digital string of purposeful binary choice commands represented by either a "1" or a "0." The sequencing or syntax of these choice-contingent commands provides a growing hierarchy of computational functionality.

Of course, computer programming is about creating semantic constructs that can be translated/compiled to run on a given computational system. The system could just as easily be quaternary rather than binary, as is the case with DNA base 4 prescription of biofunction.

Digital means each unit is discrete and definite. Programming choices have an "excluded middle." Switches must be turned either on or off. There is no in-between. A definiteness and clarity exists with each chosen command. No gray zone exists. Each selection is black or white.

As explained in Chapter 4, section 8, a bona fide "system" is an abstract, conceptual organization generated by choice contingency, not chance or necessity, that typically generates formal processes or procedures with pragmatic results. A "weather system" is not a true system. It is merely a physicodynamic interface of wind, temperature and atmospheric pressure differential. A weather front may involve phase changes and manifest self-ordering; but it is not organized. It manifests no choice contingency, no purposes or goals, no accomplishment of function or utility. Weather fronts have no formal components, no computational achievements, and no algorithmic optimization, no intended purpose given materialistic presuppositions.

The term MSS for Material Symbol Systems was first used by Rocha in his Ph.D. thesis [10, 11]. Recorded signs, symbols and tokens outside of human minds are representational physical entities called "physical symbol vehicles." Any system of communication using these physical symbol vehicles is a material symbol system. But how can a physical symbol vehicle, or group of such physical symbol vehicles in a MSS, *represent* instructions in a purely materialistic world? The Mind-Body problem is closely related to the symbol-matter problem. The symbol-matter problem is known in philosophy as the "problem of reference." How do symbols form to "stand for" or represent material structures? [12-14, 15, pg. 11, 16]

These problems are, in turn, closely related to the measurement problem not only in quantum physics, but in Newtonian physics as well. As physicist Howard Pattee has pointed out in many publications, the measurements of initial conditions used in the laws of physics are formal representations (mathematical symbols) of physicality, not physicality itself [15].

The first problem encountered by semiotics in a MSS is the *nature* of symbols themselves. Charles Sanders Peirce proposed the triadic interpretation of signs. Meaning is created mentally through consideration of recursive relationships [17]. Peirce's interpretation of signs and semiotics involving representamen, object and interpretant is inseparable from human cognition and agency. Representations are necessarily abstract, conceptual and formal. Peirce's triadic relation work incorporates abundant human psychological and epistemological components. Under no circumstances are such representations "natural" (purely physicodynamic). Representations are never physical. Representations can be arbitrarily assigned to physical tokens in a MSS. But the representations assigned to those physical tokens are always agent-chosen according to formal rules, not physical laws.

No justification exists for trying to circumvent the fact of "volition" using Peirce's category of thirdness (mere "habit formation") [18]. Habits are nothing more than redundant patterns of volitional social behavior. If a pattern does not originate out of true behavior—volitional tendencies—then that pattern is simply reflective of physicodynamic necessity. Such cause-and-effect determinism is ordered by the regularities of nature described by physical law as further refined by statistical distribution curves. Patterns caused by physicodynamic necessity have no formal prescriptive significance and produce no sophisticated utility.

Semiotic and cybernetic functions both employ formal symbolization according to previously agreed-upon, arbitrary rules (not physical laws) in order to convey meaning. Neither cybernetics nor semiosis can be reduced to the mere physicality of its switches or physical symbol vehicles. The uncoerced choice contingency that selects those symbols or that sets those configurable switches is the key.

To ascribe semantic value to physical entities requires both *contingency* and *volition*. Neither necessity (forced, law-like, cause-and-effect determinism) nor chance contingency can generate meaning. Choice contingency is required [1, 4, 6]. Semantics entails "aboutness." Aboutness and meaning are absent from the category of inanimate physicodynamic interactions.

Rosen [19] regarded sign systems as "anticipatory." He argued that conventional physicodynamic theory cannot possibly model a sign system's descriptive behavior. But the problem extends far beyond having to explain the phenomenon of description. Far more important is the function of symbol systems to *prescribe*—to indicate *determinative choices and controls* that will be efficacious in producing utility *in the future* [6]. Undirected natural selection cannot *select for* not-yet-existent function (The GS Principle [5, 20]). Sophisticated utility only comes into existence via integrated pre-programmed decision nodes, logic gates, and configurable switch-settings. Choice contingency's unique ability to generate pragmatic controls alone accomplishes this.

1. What's the difference between signs, symbols and tokens?

Attempts have been made within the semiotic community to clarify the difference between signs and symbols, the most specific and recent being [21]. In this author's opinion, the latter paper only confuses the distinction rather than clarifying it.

A *sign* is typically a two-dimensional picture or drawing conveying representational meaning to one's senses. The picture or drawing is self-explanatory because we recognize by sight what is being depicted from our every-day empirical world. A visual image of real world objects is delivered by the sign. Our consciousness links the two-dimensional picture with our experience of and with that object. See Figure 1.

A *symbol*, on the other hand, is an arbitrarily-shaped/generated character representing some assigned meaning by definition. A symbol, unlike a sign, conjures no meaning from one's sight memory of physical objects. The letters of most language alphabets are not signs, but symbols. Strings of such symbol characters spell words leading to lexicons of words. Hierarchies of phrases, clauses, sentences, and paragraphs can be constructed from the lexicon of words according to syntactical rules. Sometimes only one letter symbol, such as "H" or "C" on a faucet handle, conveys meaning.

Mathematical symbols such as π, Ω, ξ, Δ, $=$, and $\neq$ are symbols, not signs. We cannot ascertain the meaning of these symbols from the symbol itself, except that we sometimes become so familiar with a certain symbol that it begins to take on a function similar to a picture or drawing, thereby having a sign-effect from our sight memory (e.g., the symbol " $=$ " begins to be recognized visually as the a sign of equality). Such symbols are not pictures or drawings of real-world physical objects that we have previously observed. The meaning of these arbitrary "strokes of pen" is just assigned and agreed to by source and recipient. Otherwise, the message will not have meaning or function at its destination.

Figure 1. An example of a common *sign* (drawing of a cigarette) with over-laid *symbol* (an abstract representation of "No!") conveying the composite meaning of "No smoking."

In Figure 1, the drawing of the cigarette is a sign. The universally under-stood slash through a picture is, by convention, a symbol representing "No!" to whatever is being pictured by the sign beneath the symbol. The cigarette is a physical object. The symbol meaning "No" does not depict a physical object, but a formal abstract concept of prohibition.

Semaphore is technically a symbol system, not a sign system. Each flag position abstractly represents a letter of the alphabet, although our minds quickly begin associating each flag position symbol with a mental picture of the letter represented. The abstract symbols thus begin to function as iconic signs in our minds.

No signs exist within cells. Molecular biology does not create pictures, drawings, or blueprints. But interestingly enough, representational symbols *do* exist within cells. As we shall see below, physical symbol vehicles and mate-rial symbol systems are undeniably employed by living cells. Representation-alism is a formal function, not a physicodynamic interaction. This fact goes a long way toward addressing the age-old question of whether life can be re-duced to inanimate physicality. Mass/energy interactions cannot generate con-ceptual representationalism. But, in molecular biology, we seem to be talking about *material* symbol systems. Aren't they obviously physical? The answer is "NO!" MSS's *use* physical symbol vehicles, to be sure. But their represen-tational and symbolic function is purely formal. The triplet codon table of molecular biology is purely formal. To understand this reality we must proceed from a discussion of signs and symbols to a discussion of "tokens."

We must clearly distinguish between *symbols* and *physical symbol vehicles (tokens)*. Physical symbol *vehicles* are physical. Symbols are not. Symbols are *conceptual representations of meaning*. The symbol π represents a formal mathematical idea in our minds when referenced in the domain of geometry, for example. We can instantiate this abstract symbol with its meaning into a physical symbol vehicle through handwriting π onto paper with physical ink, converting the mental idea into physical sound waves in conversation, typing it onto paper or into a word processor physical system. But the recordation and transmission of physical symbol vehicles does not change the fact that the symbols being represented are abstract ideas with arbitrarily assigned meaning. No physicodynamic constraints or causation can explain cognitive representationalism and symbolization.

Cybernetic function requires deliberate selection. First, the actual uncoerced and nonrandom selection must be made. Then, that choice must be formally represented using a mentally-derived symbol. Finally, that cognitive symbol can be instantiated into physicality by selecting a certain physical symbol vehicle (token) in a MSS. Alternatively, choice contingency can be instantiated into the setting of a physical configurable switch to achieve formal pragmatism. We can also choose how to arrange physical parts into a holistic functional device (e.g., a machine). A machine may be physical, but the organization of its physical parts to achieve nontrivial functional capacity is purely formal.

A *token* is merely the physical vehicle *of* a sign or symbol. A Scrabble piece is a token. It is a physical block of wood with a symbol drawn onto or etched into its surface. Such "physical symbol vehicles" are used in MSSs [10, 11] to spell meaningful words of a language, depict computations, or (in the case of sign tokens) to portray larger composite pictures (e.g., holograms). It is all-too-easy for us to forget that the meaning of a string or cluster of tokens has absolutely nothing to do with the physicality of those tokens. The physical tokens are just instantiations of formal arbitrary choices. Each token must be purposefully selected from a phase space—a pool of physical objects, each with an abstract symbol recorded on it. Only then can a MSS be generated to convey formal meaning or achieve formal function at its destination.

Like meaning, functionality is a formal concept, not a physicodynamic interaction. The ability of a machine to perform useful work is formal, not physical. If we redefine work to mean nothing more than heat exchange between two objects, then maybe work can be purely physicodynamic. But that is not the kind of work that makes reality interesting. Usefulness—

pragmatism—is what matters. Such functionality is an abstract formal concept.

Messages are entirely formal, not physical, even though they can be instantiated into a MSS of physical symbol vehicles. Smoke signals can be used to send a message. But that does not mean that the message itself is physical. The same is true of formal programming using physical configurable-switch and logic-gate settings. Language, mathematical computation and programming all traverse The Cybernetic Cut across the one-way-only CS Bridge from the abstract conceptual world into physical manifestations of formal organization [4](see Chapter 3). Thus instantiation of message *meaning* into a physical matrix of retention and transmission is still fundamentally nonphysical.

2. Blueprints vs. linear digital prescription

A blueprint is a two-dimensional drawing (pictorial representation, similar to a sign) of a potential physical construction such as a building or other functional structure (e.g., a bridge, automobile or airplane). Blueprints and schematic diagrams consists of a composite of many signs. A blueprint has little or nothing to do with linear digital symbol systems. The latter consist of a string of symbols, not signs, that conveys meaning and function according to arbitrarily agreed-upon rules (not physicodynamic laws).

It is grossly inaccurate to refer to genetics as a "genetic blueprint." Genetic prescription is non-pictorial. Although genomics is ultimately multi-dimensional to extraordinary degrees, it is first and foremost a uni-dimensional linear digital symbol system, not a two-dimensional drawing. All of the other dimensions of genomic Prescriptive Information (PI) follow in secondary, tertiary and quaternary layers of conceptual complexity that emerge from the initial primary structure (sequencing) of monomers (tokens) in a string. All 3'5' phosphodiester bonds between nucleotide monomeric tokens in nucleic acid molecules are the same. All peptide bonds between amino acid monomeric tokens in proteins are the same. The particular selections and sequencing of nucleotides in coding regions, and of the resulting amino acids, are physicodynamically inert or indeterminate—undetermined by physics and chemistry. The opportunity to arbitrarily select each coding nucleotide token to be polymerized next to an existing positive strand provides programming freedom to "spell" meaningful messages, program, compute and prescribe eventual specific three-dimensional molecular machines and catalysts.

3. Signals vs. messages?

Messages could be viewed as a special subset of signals. From this perspective, all messages would be considered signals. But not all signals are messages.

A message is a meaningful transmission intended to relay Functional Information (FI) from a source to a receiver and destination. Often the FI provided by a message provides specific Prescriptive Information (PI) rather than just Descriptive Information (DI). An example of a PI message would be the sending and receipt of instructions for how to accomplish some multi-step task. The successful receipt and utilization of the message requires prior agreement between source and recipient as to what arbitrary "language" (symbol system) will be used. Both parties must use the same symbols, syntactical and grammatical rules in order for the meaning and prescription to be efficacious.

A signal can be nothing more than a string of inanimately generated regular pulses. But such a string of impulses would contain near zero meaning or function. A pulsar sends out a signal. That signal contains almost no Shannon uncertainty.   Such redundant, constant-frequency, high-probability, low-bit-content pulses contain almost no potential for information instantiation and transmission. The entire signal can be reduced to an exceedingly-short compression algorithm (e.g., "Emit one electromagnetic pulse of near equal energy per nanosecond, repeat x times." Virtually no meaningful message could be instantiated into such a physicodynamically militated ("necessary") and "certain" physical "matrix" (photon stream). A pulsar signal is much too highly ordered to have significant FI instantiated into it.

To send a message requires freedom from physicodynamic determinacy. Uncertainty in the physical matrix is required for instantiation of language and programming controls. Necessity theoretically eliminates uncertainty. The sender of any message must have the ability to exercise choice determinism at each decision node. Symbols and physical tokens must be freely selectable from among multiple options. The sender must have full control over these *arbitrary* selections. By arbitrary we do not mean random. We simply mean "freedom from fixed law-like determinism and the ability to freely choose from among real options." The receiver must perceive the same freedom from forced order so as to attach any meaning to the symbol selections within the message.

In molecular biology, arbitrariness of selection is made possible by the fact that all monomeric bonds are the same despite varying monomers. A nagging problem for philosophic naturalism is that the physicodynamic inertness that generates the needed combinatorial uncertainty for PI instantiation is the same physicodynamic inertness that makes any physicodynamic explanation for code-origin impossible. If the selection of each particular nucleoside in a single positive strand is not determined by physics and chemistry, how is any meaningful message spelled? We are not talking about base-pairing replication here. We are talking about the origin of initial message meaning in the single positive informational strand.

4. Proper use of the word "code"

For our purposes, the word "Code" is a representational symbol system used to assign associations (e.g. via a codon table), or to convey meaningful/functional messages (e.g., messenger molecules). In an everyday connotation, coding signs and symbols are usually substituted for letters or words. Most codes (e.g., ASCII, Zip code) are "open," (non-encrypted) with arbitrary meaning to communicate between two otherwise independent worlds. The codon/amino acid code is the most widely known code in life, but more than 20 other semiotic codes have been discovered in the past decade, each with no known physicochemical "cause." [22-25].

A peer reviewer questioned whether the "code" produced by a digital voice encoder (vocoder) would fit the above definition. Since numbers can be used as symbols, and measurements and calculations can also be used as formal representations of physicality, the answer is yes. For, a vocoder comprises a built-in mathematical (formal) model of voice (physical sound waves), which includes parameters which vary instantaneously with the voice. These instantaneous parametric values are sampled at some regular interval and the values thus obtained are what actually get encoded and transmitted, thus allowing reconstitution of the voice at the receiving end. In this case, the encoded message is a formal representation of physicality using numerical symbols, measurements, calculations, and a set of formal governing rules, not fixed physicodynamic laws.

The word "code" can also be used to describe a conversionary algorithm that translates one symbol system into another. In this context, a code is a set of rules that governs bijection (substitution or mapping) of one symbol for another between two different symbol systems. Sometimes we can have a two-to-one or a three-to-one bijection (e.g., triplet codons) where multiple symbols in one symbol system prescribe only one symbol in another symbol system. This reduces the likelihood of noise pollution and error in translation. But any translative coding requires knowing and following formal algorithmic rules that relate one arbitrary symbol system to the other. Meaning and function in the message are retained despite a complete change of language symbols and rules.

In molecular biology, genetic code is specifically used for:

- instantiation of formal, immaterial programming choices into physicality
- efficiency in translation between two different material symbol systems where molecules serve as "physical symbol vehicles" (tokens) in two different material symbol system (MSS) rather than being mere physicochemical interactants/reactants
- ease-of-transmission
- noise pollution prevention in the Shannon channel (e.g., redundancy block coding)
- proof reading and error correction (e.g. the processing of parity bit coding to detect noise pollution)

Shannon's channel capacity theorem precludes a one-to-two or a one-to-three bijection [25-27]. This mathematical fact of reality immediately falsifies any code-origin theory suggesting the slow growth from simpler bijection rules into the current translative system summarized by the codon table [25-27]. Insufficient PI, and even insufficient Shannon uncertainty, is contained in the simpler 1:1 coding system to be able to map to a 1:2 or 1:3 system. In one sense, the sequence of nucleotides in a ribozyme is not itself a code. It is just a linear digital string of ribonucleotide tokens. That string could be random (a stochastic ensemble). But no nontrivial protometabolism, let alone metabolism, has ever been observed to arise from stochastic ensembles.

The word "code" is often used in molecular biology to refer to a specifically selected syntax of many monomeric tokens (nucleosides). A particular functional sequence, all with the required right-handed sugars and correct 3'5' phosphodiester bonds, can successfully prescribe the secondary and tertiary structure of a functionally-folded ribozyme. But, for a stochastic ensemble to produce a single ribozyme is statistically prohibitive without behind the scenes experimenter steering of supposedly-random Markov chains (drunken walks). Ribonucleotides are too difficult to make and activate even with extensive investigator involvement. They are also very unstable. Even the best human-engineered ribozymes have very limited function. Thousands of ribozymes might be needed to substitute for a few dozen protein molecules. Even then the sophistication of function is far less than what proteins catalysts accomplish. The simplest bacteria normally code for two to three thousand proteins. The requirements for any protocell consisting only of ribozymes to spontaneously organize and come to life yields a calculable Universal Plausibility Metric that far exceeds what is allowed by the Universal Plausibility Principle [28]. The latter Principle definitely falsifies any such spontaneous generation-based hypothetical scenario.

But in the sequencing of monomers required for functional ribozymal folding we discover an initial inherent coding feature. Linear digital representational sequencing is "translated" in a very unique way into functional secondary and tertiary physical folding. Thermodynamic tendencies are "used" by this inherent "code" to instruct three-dimensional function. New added dimensions of PI arise from the original primary structure (the linear digital sequence of ribonucleotide tokens). In that sense, the sequencing of ribonucleotides in RNA achieves a unique coding and translative status. The linear digital Prescriptive Information (PI) found in ribonucleotide sequencing makes use of the yet-to-be-realized physicodynamic properties of minimum Gibbs-free energy sinks to instruct the manufacture of the sophisticated molecular machines known as ribozymes and ribosomes (RNA-protein complexes). Highly integrated biological factories contribute to ever more hierarchically organized holistic metabolism.

In the case of DNA, when we introduce the reality of a constant grouping of three nucleotides to represent each amino acid prescription, we have introduced a noise-pollution-reducing Hamming redundancy "block code." The codon "table" is a translative map (a three-to-one bijection) used by the ribosome/tRNA/tRNA-aminoacyl synthetase translative system. The algorithmic processing performed by this system links the nucleotide linear digital prescription of meaning and function to another polyamino acid "language." This is a true "encoding" function where mRNA fragments (triplet codon syntax) through ribosomes cause a "request" of each tRNA molecule through the use of elongation factor protein sets. Human mentation did not devise this noise-reducing translative system. We merely discovered it. It cannot be reduced to human consciousness or epistemology. It prescribed not only us, but every other species before we arrived on the scene to investigate the reality of this coded prescription, bijection of codon to amino acid, and decoding of the formal genomic PI into potential physical metabolic achievements.

In both cases, RNA and DNA, Prescriptive Information (PI) is provided initially as a linear digital string of symbols each selected from an alphabet of symbols (four nucleotide options). The syntax of these symbols represents higher levels of meaning and function through inherent code and translation. The nucleotide (token) sequence must be selected in advance with rigid covalent bonds prior to translation into functional three-dimensional structures. It is incumbent upon materialists, physicalists, and naturalists to explain how chance and/or necessity could possibly have made these *functional selections* at the molecular/genetic level prior to the existence of any physical folding and phenotypic (organismal) fitness (The GS Principle [5, 20]). Phenotypic fitness is essential for any Darwinian progress to occur. Until such explanation is provided, philosophic naturalism finds itself in the very compromising position of being nothing more than unsubstantiated metaphysical dogma. Justification for its incorporation into the very definition of the science should be called into question. Chance and necessity cannot program (cannot traverse The Cybernetic Cut (Chapter 3) or decode nucleic acid programs into the alternate "language" of polyamino acid [protein] sequencing). See The GS Principle in Chapter 7 to understand why natural selection cannot possibly explain the linear digital PI of genetic/genomic/epigenomic programming and instruction.

No biopolymer could possibly function as a reliable "messenger molecule" without selection of functional base sequences that only later contribute to integrated formal function. Stochastic ensembles do not prescribe contributions to holistic metabolism. Random strings contain zero PI.

In the case of a game of Scrabble, even if stochastic ensembles of Scrabble tokens happenstantially *appear* to spell meaningful words, it would only be our minds ascribing meaningful sequence by association. The stochastic ensemble would still be a random string despite the *appearance* of a meaningful message. In addition, a formal decoding system would still have to be in place to interpret and translate the apparent functional string. Otherwise, that happenstantial string *resembling* PI would be unintelligible at the receiver and destination. The recipient would have to know and exercise the rules and algorithms of that language convention to derive any function from nucleotide sequencing. In short, the appearance of meaning in a random string does not provide meaning or function.

5. Semiotics vs. Cybernetics

Symbol systems allow representation, recordation and transmission of formal choices [29-34]. Symbols represent specific selections from among real options. These selections then become determinative of language and message meaning, of programming function and of computational success. All of these functions can be transmitted as instructions containing PI [6]. Choices of signs/symbols/tokens are always fundamentally cybernetic (controlling), even when only Descriptive Information (DI) is sent. Symbol selection is not made randomly or physicochemically [35]. Symbol selection, if it is to have sophisticated utility at the message's destination, must be made freely and deliberately by the source [2, 36].

Selection of the symbol "1" or "0" represents the simplest binary control decision. Each such purposeful choice is the fundamental unit of PI and instruction [6, 37]. If the "1" or the "0" is selected randomly, one bit of *potential control* is immediately lost. Or, if a single "1" or "0" is determined by prior cause-and-effect chains of physicodynamic necessity, one bit of potential control is also immediately lost. The ability to steer events through many decision nodes toward computational success quickly deteriorates and dies with each new denied binary control choice. When every "choice" is determined the same way by "necessity," the resulting "program" consists of all "1's" (OR all "0's"). Neither chance nor constraints can select the path with greatest function potential. Neither chance nor constraints can program or compute. Constraints exert their physicodynamic influence independent of formal pragmatic considerations. Controls, on the other hand, program pragmatic success at the foundational binary decision-node level of "Yes, No," "Open, Closed," "On, Off." Controls select each ideal configurable switch-setting prior to the realization of any function.

The biological scientific community often seems blind to the fact that selection *for potential* function is something that undirected natural selection cannot do [4-6, 37, 38]. Absolutely no selection pressure exists at the genetic/genomic programming level. The GS (Genetic Selection) Principle reigns at the level of nucleotide selection in forming positive informational strands of nucleic acid [5, 20].

To communicate a meaningful or functional message, first, we must arbitrarily assign an alphabet of usable symbols. Next, we must again arbitrarily assign meaning to letters or small groups of alphabetical characters, the equivalent of words. This is done according to arbitrarily defined rules, not constraints or laws. The rules are freely selectable, not constrained by physicodynamics. In short, symbol systems are entirely free, formal and cybernetic. Each choice of symbol represents a discrete unit of control.

The above cybernetic realities in no way deny, of course, that the execution of the running systems presupposes attention to the constraints of physicodynamics in order to function. Nothing about computer science, engineering or life entails the suspension of the laws of physics and chemistry. But the laws of physics and chemistry are grossly inadequate to explain the biocybernetics and biosemiosis that make life possible. Symbol systems, including MSS's, are fundamentally cybernetic [39-43], whether the information conveyed is descriptive or prescriptive, and whether or not the those formal decision-node choices are secondarily instantiated into physicality.

Cybernetic means steering and controlling. Signs/symbols/tokens must be purposefully chosen from a phase space or alphabet of tokens to generate any bona fide message. Message generation, therefore, requires traversing the one-way-only CS Bridge from the formalism side of The Cybernetic Cut to the physicality side in order to instantiate formal function into the physical world [4]. Sophisticated processes like language communication/ interpretation, programming and computation must be steered toward functional goals and away from non-functional dead-ends.

All applications of Decision Theory and Systems Theory require steering and control. The creation and refinement of algorithmic processes requires more than mere inanimate physicodynamic constraints. At the very least, *particular constraints must be deliberately chosen* and others rejected to steer a cause-and-effect chain towards formal pragmatic worth [6].

Choice contingency is always formal rather than physicodynamic. The biosemiotic research community must come to terms with the simple reality that chance and/or necessity cannot choose, steer or control. Chance and/or necessity, therefore, *cannot generate meaningful/functional symbol selection.* Semiosis, including biosemiosis, is impossible without choice-with-intent at bona fide decision nodes. This one fact alone should falsify any purely physicalistic notion of spontaneous generation of life. All known life is cybernetic and biosemiotic. Metabolism is steered and regulated by programming choices and linguistic like instructions found in linear digital genetic Prescriptive Information.

6. Could life exist without controls and messaging (biosemiosis)?

Myriad messages are continuously sent between components within a living cell as well as between living cells. In the larger field of biosemiotics, endosemiosis refers to message sending within the same organism. Exosemiosis refers to messages sent between organisms. But in life-origin science we encounter a special circumstance that requires alternate usage of these terms. The first organisms being investigated were unicellular. In researching the derivation of embryonic messages within the first protocell, we use the term endosemiosis to refer to messages sent within that first primordial cell. Exosemiosis then refers to messages sent between unicellular organisms leading to perhaps the first filamentous or colonial type relationships between primitive cells (e.g., cyanobacteria).

Every aspect of metabolism with a single cell depends upon programmed instructions, the messaging of those instructions, and feed-back messaging about how well the initial messages were received and carried out. Messages deliver the PI that controls and regulates metabolism. PI determines the amount and time for small RNA and protein manufacture at the right site within the cell. Availability of each needed metabolite must be communicated back to the system (feedback) to know when to start and stop production. Without this communication, cellular activity would not only quickly become chaotic, it would be incompatible with life. Spontaneous positive and negative feedback, however, must never be confused with formal controls. Feedback in non-cybernetic life is nothing more than circular constraint. More (positive) or less (negative) of the same can be constrained. But this constraint is never pragmatically adjustable apart from formal PI. It is merely forced physicodynamically without regard to function. Constraints in a prebiotic environment that is devoid of formal pursuits cannot generate feedback *controls*.

The only factor that allows metabolism to move "far from equilibrium" is organization, not inanimate constraints and redundant order. Crystals are highly ordered; crystals are not alive. Formal organization is the key to locally and temporarily circumventing the 2nd Law. But organization is impossible without programming, steering, and the transmission of controlling messages. Apart from organizational choices (e.g., Maxwell's demon choosing when to open and close the trap door), the 2nd Law rather than formal regulation would prevail within any theorized protocell. Without programming and the bio-semiosis of those instructions, no progress could be made within any micelle, vesicle or protocell toward eventual life in a true cell.

7. Do symbol systems exist outside of human minds?

Genetic instruction uses a formal linear digital MSS. We tend to assume that symbol systems exist only within human consciousness. Yet MSSs clearly exist apart from human consciousness at the molecular biological/genetic and genomic level. They pre-date human existence. No living organism, including such organisms as non-free-living Mycoplasmas, is known to exist that does not depend upon MSS programming. Human brains themselves are prescribed by molecular/genetic MSSs.

Genetics and genomics not only utilize a linear digital symbol system, but also an abstract Hamming block coding to reduce noise pollution in the Shannon channel (triplet codons prescribing each amino acid). Anti-codons are at opposite ends of t-RNA molecules from amino acids. The linking of each t-RNA with the correct amino acid depends entirely upon on a completely independent family of tRNA aminoacyl synthetase proteins. Each of these synthetases must be specifically prescribed by separate linear digital programming using themselves the same MSS and ribosome processing as they help prescribe for other proteins. These symbol and coding systems not only predate human existence, they *produced* humans along with their anthropocentric minds. The nucleotide and codon syntax of DNA linear digital prescription has no physicochemical explanation. All nucleotides are bound with the same rigid 3'5' phosphodiester bonds. The codon table is arbitrary and formal, not physicodynamically determined. The semantic/semiotic/bioengineering function required to make proteins requires dynamically inert configurable switch-settings and resortable physical symbol vehicles. Codon syntax communicates time-independent, non-physicodynamic "meaning" (prescription of biofunction). This meaning is realized only after abstract translation via a conceptual codon table. To insist that codon syntax only *appears* to represent amino acid sequence in our human minds is not logically tenable.

In "The Biosemiosis of Prescriptive Information"[6], we asked the question, "Exactly how do the sign/symbol/token systems of endo- and exo-biosemiosis differ from those of cognitive semiosis?" Do the biological messages that control, regulate, and integrate metabolism have conceptual meaning? "Meaning" almost invariably relates to achieving function. The purpose of messages is to convey useful information. That information can be descriptive (DI) or prescriptive (PI). Both are subsets of Functional Information (FI). What makes FI intuitive or semantic information is that it imparts pragmatic value to the recipient. We call this meaning. "Messenger molecules" impart such meaning and potential function to their targets within the cell or in neighboring cells. No fundamental difference exists between the use of MSSs within cells at the molecular biological level and the use of MSSs to convey language or cybernetic programming by human minds. According to Chomsky, human consciousness cannot even take credit for human innate language [44].

Metabolism employs primarily proteins. The nucleotide sequences in mRNA prescribe the amino acid sequences that determine protein identity. DNA is largely inert. It plays a minimal direct physicochemical role in protein binding, transport and catalysis. Molecular biology's two-dimensional complexity (secondary biopolymeric structure) and three-dimensional complexity (tertiary biopolymeric structure) are both ultimately determined by linear sequence complexity (primary structure; functional sequence complexity, FSC). The chaperone proteins that aid polyamino acid folding are also prescribed by the linear digital genetic programming instantiated into DNA sequencing.

Figure 2 shows the prescriptive coding of a section of DNA. Each letter represents a choice from an alphabet of four options. The particular sequencing of letter choices prescribes the sequence of triplet codons and ultimately the translated sequencing of amino acid building blocks into protein strings. The sequencing of amino acid monomers (basically the sequencing of their R groups) determines minimum Gibbs-free-energy folding into secondary and tertiary protein structure. It is this three-dimensional structure that provides "lock-and-key" binding fits, catalysis, and other molecular machine formal functions. The sequencing of nucleotides in DNA also prescribes highly specific regulatory micro RNAs and other epigenetic factors. Thus linear digital instructions program cooperative and holistic metabolic proficiency.

Not only are symbol systems used, but a bijection must occur between two independent symbol systems. Bijection (translation; a symbol system to symbol system correspondence or mapping) is rule-based, not physical law-based. No cause-and-effect necessity exists in the linking of anticodons, amino acids, tRNAs, and amino acyl tRNA synthetases with codons. The correspondence between the two languages is arbitrary and abstract. By arbitrary, we do not mean random. Arbitrary means freely chosen—free from physicodynamic determinism. Bijection (mapping) rules are uncoerced by cause-and-effect physicochemical chains. Translation of this linear digital prescription into functionally specific polyamino acid chains cannot be explained by physico-dynamics. It is not law-based, and it certainly is not random. If this were an empirical/inductive contention, "cannot" would have to be replaced with "has not yet been." But the statement is a valid deduction.

```
   1 gctagtgtag cttaagcaaa gcataacact gaagatgtta agatgggccg tagaaagccc
  61 cacgggcaca aaggtttggt cctgacttta ttatcagctt taacccaatt tacacatgca
 121 agcctccgca cccctgtgag gatgccctca atccccgtc cggggacgag gagccggtat
 181 caggcacact ttttagccca agacgccttg cttagccaca cccccaaggg aattcagcag
 241 tgatagacat taagccataa gtgaaaactt gacttagtca gggttaagag ggccggtaaa
 301 actcgtgcca gccaccgcgg ttatacgaga ggccctagtt gattcactcg gcgtaaagag
 361 tggttatgga gaataaaata ctaaagccga agacccctta ggccgtcata cgcacctagg
 421 ggctcgaatt atagacacga aagtagcttt accccttccc accagaaccc acgacagctg
 481 ggacacaaac tgggattaga taccccacta tgccccgccg taaacttaga tattccagta
 541 caacaaatat ccgccagggg actacgagcg ccagcttaaa acccaaagga cttggcggtg
 601 cttcagaccc ccctagagga gcctgttcta gaaccgataa cccccgttca acctcactac
 661 tccttgcttt tcccgcctat ataccaccgt cgccagctta ccctgtgaag gtactacagt
 721 aagcagaatg agtaatactc aaaacgtcag gtcgaggtgt agcgtacgaa gtaggaagaa
 781 atgggctaca ttatctgatc cagattattc acggaaggtt gtctgaaacg acaatccgaa
 841 ggtggattta gcagtaaagg gggaatagag tgcccccttg aagccggctc tgaagcgcgc
 901 acacaccgcc cgtcactctc cccaacaacc gcctacacca aggtaaataa cacaacatcc
 961 gtcacaaggg gaggcaagtc gtaacatggt aagtgtaccg gaaggtgcac ttggaataat
1021 cagggtgtgg ctgagacagt taagcgactc ccttacaccg agaagacatc catgcaagtt
1081 ggatcaccct gaactaaaca gctagctcaa actataaaaa ccaaattaat gatatagata
```

Figure 2. A section of *Alosa pseudoharengus* (a fish) mitochondrion DNA. This reference sequence continues on all the way up to 16,621 "letters." Each nucleotide is a physical symbol vehicle in a material symbol system. The specific selection of symbols and their syntax (particular sequencing) prescribes needed three-dimensional molecular structures and metabolic cooperative function *prior to* natural selection's participation.(Source: http://www.genome.jp/dbget-bin/www_bget?refseq+NC_009576)

The conclusion is as unequivocal as that produced by balanced mathematical manipulations of any equation. Neither fixed/forced laws nor chance can logically make nontrivial computationally halting programming decisions. It is a logical impossibility for chance and/or necessity to exercise bona fide choice contingency. They are in isolated categories (see Section 8). Neither unaided Markov chains nor physicodynamic determinism can select for *potential* formal function.

The noise-reducing Hamming "block coding" of triplets of nucleotides to prescribe each specific amino acid is all the more abstract and formally conceptual. The triplet codon/amino acid coding table has been shown to be conceptually ideal in a formal sense [45]. Block-coding greatly reduces the ill effects of a noisy channel on transmitted messages. Fewer prescriptive reading errors occur. Translation between the nucleotide and amino acid symbol systems is extraordinarily reliable. In addition, organisms possess amazing repair mechanisms to undo what noise pollution effects do compromise biomessages. Physics and chemistry provide no mechanisms to explain any of these sophisticated formal control and correction capabilities. They clearly traverse The Cybernetic Cut [4]—a great divide in nature between those phenomena that can be explained through the chance and necessity of natural process vs. those phenomena that can only be explained through formal steering and controls.

But the peculiarity of life over inanimate physics extends far beyond the above discussion. DNA requires editing in the course of its transcription to coding mRNA. And we have not even touched on the roles of many other independent players in the formal integration of transcription, translation, regulation, metabolism, and development. Most of DNA's Prescriptive Information is found in its non-protein-coding (prescribing) regions that instruct small regulatory RNA production. In addition, some supposedly non-protein-coding RNA's have been found to prescribe functional peptides and very short proteins [46, 47]. Epigenetic factors controlling differentiation and development are a large part of overall holistic true organization [48-53]. Post-translational editing also plays a role [54-57]. Gene overlaps, anti-sense transcriptions, genes assembled from multiple chromosomes are just a few of the growing list of layers of extraordinary formal PI instructing life.

"Semantic/semiotic/bioengineering function requires dynamically inert, resortable, physical symbol vehicles that represent time-independent, non-dynamic "meaning." (e.g., codons)." [8] No empirical or rational basis exists for granting to physics or chemistry such non-dynamic capabilities of functional sequencing. Neither chance nor necessity (fixed law) can program configurable switches to integrate circuits or organize formal utility.

Linear digital prescription in physical nucleic acid has thus far invariably been associated with life. A fully postmodern anthropocentrism cannot argue a logically consistent macroevolutionary paradigm. If naturalistic/materialistic science believes anything, it believes that an objectively real "physical brain secretes mind as the liver secretes bile" [as Pierre Jean Georges Cabanis (1757-1808), Karl Vogt and many others since have phrased it]. Jakob Moleschott (1822-1893) is generally given credit for the renal version: "The brain secretes thought as the kidney secretes urine." For macroevolution theory to fly, a very real genetic symbol system must evolve through objectively real early eukaryotes, invertebrates, vertebrates, mammals and primates. A purely subjective or solipsistic view of nucleotides and codons—trying to deny that they are real physical symbol vehicles—totally compromises macroevolutionary theory.

Macroevolution theory of necessity presupposes a literal history of progressive adaptation of millions of objectively existent species through changes in objectively existent nucleotide symbol sequencing. The formal, representational codon table not only predates human minds, but humans themselves.

8. Doesn't the physicality of MSSs prove that information is physical?

The addition of each new nucleotide to a single positive polynucleotide strand represents a decision-node *selection* from among four real contingent options. Each nucleotide selection corresponds to the equivalent of a quaternary decision node (a four-way switch) rather than a binary decision node (mere On-Off switch).

The MSS of molecular biology is unique in that the tokens do have direct three-dimensional physicodynamic effects. They are not like Scrabble pieces that are only representational and physicochemically inert in their functionality. The sequence of physical monomers determines minimum Gibbs-free-energy sinks which in turn determines folding function. But this physical determinism is secondary to sequencing (primary structure). And this sequencing of monomers is still dynamically inert (physicochemically indeterminate) in forming the positive single strand of polynucleotides. Even in the case of molecular biology, the meaning and physical function of monomeric token sequences (Prescriptive Informational polymers) is still fundamentally formal, not physicodynamic, in their origin.

We sometimes see the loss of PI in nucleic acid following critical base substitutions or other mutation events. We also see the loss of protein function from denaturization. It is tempting for some to falsely conclude from these events that PI and the function it prescribes is purely physical. But the loss of function through mutations and denaturization tells us nothing about the *source* of the message found in those strings in the first place. MSS's always have a formal origin. They are invariably derived from the purposeful choice of tokens from an alphabet of tokens. When nucleoside tokens are randomly polymerized, no sophisticated function is prescribed. When physicodynamic causation constrains token "selection," non-functional homopolymers tend to result (e.g., clay adsorption of polyadenosines). *Many scientists consistently confuse the instantiation of message meaning into physicality with physicality itself.* The inference is fallacious.

We don't make this mistake when we look at physical integrated circuits, computer chips, or robots. We unquestionably know that such physical devices resulted only from formal choice contingency causation and control (CCCC) *making use of physicality*, not from physicodynamic determinism. Artificial life models are not created and engineered by spontaneous inanimate physical interactions. They are invariably designed by human intelligence. Sophisticated machines result only from wise programming choices at bona fide decision nodes, logic gates and configurable switches. Physicality alone has never been observed to generate so much as a paper clip spontaneously.

When it comes to life, most biologists fanatically insist for purely metaphysical reasons that the cell was generated by nothing but the chance and/or necessity of physicochemical interactions. Empirical and prediction-fulfillment supports for this belief system are completely lacking. Rationality denies that anything other than gibberish can be generated by random processes. No computational program has ever been produced by a random number generator. But because of prior metaphysical commitment to the religion of physicalism, materialists insist that physicodynamics alone *HAD to have* generated the nonphysical formalisms needed to organize the simplest of living organisms.

Meaningful linear digital biopolymeric syntax is not generated by mere combinatorial uncertainty. Even when duplication of meaningful syntax occurs, there is no reason to believe that mere random variation (noise pollution) of this duplicated meaningful text would improve its meaning or function. We had no expectation of our PhD theses improving from typographical errors when given to typists unfamiliar with our fields of expertise. Why do we exercise such blind faith in the power of random mutations to improve duplications of the PI in genomes? We have no evidence of semantics and pragmatics being generated by chance or necessity, or any combination of the two [58].

Just because formalisms have been instantiated into physical symbol vehicles in a MSS does not change the fact that that the symbols, symbol system and meaning are purely formal. If this page burns up in a fire, we cannot conclude that the functional information recorded with physical molecules on physical paper was purely physical. The "page" still exists on electronic storage at the book publisher's server. If that physical medium were also lost in a fire, the ideas still remain in the head of this page's author, and in the minds of many who have already read this page. Symbol systems will always be formal, even when instantiated into a MSS.

We are also confused by the fact that linear digital token (amino acid monomer) sequences fold into physical three-dimensional catalysts and physical structures. Binding seems purely physical. We forget that binding depends upon globular tertiary structure. Globular structure depends upon the minimum Gibbs-free-energy sinks determined by monomeric sequencing. Even functional peptides, small and large regulatory proteins and chaperones are themselves prescribed by linear digital semiosis. The nucleotide and codon sequencing that prescribes all of these is physicodynamically inert. Selection of each monomer in the positive informational strand is formal, not physicodynamic. No physical causation exists to explain the particular PI sequence of each positive instructional strand. The chance and necessity of physicalism cannot explain the functional sequencing of the primary structure that determines folding.

Metaphysically disallowing formalism in one's model of reality precludes not only Hamming redundancy coding (codon to amino acid bijection), it precludes semiosis. A purely physical semiotic system cannot exist or function as a messaging system. "Representationalism requires both combinatorial uncertainty and freedom to purposefully select tokens. Naturalistic physical ISness cannot generate representationalism. Formalism alone can send and interpret linear digital messages. This remains true even when a material symbol system with physical symbol vehicles is used by formalism. Polynucleotide genes are such an MSS." [6]

Neither computer programs nor genetic instructions can be written by physicodynamic determinism. We cannot conclude that mathematics is physical just because the equations are written with physical chalk on a physical blackboard. Purposeful choices must be instantiated into computer hardware and software for computation to succeed. The same is true of genetic instruction. "Both mathematics and life are fundamentally formal. Even most epigenetic factors can be shown to be formally produced and integrated into a conceptual, cooperative, computational scheme of holistic metabolism. Life cannot exist without sophisticated, formal, genetic PI." [6]

The functional sequencing of miRNAs and base-paired negative (yet highly informational) strands points to even more sophisticated dimensions of formal programming. Intron sequencing contains abundant redundancy. Yet the redundancy is clearly highly functional, and is an integral part of the programming of miRNA folding and higher regulatory function. High order is not always the product of physicodynamic necessity. Sometimes repeated selections of the same tokens are deliberate and a vital part of formal programming.   What was thought to be junk DNA resulting from pointless duplications in introns is now known to be sophisticated regulatory programming.

9. The genetic code is conceptually ideal

 It is widely appreciated that not only the genetic code, but the genomic and epigenomic integrated systems are incredibly optimized [45, 59-61]. Undirected natural selection is generally given credit for having achieved these algorithmic optimizations. A plausible scientific mechanism or model is never provided for how physicality achieved formal optimization. As we shall see in chapter 7, undirected natural selection cannot possibly account for code origin and algorithmic optimization at the molecular/genetic level. Evolution works only on already-programmed, already-living phenotypic organisms through differential survival and reproduction. Random variation of duplicated PI cannot possibly optimize symbol system rules, make purposeful choices of symbols according to those rules, or pursue the goal of integrating systems or achieving potential functionality. All of these are formal functions, not physicodynamic interactions. Some other explanation than differential survival of the fittest already-programmed, already-living organisms is needed.

The source of genetic programming lies in the free selection of nucleotides, and in the unconstrained sequencing of those particular nucleotide selections. Says Fontana and Schuster, "Understanding which phenotypes are accessible from which genotypes is fundamental for understanding the evolutionary process." [62] The sequencing of DNA nucleotides has no meaning or function independent of an overarching formal system of arbitrary (could have been otherwise) symbol assignments to each amino acid.

A representational symbol system is clearly employed in the triplet codon table of amino acid prescription. Codons are a form of Hamming "block code" wherein consistent groups of three symbols are used to represent each single amino acid prescription. Block coding is a form of redundancy coding used to reduce noise pollution in the transmission channel. These arbitrary assignments have been shown to be conceptually ideal in reducing noise pollution in the Shannon channel [45, 60]. The largest number of redundant codons for the same amino acid "just happens to be" assigned to the most important (frequency-wise) amino acids. Despite wobbles and point mutations, codons are often still able to prescribe the correct amino acid because of this extraordinary redundancy coding. The use of block coding prevents frame shift problems that would occur with a redundancy code of variable characters (nucleo-tide tokens).

Life-origin models cannot reduce these highly optimized phenomena to human epistemology. They are objective phenomena, not merely heuristic tools of our mental construction. Biosemiosis and biocybernetic management was integrating and engineering life's processes long before *Homo sapiens* appeared on the scene to ascribe their linguistic and cybernetic analogies to molecular biology. How would inanimate chance and necessity have conceived such an effective, formal, noise-reducing scheme?

Additional layers of ideal coding sophistication also exist. Independent coding overlays the genetic code in DNA [61]. A separate set of rules controls the binding of transcription factors and histone proteins to DNA. These additional rules control messenger RNA splicing and folding. The later contribute to regulating protein manufacture. The two coding systems are independent, but they are also coordinated. The two codes jointly control metabolism [61]. The genomic code is far more vast than the genetic code, as if we weren't already burdened trying to explain the genetic code alone through natural process. The genomic code includes the three-dimensional structure of DNA and many additional overlaid codings in molecular biology [60]. Chromosomes are grouped by centromeres into radial clusters that juxtaposition certain segments of different chromosomes so that they can cooperatively interact spatially [63]. Wistar researchers found 465 groups of genes that contribute to related structural or metabolic purposes in fission yeast cells. While linear digital prescription is fundamental to genetics and genomics, multiple 3-dimensional layers of information also prescribe function. Incredibly-optimized spatial organization also integrates virtually every cell function.

The underlying cause of such integration is formal even though physical tokens are used. All of these formally integrated systems require selection contingency, not chance contingency or fixed law, to organize [35]. Selection must take place at the genetic level of nucleotide selection for any phenotype to come into existence, let alone the fittest phenotype. This fact of reality constitutes "The GS Principle" is discussed in Chapter 7.

10. Conclusions

Nonphysical, formal, linear digital symbol systems can be instantiated into physicality using physical symbol vehicles (tokens) in a MSS. Genetics and genomics employ an MSS, not a two-dimensional pictorial "blueprint." Highly functional molecular biological MSS's existed prior to human consciousness in tens of millions of species. Genetic code is conceptually ideal. Not all signals are messages. Code is a conversionary algorithm that translates, bijects or decrypts one symbol system into another. Encode/decode is formal, not physicodynamic. Biosemiosis is cybernetic in that symbols must be purposefully chosen from alphabets of symbols. This is a form of control. All known life is cybernetic. Even protocells would require biosemiosis, controls and regulation to become alive. Neither law nor random variation of duplications can generate a meaningful/functional MSS. Even most epigenetic players are produced by MSS's.

References

1. Abel, D.L. 2009, The capabilities of chaos and complexity, Int. J. Mol. Sci., 10, (Special Issue on Life Origin) 247-291 Open access at http://mdpi.com/1422-0067/10/1/247

2. Abel, D.L.; Trevors, J.T. 2006, More than metaphor: Genomes are objective sign systems, Journal of BioSemiotics, 1, (2) 253-267.

3. Abel, D.L. 2007, Complexity, self-organization, and emergence at the edge of chaos in life-origin models, Journal of the Washington Academy of Sciences, 93, (4) 1-20.

4. Abel, D.L. 2008, 'The Cybernetic Cut': Progressing from description to prescription in systems theory, The Open Cybernetics and Systemics Journal, 2, 234-244 Open access at www.bentham.org/open/tocsj/articles/V002/252TOCSJ.pdf

5. Abel, D.L. 2009, The GS (Genetic Selection) Principle, Frontiers in Bioscience, 14, (January 1) 2959-2969 Open access at http://www.bioscience.org/2009/v14/af/3426/fulltext.htm.

6. Abel, D.L. 2009, The biosemiosis of prescriptive information, Semiotica, 2009, (174) 1-19.

7. Abel, D.L. 2010, Constraints vs. Controls, Open Cybernetics and Systemics Journal, 4, 14-27 Open Access at http://www.bentham.org/open/tocsj/articles/V004/14TOCSJ.pdf.

8. Abel, D.L.; Trevors, J.T. 2005, Three subsets of sequence complexity and their relevance to biopolymeric information., Theoretical Biology and Medical Modeling, 2, 29 Open access at http://www.tbiomed.com/content/2/1/29.

9. Abel, D.L.; Trevors, J.T. 2006, Self-Organization vs. Self-Ordering events in life-origin models, Physics of Life Reviews, 3, 211-228.

10. Rocha, L.M. 1997, *Evidence Sets and Contextual Genetic Algorithms: Exploring uncertainty, context, and embodiment in cognitive and biological systems.* . State University of New York, Binghamton.

11. Rocha, L.M. 2001, Evolution with material symbol systems, Biosystems, 60, 95-121.

12. Whitehead, A.N. 1927, *Symbolism: Its meaning and effect.* Macmillan: New York.

13. Cassirer, E. 1957, *The Philosophy of Symbolic Forms, Vol 3: The Phenomena of Knowledge.* Yale Univ. Press: New Haven, CT.

14. Harnad, S. 1990, The symbol grounding problem, Physica D, 42, 335-346.

15. Pattee, H.H. 1995, Evolving Self-Reference: Matter, Symbols, and Semantic Closure, Communication and Cognition-Artificial Intelligence, 12, 9-28.

16. Pattee, H.H.; Kull, K. 2009, A biosemiotic conversation: Between physics and semiotics, Sign Systems Studies 37, (1/2).

17. Peirce, C.S. 1991, *Peirce on Signs: Writings of Semiotic.* University of North Carolina Press: Chapel Hill, N.C.

18. Peirce, C.S. 1998, The Essential Peirce: Selected Philosophical Writings, 1893-1913 (Paperback). Indiana University Press: Bloomington.

19. Rosen, R. 1974, Planning, management, policies, and strategies: four fuzzy concepts, Int. J. General Systems, 1, (4) 245-252.

20. Abel, D.L. The GS (Genetic Selection) Principle [Scirus Topic Page]. http://www.scitopics.com/The_GS_Principle_The_Genetic_Selection_Principle.html (Last accessed September, 2011).

21. Alp, K.O. 2010, A comparison of sign and symbol (their contents and boundaries), Semiotica, 2010, (182) 1-13.

22. Barbieri, M. 2003, *The Organic Codes: An Introduction to Semantic Biology.* Cambridge University Press: Cambridge.

23. Barbieri, M. 2004, Biology with information and meaning, History & Philosophy of the Life Sciences, 25, (2 (June)) 243-254.

24. Barbieri, M. 2007, The Codes of Life: The Rules of Macroevolution (Biosemiotics). Springer: Dordrecht, The Netherlands.

25. Johnson, D.E. 2010, *Programming of Life.* Big Mac Publishers: Sylacauga, Alabama.

26. Yockey, H.P. 1992, *Information Theory and Molecular Biology.* Cambridge University Press: Cambridge.

27. Johnson, D.E. 2010, *Probability's Nature and Nature's Probability (A call to scientific integrity).* Booksurge Publishing: Charleston, S.C.

28. Abel, D.L. 2009, The Universal Plausibility Metric (UPM) & Principle (UPP), Theor Biol Med Model, 6, (1) 27 Open access at http://www.tbiomed.com/content/6/1/27.

29. von Uexküll, J. 1928, *Theoretische Biologie.* Julius Springer: Berlin.

30. von Uexküll, T. 1982, Introduction: Meaning and science in Jacob von Uexkull's concept of biology, Semiotica, 42, 1-24.

31. Sebeok, T.A. 1976, *Contributions to the Doctrine of Signs.* Indiana University Press: Bloomington, IN.

32. Sebeok, T.A. 1994, *Signs: An Introduction to Semiotics.* University of Toronto Press: Toronto.

33. Hoffmeyer, J.; Emmeche, C. 2005, Code-Duality and the Semiotics of Nature, Journal of Biosemiotics, 1, 37-91.

34. Bateson, G. 1979, *Mind and Nature*. Bentam Books: New York.

35. Trevors, J.T.; Abel, D.L. 2004, Chance and necessity do not explain the origin of life, Cell Biology International, 28, 729-739.

36. Abel, D.L. 2002, Is Life Reducible to Complexity? In *Fundamentals of Life*, Palyi, G.; Zucchi, C.Caglioti, L., Eds. Elsevier: Paris, pp 57-72.

37. Abel, D.L. Prescriptive Information (PI) [Scirus Topic Page]. http://www.scitopics.com/Prescriptive_Information_PI.html (Last accessed September, 2011).

38. Abel, D.L. The Cybernetic Cut [Scirus Topic Page]. http://www.scitopics.com/The_Cybernetic_Cut.html (Last accessed Sept, 2011).

39. von Neumann, J. 1950, *Functional operators*. Princeton University Press: Princeton.

40. von Neumann, J. 1950, Letter to physicist George Gamow (first scientist to elucidate *triplet* codons) on July 25, 1950. Cited by Steve J. Heims in "John von Neumann and Norbert Wiener: *From Mathematics to the Technologies of Life and Death*," Canbridge, MA, MIT Press, 1980.

41. von Neumann, J. 1956, The general and logical theory of automata. In *The World of Mathematics Vol 4*, Newman, J. R., Ed. Simon and Schuster: New YHork.

42. von Neumann, J.; Aspray, W.; Burks, A.W. 1987, *Papers of John von Neumann on computing and computer theory*. MIT Press ; Tomash Publishers: Cambridge, Mass.

43. von Neumann, J.; Burks, A.W. 1966, *Theory of Self-Reproducing Automata*. University of Illinois Press: Urbana,.

44. Chomsky, N. 1972, *Language and Mind*. Harcourt Brace Jovanovich: New York.

45. Bradley, D. 2002, Informatics. The genome chose its alphabet with care, Science, 297, (5588) 1789-1791.

46. Ledford, H. 2010, Mystery RNA spawns gene-activating peptides: Short peptides that regulate fruitfly development are produced from 'junk' RNA. In NATURE, Vol. Published online 15 July.

47. Mattick, J. 2010, Video Q&A: Non-coding RNAs and eukaryotic evolution - a personal view, BMC Biology, 8, (1) 67.

48. Veening, J.-W.; Smits, W.K.; Kuipers, O.P. 2008, Bistability, Epigenetics, and Bet-Hedging in Bacteria, Annual Review of Microbiology, 62, (1) 193-210.

49. Allis, D.C.; Jenuwein, T.; Reinberg, D.; Wood, R.; Caparros, M.-L. 2007, Epigenetics. Cold Springs Harbor Press: Woodbury, NY.

50. Qiu, J. 2006, Epigenetics: unfinished symphony, Nature, 441, (7090) 143-5.

51. Grant-Downton, R.T.; Dickinson, H.G. 2005, Epigenetics and its implications for plant biology. 1. The epigenetic network in plants, Ann Bot (Lond), 96, (7) 1143-64.

52. Jablonka, E.; Lamb, M.J. 2002, The changing concept of epigenetics, Ann N Y Acad Sci, 981, 82-96.

53. Griesemer, J. 2002, What is "epi" about epigenetics?, Ann N Y Acad Sci, 981, 97-110.

54. Bachmair, A.; Novatchkova, M.; Potuschak, T.; Eisenhaber, F. 2001, Ubiquitylation in plants: a post-genomic look at a post-translational modification, Trends Plant Sci, 6, (10) 463-70.

55. Eisenhaber, B.; Bork, P.; Eisenhaber, F. 2001, Post-translational GPI lipid anchor modification of proteins in kingdoms of life: analysis of protein sequence data from complete genomes, Protein Eng, 14, (1) 17-25.

56. Vaish, N.K.; Dong, F.; Andrews, L.; Schweppe, R.E.; Ahn, N.G.; Blatt, L.; Seiwert, S.D. 2002, Monitoring post-translational modification of proteins with allosteric ribozymes, Nat Biotechnol, 20, (8) 810-815.

57. Mata, J.; Marguerat, S.; Bahler, J. 2005, Post-transcriptional control of gene expression: a genome-wide perspective, Trends Biochem Sci, 30, (9) 506-14.

58. Luisi, P.L. 2007, The problem of macromolecular sequences: the forgotten stumbling block, Orig Life Evol Biosph, 37, (4-5) 363-5.

59. Freeland, S.J.; Hurst, L.D. 1998, The genetic code is one in a million, Journal of Molecular Evolution, 47, 238-248.

60. Itzkovitz, S.; Alon, U. 2007, The genetic code is nearly optimal for allowing additional information within protein-coding sequences, Genome Res, 17, (4) 405-12.

61. Segal, E.; Fondufe-Mittendorf, Y.; Chen, L.; Thastrom, A.; Field, Y.; Moore, I.K.; Wang, J.P.; Widom, J. 2006, A genomic code for nucleosome positioning, Nature, 442, (7104) 772-8.

62. Fontana, W.; Schuster, P. 1998, Shaping space: the possible and the attainable in RNA genotype-phenotype mapping, J Theor Biol, 194, (4) 491-515.

63. Tanizawa, H.; Et_al 2010, Mapping of long-range associations throughout the fission yeast genome reveals global genome organization linked to transcriptional regulation. , Nucleic Acids Research. Published online before print October 28, 2010.

The First Gene, David L. Abel, Editor 2011, pp 161-188 ISBN: 978-0-9657988-9-1

7. The Genetic Selection (GS) Principle*

David L. Abel

Department of ProtoBioCybernetics/ProtoBioSemiotics
Director, The Gene Emergence Project
The Origin-of-Life Science Foundation, Inc.
113 Hedgewood Dr. Greenbelt, MD 20770-1610 USA

Abstract. The GS (Genetic Selection) Principle states that biological selection must occur at the nucleotide-sequencing molecular-genetic level of 3'5' phosphodiester bond formation. After-the-fact differential survival and reproduction of already-programmed, already-living phenotypic organisms (natural selection) does not explain polynucleotide sequence prescription and coding. All life forms depend upon exceedingly-optimized genetic algorithms. Biological control requires selection of particular physicodynamically indeterminate configurable switch settings to achieve potential function. This occurs largely at the level of nucleotide selection, prior to the realization of any isolated or integrated biofunction. Each selection of a nucleotide corresponds to a quaternary (four-way) switch setting. Formal logic gates must be set initially that will only later determine folding and binding function through minimum Gibbs-free-energy sinks. The fittest living organisms cannot be favored until they are first programmed and computed. The GS Principle distinguishes selection *of existing* function (undirected natural selection) from selection *for potential* function (formal selection at decision nodes, logic gates and configurable switch-settings).

Correspondence/Reprint request: Dr. David L. Abel, Department of ProtoBioCybernetics/ProtoBioSemiotics, The Origin-of-Life Science Foundation, Inc., 113 Hedgewood Dr. Greenbelt, MD 20770-1610 USA E-mail: life@us.net

*This chapter is largely an update of Abel, D.L. 2009, The GS (Genetic Selection) Principle, Frontiers in Bioscience, 14, (January 1) 2959-2969, but may also contain short sections from these other publications [1-9]

Introduction: What is The Genetic Selection (GS) Principle?

"The GS Principle states that *biological selection must occur at the nucleotide-sequencing molecular-genetic level of 3'5' phosphodiester bond formation.* After-the-fact differential survival and reproduction of already-programmed, already-living phenotypic organisms (undirected natural selection) does not explain the origin of the genome or epigenome. Environmental selection provides no mechanism for the *generation* of Prescriptive Information (PI) [4] using linear digital polynucleotide syntax (nucleotide sequencing). Survival of the fittest provides no mechanism for the origin of any of the following requirements:

- Piecing together various distant genetic segments into many different local functional gene combinations [10]
- Regulation of transcription by negative-strand (formerly called "antisense"-strand) microRNAs from the same double helix as the protein-coding positive strand [11-13]
- Gene overlaps, including RNAs that can function both as coding messenger RNA and intrinsically as functional non-coding RNAs [14]
- Regulatory proteins produced by one gene that control transcription of different genes
- Selective editing of the same transcriptome to meet different functional needs [10]
- Linear digital PI simultaneously running in both directions [10]
- Formal translation from one symbol system to another,
- Noise-reducing "Hamming block coding" using 3 to 1 symbol bijection (triplet codons prescribing each amino acid)
- Optimized redundancy (the greatest number of different codons prescribing the most important amino acids),
- Bit parity and other post facto error-correction mechanisms that protect messages from noise pollution in the Shannon channel (99.9% of all mutations are neutral [no basis for environmental selection] or deleterious)
- Chromatin coiling and other structural three-dimensional factors providing specific PI of its own
- Functional genomic regions are very often not conserved, calling into question the belief that biological function requires evolutionary constraint [10].
- Symmetrically distributed regulatory sequences "just happen" to

surround transcription start sites.

- Integration of pathways and cycles into holistic metabolism,
- Purposive pursuit of cellular survival itself (what would an inanimate environment care whether any life came into existence, survived, or improved?)

Differential survival and reproduction does not provide explanation for a single one of these above phenomena. Each nucleotide must be selected at the point of polymerization with strong covalent bonds. Non-controlled constraints (e.g., environmental stresses), semi-controlled constraints (e.g., the corrupted Prescriptive Information of prion misfoldings), and controlled constraints (e.g., programmed chaperone proteins that steer (control, not merely constrain) the folding of other polyamino acid polymers into the needed functional shape. But by far the main determinant of conformational structure and function is the physicodynamically indeterminate primary structure (the sequencing) of the polyamino acid chain itself. Stochastic ensembles from sequence space do not form functional folds [15-17] and cannot begin to explain the transcription and efficaciousness of highly instructive cellular RNAs [14].

Chemical evolution cannot be grouped into the same category as Darwinian natural selection. Chang et al [18] state, "'Chemical evolution' should not be confused with Darwinian evolution with its requirements for reproduction, mutation and natural selection. These did not occur before the development of the first living organism, and so chemical evolution and Darwinian evolution are quite different processes." But far more serious a problem is that no satisfactory mechanism for chemical evolution has ever been presented. In the absence of even a theoretical mechanism, it remains to be seen whether "chemical evolution" is real at all. It certainly has never been observed. Artificial selection has been used in a number of ribozyme engineering experiments, for example. But no spontaneous natural selection of inanimate chemicals has ever been observed to generate nontrivial function or formal metabolic subsystems.

"The GS Principle" was first introduced in the literature in 2006 in a paper and later a book chapter entitled, "More than Metaphor: genomes are objective symbol systems" [9, 19]. Other papers have alluded to it [1, 2, 4, 5, 20]. But the full formal presentation and defense of the Principle did not occur until 2009 [21].

All known organisms are prescribed and largely controlled by semantic, functional information [4, 7, 19, 22-40]. Most biological Prescriptive Information (PI) [4] presents first as linear digital programming [41-44]. Additional multi-dimensional layers of information are built upon this linear digital foundation. Living organisms and their programmed metabolomes (holistic, integrated, metabolic schemes and parallel processes) arise only out of cybernetic/computational success. Von Neumann, Turing and Wiener all got their computer design and engineering ideas from the linear digital genetic programming employed by life itself [45-50]. All known life is cybernetic [1, 8, 51]. Regulatory proteins, microRNAs and most epigenetic factors are digitally prescribed [24]. MicroRNAs can serve as master regulators of gene expression [52-54]. One microRNA can control multiple genes. One gene can be controlled by multiple microRNAs.

The challenge of finding a natural mechanism for linear digital programming extends from primordial genetics into the much larger realm of semantics and semiotics in general. Says Barham: "The main challenge for information science is to naturalize the semantic content of information. This can only be achieved in the context of a naturalized teleology (by 'teleology' is meant the coherence and the coordination of the physical forces which constitute the living state)."[55] The alternative term "teleonomy" has been used in an effort to attribute to natural process "the appearance of teleology" [56-58]. Either way, the bottom line of such phenomena is *selection for* (not of) *potential* higher function *at the logic-gate programming level.* Programming choices must be made and physically instantiated (recorded) in pursuit of a utility that does not yet exist at the time of programming. This is the mammoth problem that continues to dog macro-evolutionary and naturalistic metaphysical belief, particularly as it attempts to claim scientific certitude (proven fact).

1. Natural selection cannot operate at the genetic programming level

Linear digital prescription requires selection of monomers *at the point of polymerization* of the initial positive informational strand. Primary structure (sequencing) instructs folding into secondary (two-dimensional) and tertiary structure (three-dimensional shape). While chaperones and other factors also affect folding, the sequencing of the polyamino acid polymer itself is by far the biggest determinant of shape, electrostatic charge, grooves, knobs, tunnels, hydrophobicities, and lock-and-key binding of the globular protein molecular machine or enzyme to its substrate. Folding proceeds according to minimum Gibbs free-energy sinks [59-67]. But rigidly-bounded monomeric sequencing largely determines what these thermodynamic and kinetic tendencies will be.

Self-replication tends to "get all the press" in life-origin literature. But the real issue of life origin lies in answering how the initial single positive strands of RNA instructions got sequenced so as to prescribe microRNA regulation, amino acid sequencing and eventual folding function. No new information is generated in base-pairing replications. Base-pairing has nothing to do with the generation of genetic information or coding. Base-pairing is purely physicodynamic, and quite secondary to the already-programmed, formal, linear digital instructions of the single positive strand.

The inanimate environment cannot program. The laws of physics and chemistry cannot program. Environmental selection cannot choose logic-gate settings at the nucleotide polymerization level of organization prior to computational halting. Undirected natural selection cannot pursue any goals. It cannot select for potential function. It cannot generate linear digital instructions to achieve eventual formal organization and function.

Natural selection (NS) is eliminative, not creative [68, pg 70]. Inferior specimens are only indirectly crowded out. NS does not actively "choose" at decision nodes.

Contrary to the notion of "selection pressure," NS is incapable of *pressuring* any existing organisms to improve themselves. Any improvement would have to take place first at the genetic, genomic, epigenetic and epigenomic molecular level. Improvement would only be secondarily realized phenotypically as a result of superior formal programming and regulation. NS embodies no mechanism to generate superior formal programming of the phenotype at the genetic and genomic level.

In addition, the environment cannot select for isolated function. The environment is blind to functionality. It has no preference for function over nonfunction. The environment has no capacity for caring whether the fittest organisms survive better. They just do, but only because their programming and hardware are superior. Under no circumstances can the environment take credit for these wise programming decisions. Undirected NS is not an active chooser of logic gate settings; NS is totally passive.

2. Natural selection can only favor the best already-living phenotypes

Differential survival and reproduction is realized only after the fact of already existing superiority. The environment can only select for the fittest already-programmed, already-living organisms. Undirected NS can only prefer living organisms that are superior in their hardware, programming, control, and regulatory mechanisms. Organisms and small groups of organisms survive best because their systems are programmed and engineered best. Evolution cannot be given the credit for programming or engineering talents.

No motivation exists in any inanimate environment to pursue, realize or preserve formal utility. Undirected natural selection is nothing more than the differential survival and reproduction of the fittest pre-programmed, pre-computed, already-living small groups of organisms.

The fittest organisms survive and reproduce better. Natural selection (NS) can only indirectly eliminate inferior living organisms. Less fit phenotypes die off. NS has nothing to do with generating life from non-life. In addition, NS has nothing to do with programming new more sophisticated PI.

Organisms cannot come into existence without many purposefully cooperative computational successes occurring across many different levels. None of these computational haltings will occur without selection of appropriate symbols and switch-settings according to formal rules so as to generate efficacious programming sequences. In addition, all these programming sequences must be integrated into a holistic operating system in order to organize even the simplest protometabolism. Such organization is mediated using multiple layers of material symbol systems. Physical symbol vehicles (nucleotide "tokens") are used to represent formal quaternary (four-way) switch-setting "choices." These in turn determine higher conceptual levels of transcriptional regulatory networks, multilayer hierarchical structures, transcript turnover regulation, and three-dimensional information retention in genomes [69-71]. The multidimensional layers of information realized through DNA coiling and juxtaposition of chromosome segments could not occur unless nucleotide and codon sequencing were first in place.

Evolution theory does not provide any explanation for how the phenotype and its superior fitness was programmed in primary structure sequences prior to folding. Nucleotide selections must be made and are rigidly bound with covalent bonds prior to the realization of any metabolic function or organismal fitness.

Lamarckism has some validity in immunology [72, 73], but it cannot explain the formal genetic and genomic programming that makes organismic existence and superior fitness possible.

3. The requirements of selection

Several requirements exist for efficacious selection. First, selection requires categorization. A clear differentiation of real options must exist. Option A must be discrete and cannot simultaneously be option Non A. With binary switches, an "excluded middle" exists. Either A or B, 0 or 1, must be chosen. No in-between gray zone is permitted. With multiple options (e.g. four-way quaternary configurable switches), option A cannot simultaneously be option B, C, D. To select B, C or D is not to select option A. Every option chosen excludes the other three.

Second, one category must be preferential or superior in its functionality compared to the others for selection to be worthwhile. Without such motivation and desire for utility, selection has no point, and might as well be random.

Third, freedom must exist to select a certain option from among other options. If nothing but cause-and-effect deterministic chains prevail according to fixed law, little or no possibility exists to select for function over non-function.

Fourth, impetus (the equivalent of motive) is needed to drive the selection process.

Fifth, selection *in pursuit of function that does not yet exist* must be possible. The selecting agent must be able to look ahead and anticipate what programming decisions will lead to usefulness prior to the existence of that usefulness. Undirected natural selection cannot pursue *potential* function. Sophisticated function usually requires multiple integrated choices prior to the realization of that function.

Sixth, a means of selection must exist. How will each specific selection be made? What enables the process of choosing one option over another once the intent to select a particular option exists?

Seventh, choices must be recordable and additive in order to program nontrivial computation.

How many of these criteria are met by undirected natural selection? The fittest organisms are clearly categorized from less fit organisms independent of human knowledge and description. Superior fitness also readily meets the criterion of functional superiority. Differential survival and reproduction provides a degree of freedom from fixed law to allow passive "selection." What about the *means* of selection? The means of selection is provided by gradual extinction of less successful competing organisms. The impetus to select is automatic. Differential survival and reproduction indirectly drive the selection process to its endpoint of maximum utility through time with no requirement of any external unnatural or supernatural force. No nonphysical formal component is required for environmental selection of superior organisms to occur. Thus far, natural selection seems to fit the bill in its selection capabilities.

Problems arise, however, in natural selection being able to explain how any organism, fit or unfit, came into existence in the first place. No impetus or motive exists in inanimate nature to stimulate the pursuit of any goal, function included. Many programming choices are required at the genetic, genomic, epigenetic and epigenomic level to generate the simplest metabolism. These choices must be made prior to the existence of the organism. Choice contingency manifests *the ability to voluntarily pursue and choose for potential function.* Potential means "not yet existent." "Not yet existent" means that the environment has nothing yet to select or favor. Natural selection cannot operate on not-yet-existent phenotypes. Large numbers of programming choices in pursuit of function that does not yet exist must be made for any organism to come into existence. Environmental selection cannot pursue the goal of potential function.

The next problem for natural selection is that selections (e.g., of codons) must be recorded using representational symbols. Codons formally *represent* amino acids. No direct physicochemical link exists between the two. The associations are arbitrary (could have been selected otherwise).

Let us now subject genetic programming at the molecular level to the same essential criteria of selection referred to above. All four nucleotides occur with near-equal frequencies, making genomes appear to be in an almost random distribution. Yet they are way too functional to be random. They are all bound with the same 3'5' phosphodiester bonds. Can single-stranded polynucleotide primary structures be distinguished and categorized by physicodynamics alone, prior to folding? Perhaps. Physicochemical, steric, electrostatic, and structural differences do exist between various stochastic ensembles of polynucleotides even before folding. But do any physicodynamic differences relating to nucleotide sequence *matter* to nature at the point of primary structure formation? In a prebiotic environment, would the environment prefer one stochastic ensemble over another? Although ribozymes and DNAzymes exist, they contribute *needed* function only in a holistic metabolic context. No reason exists for nature to *prefer* a catalytic RNA over a non-catalytic one. Nature has no goals, preferences or motives, evolution included. At the programming level of gene formation, function in an integrated metabolic scheme does not yet exist. No living phenotypic superiority exists for the environment to favor. Apart from a polynucleotide's participation in the instructional symbol system of an already-living organism, any one single strand of RNA or DNA is just as good as any other. A self-replicating RNA could theoretically form spontaneously. But as we will discuss later, it is not at all clear what such a self-replicating RNA would contribute to any potential *metabolic* scheme. It would also consume so many resources few would be left for thousands of other needed metabolites to form out of a severely depleted sequence space. *Functional superiority* of one sequence option (primary structure) over another is completely lacking in an abiotic environment. Mere self-replication is of no value to inanimate nature. If a self-replicative sequence occurred happenstantially, it would self-optimize its self-replicative function to the exclusion of other potentially metabolic functions and consume all resources. Self-replication of that one molecule would indeed be "selfish" in a purely materialistic sense [74-76]. Nothing would be "preferred" other than self-replication of that one molecule and its one "function" of self-replication. But self-replication has nothing to do with contributing to a cooperative metabolic scheme with many other players. But the immediate point here is that undirected natural selection fails the test of selection at the genetic programming level. The environment cannot yet adequately categorize molecules into ones that will eventually contribute to holistic metabolism and those that will not.

What about the next criterion of selection of *means*? With natural selection, the means is differential survival and reproduction of already-living organisms. But at the genetic programming level in a prebiotic world, no life or differential survival exist yet. *Means* is totally lacking for evolution to occur at the genetic programming level.

What about the essential criterion of selection of *impetus*? In environmental selection, differential survival and reproduction of small populations drives the selection process. The impetus is automatic. But at the positive strand formation level, what is the natural-process impetus for selection of one nucleotide selection or one sequence over another? Phenotypic fitness does not yet exist. Life does not yet exist. Differential survivability cannot be a factor to drive the selection process. While a self-replicating polynucleotide might differentially dominate the sequence space, a sequence optimized for self-replication would not have the ideal sequencing for almost any other metabolic function. The self-replicating strand would merely consume all the resources required for mass-producing itself. But with respect to prescribing all of the metabolic functions needed for life, the mass-produced strand would be gibberish.

"Natural selection" and "environmental selection" both include the term "selection" for good reason. Evolution is impossible without selection. "Selection" in our naturalistic paradigms replaces the more subjective and purposeful "choice with intent" of our cognition. The best phenotypes in any environment are selected indirectly through *differential reproductive success and differential survival*. The fittest phenotypes are those that perform the most utilitarian processes and functions and provide the most heritable and biofunctional cybernetic controls. But evolution cannot generate any of these formal functions.

Thus, environmental selection can play no role whatsoever in the selection of nucleotide or codon linear digital prescription. Nothing exists in inanimate nature to steer sequencing *toward* meaning or function. Yet these selections constitute the setting of critical logic gates. Nucleotide selections clearly constitute the programming of configurable switches. If the switches are not set properly, no life will come into existence to be favored. Yet at the point of polymerization of any certain sequence, no physicochemical superiority exists for the environment to favor.

It follows that a purely physicalistic nucleotide polymerization of single positive strands in solution fails to manifest any of the above essential criteria of a selection process. No basis exists for natural-process programming of the covalently-bound strand. Under these conditions, linear digital programming by environmental selection is impossible. Natural selection cannot occur at the programming level of configurable switch-setting (the choice of which nucleotide to polymerize next). Thus environmental selection cannot program computational linear digital programs. Yet environmental selection is the only kind of selection known to natural process.

Programming selections at successive decision nodes requires anticipation of what selections and what sequences *would be* functional. Selection must be made *for potential* function. Nature cannot anticipate—let alone plan or pursue—formal function. Natural selection can only preserve the fittest already-existing holistic life.

The specific selection of one ribonucleotide from among four real options functions as a quaternary (four-way) configurable switch-setting. Configurable switches *control and prescribe*, not merely describe, translated metabolic utility. Covalently-bound selection commitments are rigid by comparison to weaker H-bonded secondary folding. Primary structure is the primary determinant of what H-bonding and van der Waals forces can accomplish. Configurable switch-settings in the form of specific nucleotide selections constrain minimum-free-energy folding space. Both ribonucleotide polymerization reactions and folding are subject to the laws of motion and to dynamic constraints. But the cybernetic function of the genetic material symbol system *controls* those constraints. The sequencing of nucleotides as physical symbol vehicles is not determined by physicodynamics. It is *dynamically inert* (physicodynamically incoherent; decoupled from physical determinism) [77, 78]. Once instantiated into a Material Symbol System, however, this dynamically-inert programming physically constrains dynamic folding space. The GS Principle attributes the main *control* of folding constraints to *formal nucleotide sequencing selections.* Notice that when constraints are selected by cybernetic determinism, not by physicodynamic determinism, they become bona fide formal controls rather than mere physicodynamic constraints [2, 5].

All life depends upon literal, objective, molecular biological genetic algorithms—not the applied science artificial ones. Fittest organisms cannot be favored until they are first computed. The GS Principle elucidates the source of messenger molecules' representational prescription of biofunction, a phenomenon unique in nature to life [44]. The GS Principle distinguishes selection *of existing* function (natural selection) from selection *for potential* function (formal selection at decision nodes, logic gates and configurable switch-settings).

Symbol systems employ alphabetical characters, signs, and physical symbol vehicles (tokens such as nucleotide options) to represent meaning or function. Selections must be made from an option space of real, uncoerced alternatives. Each selection represents the setting of a logic gate or configurable switch that computes or integrates a circuit. The purposeful selection of letters alone constructs words, sentences and paragraphs. Any denial of "choice with intent" reduces language to gibberish. Any attempt to replace arbitrary rules of convention with fixed law destroys information potential. Rules are voluntarily followed, not forced. Only when the destination voluntarily applies the same free and arbitrary rules of the source can the destination successfully interpret the source's intended meaning. By arbitrary, we do not mean random. We mean, "could have been otherwise" within the constraints of nature. But abiding by arbitrary rules also embodies selection in accordance with those rules at the decision-node level. The programming of polynucleotide prescription of biofunction cannot be reduced to mere linguistic and computer science metaphor [19]. Indeed, if any analogy exists, it is in the reverse direction. Life's linear digital cybernetics predates humans, their languages and their computers.

4. Genetics is a linear digital material symbol system

Nucleotides function as physical symbol vehicles in a Material Symbol System (MSS) [77, 79, 80]. Nucleotide sequences are a little like Scrabble tokens deliberately aligned in linear sequence to spell meaningful words and sentences. This is not the same as a random sequence string of tokens that just happens to *resemble* a word. Such supposed "word recognition" is nothing more than a mental association ascribed onto the random string. The string remains fundamentally random, with no meaningful prescription of function. Such string of letters only *appear* to be a word after applying a set of formal linguistic rules not common to the playing field of stochastic ensembles or to mere physicodynamic interactions.

But unlike stochastic ensembles of aligned Scrabble tokens, certain ribonucleotide sequences can contain formal linear digital prescription independent of our mental ascriptions. RNA single strands can fold back onto themselves into functional structures that have direct catalytic effects. Except in the case of rare DNAzymes, DNA sequences cause few direct physicochemical causative effects. The significance of DNA primary structure lies primarily in their formal prescription of amino acid sequencing following algorithmic processing and Hamming block code decryption. And much of DNA prescribes regulatory microRNA transcriptions. But the latter are formal functions, not physicodynamic interactions. DNA sequence represents a string of formal program-

ming choices. Each nucleoside token successively polymerized has to be selected from a pool of 4 options. If polymerized by law, there would be no uncertainty, and no possibility of information retention. If polymerized randomly, there would be prescription of nontrivial function.

To be logically consistent with naturalistic metaphysical presuppositions,
we exclude purposeful choice from any model of explanation for how this programming came about. Programming has to be explained solely in terms of
chance and necessity. The problem is that no chance and/or necessity model
has been able to explain programming function and the prescription of metabolism prior to the realization of any phenotypic benefit (organismic fitness).
Each selection of a physical nucleotide corresponds to pushing a quaternary
(four-way) switch knob in one of four possible directions. The most perplexing problem for evolutionary biology is to provide a natural mechanism for
setting so many functional configurable switch-settings at the genetic level.
These logic gates must be locked in open or closed positions prior to the existence of any living organism. Each selected nucleotide token out of four possibilities is bound with strong 3'5' phosphodiester bonds. Even after translation,
amino acid sequencing is "written in stone" with strong covalent bonds prior to
any folding. Amino acid sequencing is the primary determinate of molecular
machine shape and function. At the point of polymerization of informational
positive DNA single strands, no immediate function exists. No selectable
three-dimensional shape exists for the environment to favor when polyamino
acid strands form, either. The environment does not select for isolated functions, anyway. It could care less whether anything functions. The environment is blind to "usefulness." It has no goals, values or motives. The environment only selects very indirectly for best surviving, best-reproducing organisms. Evolution plays no role in programming decisions in pursuit of the
goal of successful computation and formal utility.

Primary structure (sequencing) of polyamino acid strings serves the same
relative purpose as primary structure of nucleic acid prescription. Amino acid
sequencing instructs secondary and tertiary structure (three-dimensional
shape). While chaperones and other factors affect folding, the sequencing of
the polyamino acid itself is by far the biggest determinant of shape, electrostatic charge, grooves, knobs, tunnels, hydrophobicities, and lock-and-key binding
of the globular protein molecular machine or enzyme to its substrate. Folding
proceeds according to minimum Gibbs free-energy sinks [59-67]. But, rigidly-
bound monomeric sequencing largely determines what these thermodynamic
and kinetic tendencies will be.

It is not sufficient for the environment to select the fittest living organisms. Organisms do not exist until after cooperative computational haltings

occur on many different levels. None of these computational haltings will occur without selection of appropriate symbols so as to generate formally efficacious programming sequences. In addition, all of these programming sequences must be integrated into a holistic operating system in order to organize even the simplest protometabolism. Organization, too, is mediated using multiple layers of material symbol systems (MSSs). Physical symbol vehicles (nucleotide "tokens") are used to represent formal quaternary (four-way) switch-setting "choices." These in turn determine higher order levels of transcriptional regulatory networks, multilayer hierarchical structures, transcript turnover regulation, and three-dimensional information retention in genomes [69-71].

5. Duplication plus variation produces no new Prescriptive Information

Self-replication tends to "get all the press" in life-origin literature. But the real issue of life origin lies in answering how the initial single positive strands of RNA instructions got sequenced so as to prescribe functional ribozymes, microRNA regulation, and eventual folding function. No new information is generated in base-pairing replications. Base-pairing has nothing to do with the generation of genetic information or coding in the initial positive informational strand. Base-pairing is purely physicodynamic, and quite secondary to the pre-programmed, formal, linear digital instructions of the single positive strand. Of course, now we know that the negative strand is also prescriptive, and that DNA is read in both directions [10, 12-14]. But this does not change the fact that base pairing itself is purely physicodynamic. The fact that the negative strand is also prescriptively informational only adds to the mystery of how either (or both) strands could have meaning and prescriptive function in their syntax.

Duplication of symbol syntax occurs frequently in cellular molecular biology. It is common for us to hypothesize extensively about the value of duplication without ever addressing the question, "Duplication of what?" A stochastic ensemble contains the most bits of Shannon uncertainty. But uncertainty is not information. What functionality would duplication of a stochastic ensemble provide? A stochastic ensemble is the equivalent of gibberish in language; it is the equivalent of "garbage in, garbage out" in programming.

If the strand being duplicated is something better than a stochastic ensemble, how did this sequence acquire its instructive functionality? The environment does not select for isolated function. It selects only for the fittest already programmed, already-living phenotypic organisms. We tend to just presuppose the functional value of the string being duplicated rather than explaining it. Presumably the value of duplication would be copying, not doubling, the amount of Prescriptive Information (PI) found in the initial strand. No new

Shannon uncertainty, let alone PI, is created in copying [81, pg 78]. So before proceeding into a discussion of duplication plus variation, it would be worth-while to reveiw that we have yet to explain how the initial positive informa-tional strand that is to be duplicated was itself programmed. How did *the object* of duplication acquire its functional value as a linear digital instructive string of logic gate settings? No naturalistic explanation seems to exist within scientific literature. The fact that that the negative base-paired strand instructs completely different microRNA regulatory function is nothing less than mind-boggling. We certainly have no naturalistic explanation for this additional in-credible phenomenon.

As mentioned above, no increase in the number of bits of uncertainty, let alone fits (functional bits) of Functional Sequence Complexity (FSC) [82] is achieved with duplication of an existing string. So the number of bits of po-tential information and fits of FSC remains the same with base-pairing dupli-cation, including any self-replication by ribozymes. If the number of bits of potential information remains the same with duplication, no new Prescriptive Information (PI) [4], either, could be generated by a self-replicating ribozyme from duplication alone.

But what about variation? Surely duplication *plus* variation can generate new PI, can't it? The answer is, "No!" If variation is random, it is noise pollu-tion of whatever PI previously existed in the string that was faithfully being duplicated. Information theory goes to great lengths to protect existing infor-mation in the Shannon channel from deterioration due to random noise. Typo-graphical errors never increase the functional information and meaning of any text. No scientific justification exists for our love of random variation as a supposed source of new PI.

Well, what if the variation is *not* random? What if the variation is somewhat militated by law-like orderliness and regularity? Can't that in-crease the PI content of the duplicated strand? The number of bits of uncer-tainty (potential information) is always reduced by any self-ordering tenden-cies of law-like cause-and-effect determinism. Necessity increases the proba-bility of events and spells the loss of freedom and uncertainty. Uncertainty is related inversely to probability. Any form of self-ordering in the variation of the duplication only increases the probability of redundant law-like interac-tions and decreases the bit content that measures uncertainty. The decrease in uncertainty results in a decrease in potential information retention, and there-fore a decrease in actual PI retention.

So, let's review where we are. We attribute nearly all sophisticated bio-logical programming to duplication plus variation of simpler instructions. But we are forced to note first that no new uncertainty or PI is produced in duplication itself. We note further that variation, whether totally random or partially cause-and-effect induced, only decreases PI, not increases it. Totally random sequencing produces virtually no nontrivial function. Physicodynamic necessity only increases probability, and therefore decreases uncertainty and information retention possibility. If variation's bits of uncertainty decrease, it's potential information content must and will simultaneously decrease. There can be no increase in PI with duplication plus variation. So whether variation is random, ordered, or mixed, it cannot possibly increase the PI of any symbol string, whether that symbol string is a stochastic ensemble or was originally somehow programmed. Variation, if random, can at best only produce increased bits of uncertainty with no nontrivial function. Randomness never produces purposefully considered programming choices. There is no reason to even hope for an increase in PI from random variation *or* law-like physicodynamic determinism.

There is only one kind of variation that can possibly increase the PI in a duplicated strand: choice-contingent variation. Since no new information was generated in the duplication process itself, the nontrivial PI of any strand subjected to duplication plus variation will only decrease absent efficacious choice-contingent intervention in that variation. Programming (Choice contingent causation and control [CCCC]) for potential function must occur to produce formal computation.

6. No observations of the spontaneous generation of formal prescription

There's a very good reason why we have never observed the spontaneous generation of life in accordance with Virchow's and Pasteur's First Law of Biology (all life must come from previously existing life). We have never seen the spontaneous generation of *any* Material Symbol System (MSS). Life is wholly dependent upon MSSs. All known life is cybernetic and semiotic. The key to life is sophisticated formal controls and the messaging of those controls, not boring, unimaginative, redundant, physicodynamic constraints.

The logic gate settings that program life are recorded into the 2-bit token selections (one of four equiprobable, independent nucleotide selections) of a genetic MSS. While genetic programming is fundamentally nonphysical (physicodynamically indeterminate or inert), that programming is instantiated into physicality using purposeful token choices and linear digital sequences of those token selections. The compounding multi-dimensional layers of PI in molecular biology do not change this. They only add to the extraordinary sophistication of genomic prescription.

No selectable phenotypic function exists at the time of each token selection (polymerization of the positive informational single strand of nucleic acid)[21, 83]. The inanimate environment has no perception of or interest in isolated function over non-function even if functionality were immediate with each nucleotide selection. But the usefulness of any nucleotide sequence is always only *potential*. Selection of each nucleotide must be *for* potential utility, not *of* current utility. The environment's variability can play no role in making purposeful programming choices (token selection) at the molecular/genetic level of positive strand nucleotide sequencing (The GS Principle [21].

Formal prescription of metabolic activity—life's PI—is primarily recorded in the sequencing of nucleotide tokens, three at a time using structural pairing, in the genetic MSS. Corruption of this PI is formally prevented to a large degree by use of an ideal Hamming block code that uses three nucleotide tokens to represent each prescribed amino acid. No naturalistic explanation exists for the origin of life's PI using a MSS and conceptually ideal redundancy block code to prevent frame-shift and other noise-pollution errors in the Shannon channel. Sophisticated repair mechanisms also exist in the cell that employ parity bits and numerous other formal functions to correct errors [84, 85].

In addition to the lack of any plausible hypothetical model for PI generation by inanimate nature, no empirical observations or prediction fulfillments of spontaneous PI generation have occurred. Chance and necessity logically, deductively, absolutely cannot make programming choices at bona fide decision nodes. In short, belief in the spontaneous generation of MSS's, or life, through natural process is without rational, empirical or predictive support. The notion is not even falsifiable. It violates both the First (Virchow and Pasteur) and Second Law of Biology (The GS Principle). Belief in the spontaneous generation of MSS's and life are in fact a blind faith more akin to adherence to a materialistic worldview ideology than to science.

7. Chance and necessity cannot program, steer, or regulate

Chance and necessity provide no plausible mechanism for programming, steering, regulation or control via genetics, genomics, epigenetics or epigenomics. Control is just metaphysically presupposed in every naturalistic model without any real explanation or scientific justification.

Control requires the exercise of purposeful choice contingency at bona fide decision nodes, logic gates, and configurable switch-settings. Neither fixed law nor chance can steer events toward formal utility. No combination of chance and necessity can write the formal rules for or actually set up a symbol system. To steer inanimate events toward potential usefulness and to integrate those functions into a cooperative effort require traversing The Cybernetic Cut across the one-way-only Configurable Switch (CS) Bridge [2] (See Chapter 3).

No physicochemical factors determine monomeric sequencing. Physicodynamic determinism would severely reduce the uncertainty required for information retention in any physical matrix. Sequencing is physicodynamically inert (dynamically incoherent) [77, 78]. Sequencing is decoupled from physicodynamic causation. It is independent of cause-and-effect physical determinism. This freedom is the very key to carbon chemistry being ideal for instantiation of large amounts of genetic prescriptive information into a physical matrix.

Attempts to explain the origin of formal programming from chance and necessity, thermodynamics and kinetics [86-96], have been unconvincing. Brillouin and others have attempted to equate Shannon's "informational" uncertainty with Maxwell-Boltzmann-Gibbs entropy. Notions of negentropy abound despite Boltzmann's prohibition of a negative constant in his famous equation: $S = k \log W$ [97]. Neither the number of microstates (W) nor Boltzmann's constant ($k = 1.38065 \times 10^{-23}$ joule/Kelvin) can be negative. In addition, every probability distribution is unique. Yockey showed that the probability distribution phase space of Boltzmann's physical entropy (S) cannot be equated or synthesized with Shannon's probability distribution of "informational" uncertainty (H) despite seemingly identical S and H equations (apart from the disallowed sign reversal of S) [41].

Schneider [98] and Adami [99, 100] are correct that uncertainty is not information. But mere subtractions of "after uncertainty" from "before uncertainty" do not measure up to the formal cybernetic proficiency of genetic control. Even archaeal genomes positively program hundreds of integrated computations. These cybernetic processes are not just decreasing measurements of combinatorial probabilism. Even if they were, inanimate nature cannot measure [101-104]. Human knowledge and measurement of the change in uncertainty inserts a human mental factor into the definition of even Shannon "information" that did not exist for most of the presumed 3.5 billion years of genetic prescription. Genetic instructions stand alone in their proficiency at making life happen.

Adami is also correct that bona fide information must be *about* something [99, 100]. But no source of formal "aboutness" has ever been provided from physics, chemistry and the physical environment alone. Yes, the environment provides a context. But the environment does not program the meaningful (cybernetically functional) configurable switch-settings that enable living organisms to overcome environmental insults and challenges that will occur only in the future. Aboutness will not be found in the environment itself.

Aboutness requires choice contingency. Aboutness is generated by the particular settings of configurable switches that program computational halting and that organize integrated circuits to meet future challenges. Meaning is formal, not physical. The environment cannot exercise choice contingency at genetic logic gates. The environment cannot program configurable switches. It can only favor the best already-computed living organisms.

A long string of Nobel laureates including Niels Bohr [105] and Jacques Monod [106] have argued that chance and necessity cannot generate nontrivial linear digital biological instructions. Bohr argued that "Life is consistent with, but undecidable from physics and chemistry." [105]. Many additional first-rate biologists such as Ernst Mayr [107, 108] and Bernd-Olaf Küppers [109, pg 166] have argued that physics and chemistry do not explain life. Says Hubert Yockey, "More than any other characteristic, computational linear digital algorithms distinguish life from non-life" [110].

What about some new law that has yet to be discovered? Couldn't that explain biological programming? Absolutely not. Law-like behavior is highly probable. Any such new law would be poison to programming freedom, uncertainty, Prescriptive Information (PI) and functional prescriptions. Clay surface adsorption is also purely physicodynamic with highly redundant order. Nucleotide adsorption results in homopolymers such as poly[A] and poly[G] onto montmorillonite and kaolinite clay surfaces [111-114]. Such homopolymerizations do not even contain much Shannon information, let alone semantic information. Such homopolymers offer no potential for prescription of formal function. Inflexible laws cannot select nucleotides and codon "block codes" (many-to-one symbol assignments used to reduce noise pollution in the Shannon channel) to achieve formal programming function. Life is the most highly informational phenomenon known to humans. Physicochemical propensities only preclude the generation of new information. New information generation requires uncertainty, not forced physical causation according to fixed laws. Laws are compression algorithms of reams of data made possible because the information content of all that data is so redundant and reducible. Functional sequencing, therefore, will never be explained by law. This includes hoped-for, as-of-yet undiscovered imaginary laws. Laws are too low informationally. Laws describe redundant, monotonous, highly ordered behavior with a probability approaching 1.0. Such behavior has almost no mathematical uncertainty. Law-like behavior precludes highly informational uncertainty required for linear digital prescription of formal function.

The high probability of adenosine's occurrence at each "decision node" in clay adsorption strings approaches 1.0. The uncertainty of such an RNA strand is close to the summation of 0 bits at each locus in the string times the number of ribonucleotides (- $\log_2 p$ = - $\log_2 1.0$ = O bits of uncertainty times n loci = 0 additive bits). Physicodynamics cannot generate nontrivial Prescriptive Information (PI). The generation of programming instructions requires freedom of selection at each logic gate. More importantly, it requires selection *for potential function* at each decision node. This is the very reason that sequencing is able to become a control function, not just a physical constraint. And the sequencing is not stochastic either, given that the control function manifests too much computational utility. No empirical evidence, predictive success, or rationality exists in the history of human experience to justify believing that unaided Markov processes can generate sophisticated algorithmic programming apart from hidden investigator involvement. Although the experiment begins with a random pool, selections of particular iterations are usually employed behind the scenes in order to isolate or produce the desired potential function.

The pre-RNA and RNA World models of life origin provide simplification and ideal reductionism for the study of the birth of biocybernetics and biosemiotics. Small RNA's provide a multitude of controlling (regulatory) functions even in current life. But spontaneous RNA generation is a biochemical vertical cliff [115-119]. In addition, ribose and RNA are too unstable for the long highly informational strands needed for life to have slowly developed in small increments [116, 120]. As a result, many life-origin specialists have been forced to return to Peptide First and Metabolism First models advocated by Gánti [121], Shapiro [122, 123], Dyson [124], Kauffman[125], Wachtershauser [126], Morowitz [127], Deamer [128], Lindhal [129], Russell [130], and many others.

Yet a Metabolism First origin of life is far from a foregone conclusion [131, 132]. Few life-origin scientists have been more respected than Leslie Orgel. Wrote Orgel, "In my opinion, there is no basis in known chemistry for the belief that long sequences of reactions can organize spontaneously---and every reason to believe that they cannot." [133] Indeed, organization should never be confused with mere self-ordering phenomena in nature [1, 4, 8]."

For life to have spontaneously generated, specific nucleotides would have had to have been selected according to programming or linguistic-like formal rules to prescribe post-folding, three-dimensional, tertiary and quaternary function. The function of protein complexes, zinc fingers, etc. would have had to have been anticipated in order to organize any protometabolism and eventual life.

The supposed new first law of biology, "The Tendency for Diversity and Complexity to Increase in Evolutionary Systems" [134] is metaphysically presupposed, not scientifically established. Evidence across the board in many species continues to grow that duplications plus variation, and mutations of all kinds, lead only to the deterioration of existing Prescriptive Information, not to a spontaneous increase in sophistication of biological computations. After 160 years of intense study of tens of millions of species, no new superior genus of plant or animal has ever been observed to arise spontaneously from existing genera. We have not even been able to purposefully *engineer* superior genera in the lab.

Nothing in science has ever overturned the real First Law of Biology—Virchow's "all life must come from previously existing life" and his cell theory, "omnis cellula e cellula," which means that every living cell comes from another living cell.

All that has happened over the last 160 years is a complete takeover of biology by materialistic, philosophic pre-assumptions that sweep the reality of needed PI under the rug. The current paradigm is bankrupt when it comes to explaining the source of genetic and genomic programming and regulation. No falsification has been provided of the GS Principle. The 2nd Law of Biology is alive and well: Selection must take place at the molecular/genetic level of genetic and genomic programming. Natural selection cannot explain the origin of PI needed for *any* species to come into existence, let alone all species.

8. Additional dimensions of PI do not negate linear digital prescription

Growing elucidation of multiple layers as well as three-dimensional cellular Prescriptive Information (PI) only adds to the sophistication of life's control systems [69]. It does not detract from the major role of linear digital prescription using a material symbol system. It only adds to it.

The fact that genomic "Turing tape" instructions must be edited and algorithmically processed does not obviate the importance of the Turing tape instructions themselves. Neither do gene overlappings, multi-chromosome assemblies of gene components, anti-sense strand transcriptions, and three-dimensional coiling information. These additional layers and dimensions of PI magnify the sophistication of linear digital prescription. The positive "sense" strand of DNA can prescribe a protein while the "antisense" strand from that same DNA molecule can convey instructions through providing transcription products needed for regulation [12, 135, 136]. Both strands are a form of linear digital prescription. Gene overlapping, too, is still a form of multi-layered linear digital prescription. The cell's layered and multi-dimensional cybernetic systems are mind-boggling. No conceivable physicodynamic explanation exists for such formal organization. After-the-fact environmental selection of the best holistic systems does not explain the integrated programming and parallel processing that produces those fittest organisms [21, 83].

No disagreement exists with the contention that "program language" may not accurately describe various epigenetic developmental controls. But the major steering role is still genetic and genomic. Small RNA, peptide, and small protein regulation of genomics is considerable. "Embryonic development is an enormous informational transaction in which DNA sequence data generate and guide the system-wide spatial deployment of specific cellular functions." [137].

Non-protein-coding regulatory DNA anti-sense strand transcriptions, covalent histone modifications and DNA methylation roles are all becoming clearer. Single-molecule fluorescence-spectroscopy has helped determine the distances of single motor proteins from polymerase molecules in transcription factories in living cells [138]. All these types of regulatory mechanisms yield quantitative three-dimensional information about the dynamic organization of living cells.

The genome's three-dimensional structure also allows sections of different chromosomes in tangled clumps to "converse" with each other even in fission yeast. [139]. Fission yeast have only three pairs of chromosomes. Yet even at this phylogenetically simple level, related genes are purposefully juxtapositioned rather than randomly tangled. Multiple genes are turned off and on simultaneously depending upon their relative position it the genomic structure. Groups of genes are found to share a related purpose in metabolism. The structure does not just arise from physicodynamic attractions and interactions. The juxta-positioning is clearly related to efficiency of formal function [140]. Groups of related genes are transcribed at certain sites. Transcription factors bind to DNA in "factories" and contribute to the formation of functional three-dimensional structure.

These are just a few of the additional epigenetic and multidimensional control mechanisms that are unfolding within cellular physiology. But linear digital prescription of function and genomics still reign supreme in the cybernetics of life.

9. Conclusions

The GS Principle states that selection must occur at the molecular/genetic level, not just at the fittest phenotypic/organismic level, to explain the generation of polynucleotides into polycodon linear digital prescription. Organismic/phenotypic selection (natural selection) cannot prescribe the linear digital programming of coded genetic instructions. Environmental selection cannot set configurable switches so as to achieve potential integrated circuits. "Selection pressure" comes only after the fact of computational halting. No fittest organisms exist for the environment to favor without prior computational success on many cooperative levels. Nucleotides must be selected and rigidly bound in a certain sequence to prescribe and integrate metabolism. Sequencing (primary structure) is the major determinant of three-dimensional molecular-machine shape (tertiary structure) and biofunction. Nucleotide sequencing is covalently (rigidly) bound into a linear digital string long before that string can fold into a ribozyme or can digitally prescribe polyamino acid sequencing. Metabolism depends on holistic integration of thousands of individual prescriptions, including regulatory peptides, proteins and microRNAs. Ultimately even most supposedly epigenetic factors such as methylations are prescribed and tightly controlled by liner digital genetic instructions. Such symbol systems are fundamentally formal, not physical.

Oligoribonucleotides are physical. Their polymerization reactions are fully subject to the laws of motion and to physicodynamic constraints. But their specific sequencing is physicodynamically inert. Their cybernetic function consists of instantiations of formal selections for potential function, not already-existing physicodynamic necessity. Selection pressure plays no role in determining not-yet-computed function at the formal programming level of configurable switch-setting (polymerization of each particular nucleotide).

Genes are linear digital programs that employ a material symbol system. Even their editing is ultimately controlled by other linear digital programs. Genes only functionally exist in the context of a formal representational material symbol system using physical symbol vehicles. Their configurable switch-settings are *physicodynamically inert*. Highly informational metabolic instructions cannot be generated by low-informational laws. The GS Principle defines the kind of selection that is required to set physical configurable switches so as to compute metabolic integration. Environmental selection has never been observed to generate the simplest example of formal computational success. Worse yet, the latter is a logical impossibility. Physicodynamics is limited to chance and necessity. It cannot make programming decisions using representational symbols to signify those purposeful choices. Formal computation is abstract, conceptual and nonphysical. Formal computation cannot be generated apart from choice contingency at true decision nodes. Metabolic organization and life have never been observed to exist independent of formally integrated circuits, optimized algorithms, and computational success.

The specific selection of each nucleotide from among four real options *controls,* not merely constrains, biofunction. The selection of each nucleotide also *prescribes*, not merely describes, its metabolic contribution. The GS Principle states that natural selection of already optimized genetic algorithms (the fittest already-computed living phenotypes) is inadequate to explain the derivation of a single-stranded polynucleotide's digital programming and computational prowess. Selection for function must occur at each decision node—each logic gate—each configurable switch-setting—each nucleotide polymerization onto the programming string.

References

1. Abel, D.L. 2007, Complexity, self-organization, and emergence at the edge of chaos in life-origin models, Journal of the Washington Academy of Sciences, 93, (4) 1-20.
2. Abel, D.L. 2008, 'The Cybernetic Cut': Progressing from description to prescription in systems theory, The Open Cybernetics and Systemics Journal, 2, 234-244 Open access at www.bentham.org/open/tocsj/articles/V002/252TOCSJ.pdf
3. Abel, D.L. 2009, The capabilities of chaos and complexity, Int. J. Mol. Sci., 10, (Special Issue on Life Origin) 247-291 Open access at http://mdpi.com/1422-0067/10/1/247
4. Abel, D.L. 2009, The biosemiosis of prescriptive information, Semiotica, 2009, (174) 1-19.
5. Abel, D.L. 2010, Constraints vs. Controls, Open Cybernetics and Systemics Journal, 4, 14-27 Open Access at http://www.bentham.org/open/tocsj/articles/V004/14TOCSJ.pdf.
6. Abel, D.L. 2011, Moving 'far from equilibrium' in a prebiotoic environment: The role of Maxwell's Demon in life origin. In Genesis - In the Beginning: Precursors of Life, Chemical Models and Early Biological Evolution Seckbach, J.Gordon, R., Eds. Springer: Dordrecht.
7. Abel, D.L.; Trevors, J.T. 2005, Three subsets of sequence complexity and their relevance to biopolymeric information., Theoretical Biology and Medical Modeling, 2, 29 Open access at http://www.tbiomed.com/content/2/1/29.
8. Abel, D.L.; Trevors, J.T. 2006, Self-Organization vs. Self-Ordering events in life-origin models, Physics of Life Reviews, 3, 211-228.
9. Abel, D.L.; Trevors, J.T. 2007, More than Metaphor: Genomes are Objective Sign Systems. In BioSemiotic Research Trends, Barbieri, M., Ed. Nova Science Publishers: New York, pp 1-15
10. Encode-Project-Consortium 2007, Identification and analysis of functional elements in 1% of the human genome by the ENCODE pilot project, NATURE 447, 799-816.
11. T Beiter, E.R., R Williams, P Simon 2008, Antisense transcription: A critical look in both directions, Cellular and Molecular Life Sciences, (Sep 15) doi 10.1007/s00018-008-8381-y.
12. Beiter, T.; Reich, E.; Williams, R.; Simon, P. 2009, Antisense transcription: A critical look in both directions, Cellular and Molecular Life Sciences (CMLS).
13. Yelin, R.; Dahary, D.; Sorek, R.; Levanon, E.Y.; Goldstein, O.; Shoshan, A.; Diber, A.; Biton, S.; Tamir, Y.; Khosravi, R.; Nemzer, S.; Pinner, E.; Walach, S.; Bernstein, J.; Savitsky, K.; Rotman, G. 2003, Widespread occurrence of antisense transcription in the human genome, Nat Biotechnol, 21, (4) 379-86.
14. Dinger, M.E.; Pang, K.C.; Mercer, T.R.; Mattick, J.S. 2008, Differentiating Protein-Coding and Noncoding RNA: Challenges and Ambiguities, PLoS Computational Biology, 4, (11) e1000176.
15. Axe, D.D. 2000, Extreme functional sensitivity to conservative amino acid changes on enzyme exteriors, J Mol Biol, 301, (3) 585-95.
16. Axe, D.D. 2004, Estimating the prevalence of protein sequences adopting functional enzyme folds, J Mol Biol, 341, (5) 1295-315.
17. Axe, D.D. 2010, The case against a Darwinian origin of protein folds, BIO-complexity, 1, 1-12.
18. Chang, S.; DesMarais, D.; Mack, R.; Miller, S.L.; Streathearn, G.E. 1983, Prebiotic organic syntheses and the origin of life. In Earth's Earliest Biosphere: Its Origin and Evolution, Schopf, J. W., Ed. Princeton University Press: Princeton, NJ, pp 53-92.
19. Abel, D.L.; Trevors, J.T. 2006, More than metaphor: Genomes are objective sign systems, Journal of BioSemiotics, 1, (2) 253-267.
20. Abel, D.L. 2009, The Universal Plausibility Metric (UPM) & Principle (UPP), Theor Biol Med Model, 6, (1) 27 Open access at http://www.tbiomed.com/6/1/27.
21. Abel, D.L. 2009, The GS (Genetic Selection) Principle, Frontiers in Bioscience, 14, (January 1) 2959-2969 Open access at http://www.bioscience.org/2009/v14/af/3426/fulltext.htm.
22. Stegmann, U.E. 2005, Genetic Information as Instructional Content, Phil of Sci, 72, 425-443.
23. Jacob, Francois 1974, The Logic of Living Systems---a History of Heredity. Allen Lane: London.
24. Alberts, B.; Bray, D.; Lewis, J.; Raff, M.; Roberts, K.; Watson, J.D. 2002, Molecular Biology of the Cell. Garland Science: New York.
25. Davidson, E.H.; Rast, J.P.; Oliveri, P.; Ransick, A.; Calestani, C.; Yuh, C.H.; Minokawa, T.; Amore, G.; Hinman, V.; Arenas-Mena, C.; Otim, O.; Brown, C.T.; Livi, C.B.; Lee, P.Y.; Revilla, R.; Rust, A.G.; Pan, Z.; Schilstra, M.J.; Clarke, P.J.; Arnone, M.I.; Rowen, L.; Cameron, R.A.; McClay, D.R.; Hood, L.; Bolouri, H. 2002, A genomic regulatory network for development, Science, 295, (5560) 1669-78.
26. Dose, K. 1994, On the origin of biological information, Journal of Biological Physics, 20, 181-192.

27. Griffiths, P.E.; Sterelny, K. 1999, Sex and Death: An Introduction to Philosophy of Biology. Univ. of Chicago Press: Chicago.

28. Wolpert, L.; Smith, J.; Jessell, T.; Lawrence, P. 2002, Principles of Development. Oxford University Press: Oxford.

29. Sterelny, K.; Smith, K.; Dickison, M. 1996, The extended replicator, Biology and Philosophy, 11, 377-403.

30. Hoffmeyer, J.; Emmeche, C. 2005, Code-Duality and the Semiotics of Nature, Journal of Biosemiotics, 1, 37-91.

31. Barbieri, M. 2004, Biology with information and meaning, History & Philosophy of the Life Sciences, 25, (2 (June)) 243-254.

32. Barbieri, M. 2006, Introduction to Biosemiotics: The New Biological Synthesis. Springer-Verlag Dordrecht, The Netherlands.

33. Barbieri, M. 2007, Is the cell a semiotics system? In Introduction to Biosemiotics: The New Biological Synthesis, Barbieri, M., Ed. Springer: Dordrecht, The Netherlands, pp 179-208.

34. Barbieri, M. 2007, BioSemiotic Research Trends Nova Science Publishers, Inc.: New York.

35. Barbieri, M. 2007, The Codes of Life: The Rules of Macroevolution (Biosemiotics). Springer: Dordrecht, The Netherlands.

36. Hoffmeyer, J. 1995, The swarming cyberspace of the body, Cybernetics and Human Knowing, 3, 1-10.

37. Hoffmeyer, J. 1997, Biosemiotics: Towards a new synthesis in biology, European Journal for Semiotic Studies, 9, 355-376.

38. Hoffmeyer, J. 2000, Code-duality and the epistemic cut, Ann N Y Acad Sci, 901, 175-86.

39. Hoffmeyer, J. 2006, Semiotic scaffolding of living systems. In Introduction to Biosemiotics: The New Biological Synthesis, Barbieri, M., Ed. Springer-Verlag New York, Inc. : Dordrecht, The Netherlands; Secaucus, NJ, USA pp 149-166.

40. Abel, D.L. 2002, Is Life Reducible to Complexity? In Fundamentals of Life, Palyi, G.; Zucchi, C.Caglioti, L., Eds. Elsevier: Paris, pp 57-72.

41. Yockey, H.P. 1992, Information Theory and Molecular Biology. Cambridge University Press: Cambridge.

42. Yockey, H.P. 1995, Information in bits and bytes, BioEssays, 17, (1) 85-88.

43. Yockey, H.P. 2002, Information theory, evolution and the origin of life, Information Sciences, 141, 219-225.

44. Yockey, H.P. 2005, Information Theory, Evolution, and the Origin of Life. Second ed., Cambridge University Press: Cambridge.

45. Quastler, H., 1958, A Primer on Information Theory in Symposium on Information Theory in Biology In A Primer on Information Theory in Symposium on Information Theory in Biology, The Gatlinburg Symposium, Gatlinburg, 1958; Yockey, H. P.; Platzman, R. P.Quastler, H., Eds. Pergamon Press: Gatlinburg, p pages 37 and 360.

46. von Neumann, J.; Burks, A.W. 1966, Theory of Self-Reproducing Automata. University of Illinois Press: Urbana,.

47. Turing, A.M. 1936, On computable numbers, with an application to the *entscheidungs problem*, Proc. Roy. Soc. London Mathematical Society, 42, (Ser 2) 230-265 [correction in 43, 544-546].

48. von Neumann, J. 1961, Collected works. Pergamon Press: New York.

49. Wiener, N. 1948, Cybernetics. J. Wiley: New York.

50. Wiener, N. 1961, Cybernetics, its Control and Communication in the Animal and the Machine. 2 ed., MIT Press: Cambridge.

51. Abel, D.L. 2006 Life origin: The role of complexity at the edge of chaos, Washington Science 2006, Headquarters of the National Science Foundation, Arlington, VA

52. Ma, L.; Weinberg, R.A. 2007, MicroRNAs in malignant progression, Cell Cycle, 7, (5).

53. Morin, R.D.; O'Connor, M.D.; Griffith, M.; Kuchenbauer, F.; Delaney, A.; Prabhu, A.L.; Zhao, Y.; McDonald, H.; Zeng, T.; Hirst, M.; Eaves, C.J.; Marra, M.A. 2008, Application of massively parallel sequencing to microRNA profiling and discovery in human embryonic stem cells, Genome Res.

54. Royo, H.; Cavaille, J. 2008, Non-coding RNAs in imprinted gene clusters, Biol Cell, 100, (3) 149-66.

55. Barham, J. 1996, A dynamical model of the meaning of information, Biosystems, 38, (2-3) 235-41.

56. Pross, A. 2005, On the chemical nature and origin of teleonomy, Origins of life and evolution of the biosphere, 35, (4) 383-94.

57. Lifson, S. 1987, Chemical selection, diversity, teleonomy and the second law of thermodynamics. Reflections on Eigen's theory of self-organization of matter, Biophys Chem, 26, (2-3) 303-11.

58. Pittendrigh, C.S. 1958, Teleonomy. In Behavior and Evolution, Roe, A.Simpson, G. G., Eds. Yale University Press: New Haven, CN, pp 390-416.

59. Flamm, C.; Fontana, W.; Hofacker, I.L.; Schuster, P. 2000, RNA folding at elementary step resolution, Rna, 6, (3) 325-38.

60. Woodson, S.A. 2001, Folding mechanisms of group I ribozymes: role of stability and contact order, Biochem Soc Trans, 30, (6) 1166-9.

61. Martin, A.; Schmid, F.X. 2003, The Folding Mechanism of a Two-domain Protein: Folding Kinetics and Domain Docking of the Gene-3 Protein of Phage fd, J Mol Biol, 329, (3) 599-610.

62. Munoz, V. 2007, Conformational Dynamics and Ensembles in Protein Folding, Annual Review of Biophysics and Biomolecular Structure, 36, (1) 395-412.
63. Fontana, W.; Schuster, P. 1998, Shaping space: the possible and the attainable in RNA genotype-phenotype mapping, J Theor Biol, 194, (4) 491-515.
64. Schuster, P.; Stadler, P.F.; Renner, A. 1997, RNA structures and folding: from conventional to new issues in structure predictions, Curr Opin Struct Biol, 7, (2) 229-35.
65. Schuster, P.; Stadler, P.F. 1994, Landscapes: complex optimization problems and biopolymer structures, Comput Chem, 18, (3) 295-324.
66. Schuster, P.; Fontana, W.; Stadler, P.F.; Hofacker, I.L. 1994, From sequences to shapes and back: a case study in RNA secondary structures, Proc R Soc Lond B Biol Sci, 255, (1344) 279-84.
67. Martinez, H.M. 1984, An RNA folding rule, Nucleic Acids Res, 12, (1 Pt 1) 323-34.
68. Johnson, D.E. 2010, Probability's Nature and Nature's Probability (A call to scientific integrity). Booksurge Publishing: Charleston, S.C.
69. Cao, G.S.; Liu, A.L.; Li, N. 2004, [Exploration of the hidden layers of genome.], Yi Chuan, 26, (5) 714-20.
70. Ma, H.W.; Kumar, B.; Ditges, U.; Gunzer, F.; Buer, J.; Zeng, A.P. 2004, An extended transcriptional regulatory network of Escherichia coli and analysis of its hierarchical structure and network motifs, Nucleic Acids Res, 32, (22) 6643-9.
71. Mata, J.; Marguerat, S.; Bahler, J. 2005, Post-transcriptional control of gene expression: a genome-wide perspective, Trends Biochem Sci, 30, (9) 506-14.
72. Taylor, R.B. 1980, Lamarckism Revival in Immunology Nature, 286, 837.
73. Koenig, R. 2000, Uphill Battle to Honor Monk Who Demystified Heredity, Science, 288, 37 - 39.
74. Dawkins, R. 1976, The Selfish Gene. 2 ed., Oxford Univerisy Press: Oxford.
75. Dwyer;, D.S.; Ogata, H.; Audic, S.; Claverie, J.-M. 2001, Selfish DNA and the Origin of Genes, Science, 291, (5502) 252-253.
76. Orgel, L.E.; Crick, F.H. 1980, Selfish DNA: the ultimate parasite, Nature, 284, (5757) 604-7.
77. Rocha, L.M. 2001, Evolution with material symbol systems, Biosystems, 60, 95-121.
78. Rocha, L.M.; Hordijk, W. 2005, Material representations: from the genetic code to the evolution of cellular automata, Artif Life, 11, (1-2) 189-214.
79. Rocha, L.M. 1997, Evidence Sets and Contextual Genetic Algorithms: Exploring uncertainty, context, and embodiment in cognitive and biological systems. . State University of New York, Binghamton.
80. Rocha, L.M. 1998, Selected self-organization and the semiotics of evolutionary systems. In Evolutionary Systems: Biological and Epistemological Perspectives on Selection and Self-Organization, Salthe, S.; van de Vijver, G.Delpos, M., Eds. Kluwer: The Netherlands, pp 341-358.
81. Johnson, D.E. 2010, Programming of Life. Big Mac Publishers: Sylacauga, Alabama.
82. Durston, K.K.; Chiu, D.K.; Abel, D.L.; Trevors, J.T. 2007, Measuring the functional sequence complexity of proteins, Theor Biol Med Model, 4, 47 Free on-line access at http://www.tbiomed.com/content/4/1/47.
83. Abel, D.L. The GS (Genetic Selection) Principle [Scirus Topic Page]. http://www.scitopics.com/The_GS_Principle_The_Genetic_Selection_Principle.html (Last accessed May, 2011).
84. Bradley, D. 2002, Informatics. The genome chose its alphabet with care, Science, 297, (5588) 1789-1791.
85. Mac Donaill, D.A. 2003, Why nature chose A, C, G and U/T: an error-coding perspective of nucleotide alphabet composition, Orig Life Evol Biosph, 33, (4-5) 433-55.
86. Rojdestvenski, I.; Cottam, M.G. 2000, Mapping of statistical physics to information theory with application to biological systems, J Theor Biol, 202, (1) 43-54.
87. Bennett, C.H. 1982, The thermodyamics of computation - a review, Int. J. Theor. Phys., 21, 905-940.
88. Schneider, T.D. 2000, Evolution of biological information, Nucleic Acids Res, 28, (14) 2794-9.
89. Wallace, R.; Wallace, R.G. 1998, Information theory, scaling laws and the thermodynamics of evolution, J Theor Biol, 192, (4) 545-559.
90. Weinberger, E.D. 2002, A theory of pragmatic information and its application to the quasi-species model of biological evolution, Biosystems, 66, (3) 105-19.
91. Yokomizo, H.; Yamashita, J.; Iwasa, Y. 2003, Optimal Conservation Effort for a Population in a Stochastic Environment, J Theor Biol, 220, (2) 215-231.
92. Klir; Wierman 1998, Uncertainty-Based Information. Physica-Verlag.
93. Ricard, J. 2003, What do we mean by biological complexity?, C R Biol., Feb;326, (2) 133-40.
94. Strippoli, P.; Canaider, S.; Noferini, F.; D'Addabbo, P.; Vitale, L.; Facchin, F.; Lenzi, L.; Casadei, R.; Carinci, P.; Zannotti, M.; Frabetti, F. 2005, Uncertainty principle of genetic information in a living cell, Theor Biol Med Model, 2, 40.
95. Balanovski, E.; Beaconsfield, P. 1985, Order and disorder in biophysical systems: a study of the correlation between structure and function of DNA, J Theor Biol, 114, (1) 21-33.

96. Theise, N.D.; Harris, R. 2006, Postmodern biology: (adult) (stem) cells are plastic, stochastic, complex, and uncertain, Handbook of experimental pharmacology, (174) 389-408.

97. Boltzmann, L. 1877, Weitere Studien über das Wärmegleichgewicht unter Gasmolekulen. , Königliche Academie der Wisschenshaft (Wien), Sitzungsberichte II Abteilung, 66, 275.

98. Schneider, T.D. 2000, Information Is Not Entropy, Information Is Not Uncertainty. In.

99. Adami, C. 1998, Introduction to Artificial Life. Springer/Telos: New York.

100. Adami, C.; Cerf, N.J. 2000, Physical complexity of symbolic sequences, Physica D, 137, 62-69.

101. Pattee, H.H. 1989, The measurement problem in artificial world models, Biosystems, 23, (2-3) 281-9; discussion 290.

102. Pattee, H.H. 1995, Evolving Self-Reference: Matter, Symbols, and Semantic Closure, Communication and Cognition-Artificial Intelligence, 12, 9-28.

103. Pattee, H.H. 1995, Artificial Life Needs a Real Epistemology. In Advances in Artificial Life Moran, F., Ed. Springer: Berlin, pp 23-38.

104. Pattee, H.H. 2001, The physics of symbols: bridging the epistemic cut, Biosystems, 60, (1-3) 5-21.

105. Bohr, N. 1933, Light and life, Nature, 131, 421.

106. Monod, J. 1972, Chance and Necessity. Knopf: New York.

107. Mayr, E. 1988, Introduction, pp 1-7; Is biology an autonomous science? pp 8-23. In Toward a New Philosophy of Biology, Part 1, Mayr, E., Ed. Harvard University Press: Cambridge, MA.

108. Mayr, E. 1982, The place of biology in the sciences and its conceptional structure. In The Growth of Biological Thought: Diversity, Evolution, and Inheritance Mayr, E., Ed. Harvard University Press: Cambridge, MA, pp 21-82.

109. Küppers, B.-O. 1990, Information and the Origin of Life. MIT Press: Cambridge, MA.

110. Yockey, H.P. 2000, Origin of life on earth and Shannon's theory of communication, Comput Chem, 24, (1) 105-123.

111. Franchi, M.; Gallori, E. 2005, A surface-mediated origin of the RNA world: biogenic activities of clay-adsorbed RNA molecules, Gene, 346, 205-14.

112. Ferris, J.P. 2002, Montmorillonite catalysis of 30-50 mer oligonucleotides: laboratory demonstration of potential steps in the origin of the RNA world, Origins of Life and Evolution of the Biosphere, 32, (4) 311-32.

113. Sowerby, S.J.; Cohn, C.A.; Heckl, W.M.; Holm, N.G. 2001, Differential adsorption of nucleic acid bases: Relevance to the origin of life, Proc Natl Acad Sci U S A, 98, (3) 820-2.

114. Ertem, G.; Hazen, R.M.; Dworkin, J.P. 2007, Sequence Analysis of Trimer Isomers Formed by Montmorillonite Catalysis in the Reaction of Binary Monomer Mixtures, Astrobiology, 7, (5) 715-722.

115. Shapiro, R. 1984, The improbability of prebiotic nucleic acid synthesis, Orig Life, 14, (1-4) 565-570.

116. Shapiro, R. 1988, Prebiotic ribose synthesis: a critical analysis, Orig Life Evol Biosph, 18, (1-2) 71-85.

117. Shapiro, R. 1999, Prebiotic cytosine synthesis: a critical analysis and implications for the origin of life, Proc Natl Acad Sci U S A, 96, (8) 4396-4401.

118. Shapiro, R. 2000, A replicator was not involved in the origin of life, IUBMB Life, 49, (3) 173-176.

119. Shapiro, R. 2002, Comments on 'Concentration by Evaporation and the Prebiotic Synthesis of Cytosine', Origins Life Evol Biosph, 32, (3) 275-278.

120. Shapiro, R. 1987, Origins : A Skeptic's Guide to the Creation of Life on Earth. Bantam: New York.

121. Gánti, T. 2003, The Principles of Life. Oxford University Press: Oxford, UK.

122. Shapiro, R. 2006, Small molecule interactions were central to the origin of life, Quarterly Review of Biology, 81, 105-125.

123. Shapiro, R. 2007, A simpler origin of life, Scientific American, Feb 12.

124. Dyson, F.J. 1998, Origins of Life. 2nd ed., Cambridge University Press: Cambridge.

125. Kauffman, S.A. 2000, Investigations. Oxford University Press: New York.

126. Wächtershäuser, G. 2007, On the Chemistry and Evolution of the Pioneer Organism, Chemistry & Biodiversity, 4, (4) 584-602.

127. Morowitz, H.J.; Kostelnik, J.D.; Yang, J.; Cody, G.D. 2000, From the cover: the origin of intermediary metabolism [see comments], Proc Natl Acad Sci U S A, 97, (14) 7704-7708.

128. Deamer, D.W. 1997, The first living systems: a bioenergetic perspective, Microbiol Mol Biol Rev, 61, (2) 239-61.

129. Lindahl, P.A. 2004, Stepwise evolution of nonliving to living chemical systems, OLEB, 34, 371-389.

130. Russell, M.J.; Hall, A.J. 1997, The emergence of life from iron monosulphide bubbles at a submarine hydrothermal redox and pH front., Journal Geological Society London,, 154, 377-402.

131. Orgel, L.E. 2000, Self-organizing biochemical cycles, Proc Natl Acad Sci U S A, 97, (23) 12503-7.

132. Anet, F.A. 2004, The place of metabolism in the origin of life, Curr Opin Chem Biol, 8, (6) 654-9.

133. Orgel, L.E. 1998, The origin of life--a review of facts and speculations, Trends Biochem Sci, 23, (12) 491-5.

134. McShea, D.W.; Brandon, R.N. 2010, Biology's First Law - The Tendency for Diversity and Complexity to Increase in Evolutionary Systems. University of Chicago Press Chicago, IL.

135. Seila, A.C.; Calabrese, J.M.; Levine, S.S.; Yeo, G.W.; Rahl, P.B.; Flynn, R.A.; Young, R.A.; Sharp, P.A. 2008, Divergent transcription from active promoters, Science, 322, (5909) 1849-51.

136. Core, L.J.; Waterfall, J.J.; Lis, J.T. 2008, Nascent RNA Sequencing Reveals Widespread Pausing and Divergent Initiation at Human Promoters, Science Express, online December 4, 2008 | Science DOI: 10.1126/science.1162228.

137. Davidson, E.H. 2010, Emerging properties of animal gene regulatory networks, Nature, 468, (7326) 911-20.

138. Tinnefeld, P.; Sauer, M. 2005, Branching out of single-molecule fluorescence spectroscopy: challenges for chemistry and influence on biology, Angewandte Chemie (International ed, 44, (18) 2642-71.

139. Tanizawa, H.; Et_al 2010, Mapping of long-range associations throughout the fission yeast genome reveals global genome organization linked to transcriptional regulation. , Nucleic Acids Research. Published online before print October 28, 2010.

140. Duan, Z.; Andronescu, M.; Schutz, K.; McIlwain, S.; Kim, Y.J.; Lee, C.; Shendure, J.; Fields, S.; Blau, C.A.; Noble, W.S. 2010, A three-dimensional model of the yeast genome, Nature, 465, (7296) 363-7.

The First Gene, David L. Abel, Editor 2011, pp 189-230 ISBN: 978-0-9657988-9-1

8. The Birth of Protocells[*]

David L. Abel

Department of ProtoBioCybernetics/ProtoBioSemiotics
Director, The Gene Emergence Project
The Origin-of-Life Science Foundation, Inc.
113 Hedgewood Dr. Greenbelt, MD 20770-1610 USA

Abstract: Could a composome, chemoton, or RNA vesicular protocell come to life in the absence of formal instructions, controls and regulation? Redundant, low-informational self-ordering is not organization. Organization must be programmed. Intertwined circular constraints (e.g. complex hypercylces), even with negative and positive feedback, do not steer physicochemical reactions toward formal function or metabolic success. Complex hypercycles quickly and selfishly exhaust sequence and other phase spaces of potential metabolic resources. Unwanted cross-reactions are invariably ignored in these celebrated models. Formal rules pertain to uncoerced (physiodynamically indeterminate) voluntary behavior. Laws describe and predict invariant physicodynamic interactions. Constraints and laws cannot program or steer physicality towards conceptual organization, computational success, pragmatic benefit, the goal of integrated holistic metabolism, or life. The formal controls and regulation observed in molecular biology are unique. Only constraints, not controls, are found in the inanimate physical world. Cybernetics should be the corner stone of any definition of life. All known life utilizes a mutable linear digital material symbol system (MSS) to represent and record programming decisions made in advance of any selectable phenotypic fitness. This fact is not undone by additional epigenetic formal controls and multi-layered Prescriptive Information (PI) instantiated into diverse molecular devices and machines.

Correspondence/Reprint request: Dr. David L. Abel, Department of ProtoBioCybernetics/ProtoBioSemiotics, The Origin-of-Life Science Foundation, Inc., 113 Hedgewood Dr. Greenbelt, MD 20770-1610 USA E-mail: life@us.net

*Sections from previously published peer-reviewed science journal papers [1-9] have been incorporated with permission into this chapter:

Introduction: Would control and regulation be necessary for protometabolism?

It would be hard to imagine any molecular biologist alive today who would question life's need for extensive control and regulation. The long-overlooked regulatory role of micro RNA's [10-13], peptides and very small proteins [14], for example, has dominated research in the last few years. The phrase "junk DNA" disappeared from the literature overnight.

Even in a theoretical protocell, any hint of a protometabolism would require the steering of biochemical pathways toward contribution to a productive holistic scheme [15]. The simplest pathways are usually quite conceptually complex. Each pathway leads to indispensable players in other pathways. These vital products must be delivered to the right place at the right time in the right form. All of the biochemical pathways need integration into interconnected cycles that contribute to the fulfillment of larger metabolic goals.

Says Tsokolov, "All life today incorporates a variety of systems controlled by negative feedback loops and sometimes amplified by positive feedback loops. The first forms of life necessarily also required primitive versions of feedback, yet surprisingly little emphasis has been given to the question of how feedback emerged out of primarily chemical systems." [15] Tsokolov points to the Belousov-Zhabotinsky (BZ) reaction as a possible model for chemical "systems" that might spontaneously develop autocatalytic feedback. He argues that the metabolism of contemporary life "evolved from primitive homeostatic networks regulated by negative feedback. Because life could not exist in their absence, feedback loops should be included in definitions of life." [15]

But could mere chemical circular constraint feedback achieve formal "regulation" in the sense of fine-tuning homeosatic metabolism? We shall examine this question in great detail in section 9 of this chapter when we examine Tibor Ganti's model. For now we will just take issue with Tsokolov's use of the word "system" to describe mere circular feedback constraint. As discussed in previous chapters, mere circular constraints do not constitute formal control systems. Tsokolov's intuitive sense is quite correct that there must be negative and positive feedback *controls* that make regulation of metabolism possible. But circular constraints are not controls. Circular constraints alone cannot establish formal regulation to the end of optimizing function. Controls are needed. But controls are formal, not physicochemical. Controls must be purposefully chosen in pursuit of formal function.

The simplest protometabolic schemes are highly abstract, formally functional, and goal-oriented. The cooperation of participants is extensive, yet highly measured. The required organization for even the simplest conceivable protolife is mind-boggling.

To make our models of spontaneous life-origin work, we continue to define down life to something that empirically never seems to come close to adding up to life. Whatever life is, one thing is for certain: it depends upon controls, not just mere physicodynamic constraints. Life never violates the laws of physics and chemistry. But the laws of physics and chemistry cannot generate the Prescriptive Information (PI)[6], controls and finely-tuned feedback *regulation* needed to organize and coordinate even the simplest conceivable protometabolism. Any Metabolism First model must first address the problem that chance and necessity cannot steer events toward pragmatic success. Chance and necessity cannot generate formal controls. Chance and necessity cannot pursue "usefulness." [1, 3-9, 16-20]

1. Emergence of spontaneous controls

The difference between mere self-ordering vs. bona fide organization has been made abundantly clear in the literature [1-4, 6-9, 19]. That difference has even been made clear specifically with reference to life-origin models [9]. The logic of the very notion of "self-organization" has even been challenged [9, 21]. No physical entity can "self-organize" itself into existence. An effect cannot cause itself. Organization is the effect of choice-contingent determinism, not physicodynamic determinism or chance.

Our physical central nervous systems did not organize themselves. Nonphysical human consciousness does not even organize itself. Chomsky argued quite successfully that we are born with inherent rules of language [22], for example. Infant minds just find themselves with a certain degree of inherent, pre-existing organizational thought structure that pre-exists empirical learning [22, 23]. Logic theory, mathematical axioms and the rules of mathematical manipulations seem to predate *Homo sapiens'* consciousness altogether in the workings of cosmic physical force interactions. Mathematical laws and their governance of physical interactions predate our discovery and description of them. Underlying formalisms seems to organize every aspect of inanimate physicality even prior to any discussion of life.

Organization always requires steering a course through multiple logic gates. Chance and necessity cannot cohort to make purposeful choices [4, 7, 9, 18]. Dissipative structures can spontaneously self-order into momentary high-energy states. But dissipative structures cannot program, organize or compute sophisticated formal utility. The most highly self-ordered dissipative structure to spontaneously occur in nature is probably the tornado. Tornados do not organize anything. Tornados only destroy organization at every turn.

Tornados are themselves only self-ordered, not organized [9]. Referring to tornado formation as "self-organization" is a classic case of sloppy definition/terminology that should be altogether unacceptable in any science.

2. What specific natural *mechanisms* of emergence have been elucidated?

Before addressing specifically the self-organization of protometabolism, we need first to take a critical look at the pre-assumption of emergence of any functional "natural process mechanism." What is a "mechanism"? "Mechanism" is a directed process, programmed procedure, technique, system, or component of a machine that achieves some pragmatic goal. "Mechanism" is a formal term, not a physicodynamic term. "Mechanism," like the term "useful work," has no place in pure naturalistic physics and chemistry. The concept of mechanism was simply high-jacked by philosophic naturalism. It was then bastardized to conform to materialistic metaphysical presuppositions.

Metaphysical naturalism presupposes that mass/energy alone is sufficient to explain everything. The etiology of "mechanism" from both Latin and Greek derives from the word "machine." Metaphysical naturalism has never demonstrated the ability of physicodynamics and so-called "natural process" to produce nontrivial machines or sophisticated utilitarian mechanisms. Naturalism merely pre-assumes what it purports to have scientifically proven. In reality, no purely physicodynamic interactions have ever been able to generate a formal mechanism that yields nontrivial formal function. And no theoretical model of spontaneous emergence of machines or formal mechanism through natural process has ever been demonstrated to actually occur in nature without investigator involvement in experimental design (e.g., so-called "directed evolution" and "evolutionary algorithms," both of which are self-contradictory non-sense terms. See below.).

Every case of supposed spontaneous self-organization ever published has in fact been a case of mere self-ordered cause-and-effect determinism, not self-organization. Wherever bona fide organization has been experimentally achieved, investigator involvement in the experimental design can always be identified hidden in the background information. More often than not, the steering and artificial selection are frankly acknowledged by the authors themselves right within Materials and Methods of the paper. No natural process tendency exists in inanimate nature to self-organize any utilitarian mechanism. Only experimenter desires and formal controls produce organization.

3. What *observations* of spontaneous emergence exist in the literature?

Many people would point to the experimental evidence of "directed evolution" and "evolutionary algorithms" in answer to this question. To understand why both of these fail to provide any evidence at all of spontaneous self-organization, we must return to a point made in Chapter 2, section 6.1. The choice of particular physical constraints is a formal enterprise, not a spontaneous physicodynamic interaction. The moment that initial conditions (constraints) are *chosen* in designing an experiment, those constraints immediately become formal controls. In such instances, nonphysical formalism has been introduced as a steering and controlling factor. What was supposed to model natural selection in fact models nothing more than artificial selection. Choosing constraints can constitute a very subtle form of "experimenter interference" ("investigator involvement") in experimental design. The result creates the illusion of empirical support for self-organization and undirected achievement of utility.

The pursuit of potential function guides these choices at true decision nodes. And this pursuit takes place prior to the realization of any naturally selectable fitness. No living organism exists yet to differentially survive.

The most relevant cases of so-called "directed evolution" related to life-origin science are ribozyme engineering papers [24-33] [34, 35]. In these papers investigator involvement is apparent in the purposeful selection of which effluent to use in successive iterations. This creates the illusion of a spontaneous evolutionary pathway. But evolution has no goals or chosen pathways to goals. So-called "directed evolution" is a classic example of formal control. Directed evolution boils down to the purposeful selection of initial conditions *for each iteration* of a highly integrated experimental plan and goal. Such experiments begin with a highly touted initial random phase space of stochastic ensembles of oligoribonucleotides. But the succession of repeated runs uses only carefully selected candidates from each previous iteration [36-39]. The procedure is anything but random. And it is not just constrained by physicodynamics. It is controlled by the formal choice contingency of the experimenter who pursues his or her own formally desired catalyst or self-replicant. The situation is reminiscent of Dawkins' embarrassing "target phrase" in *The Blind Watchmaker* [40] evolutionary software. Such a process has absolutely nothing to do with evolution.

So-called "directed evolution" and "evolutionary algorithms" are both self-contradictory terms. If the process is directed, it is not evolution. Evolution has no goal. If the experiment really does model evolution, it is never directed. Similarly, algorithms are always formal processes or procedures undertaken to achieve some function. This means they cannot possibly be evolutionary because evolution is blind to function and its pursuit. Natural selection simply favors the fittest already-programmed, already-living organisms.

The very experiments that were supposed to demonstrate inanimate self-organization invariably prove the opposite—the need for purposeful steering of physicodynamic events in order to achieve the desired formal function.

The result of oligoribonucleotide "evolution" experiments is typically attributed to trial and error. We fail to realize that "trial and error" is itself a teleological process of investigating and testing for what might work (inefficient though it may be). The Markov or drunken-walk "process" is erroneously and illegitimately offered as proof of self-organization. This "evidence" is then used in support of the notion of spontaneous generation of life.

The fatal flaw in the notions of "drunken walks," "directed evolution" and "evolutionary algorithms" is that each selection made by the experimenter is artificial, not natural. Each selection is made at the programming level *in pursuit of a potential function* that does not yet exist. The GS Principle (Genetic Selection Principle) is only affirmed, not falsified, by such engineering experiments [5, 41]. Natural selection favors only the fittest already-existing function. The inanimate environment possesses no ability to select for *potential* function. The inanimate environment cannot even select for *existing* isolated functions. Natural selection is nothing more than differential survival and reproduction of the fittest *already-living organisms*. [5, 41]. No differential survival of living organisms is involved in ribozyme engineering experiments. Directed evolution is nothing more than a string of purposeful logic gate or configurable-switch settings. Directed evolution is controlled, not constrained. The terms "directed" and "process" are quite legitimate in such laboratory procedures. The term "evolution" is not. Remove the hidden experimenter involvement (investigator interference) from Materials and Methods, and nothing of interest has ever been observed to spontaneously evolve. The reason is the loss of formal steering and control. When the experimenter is denied purposeful choices of which iteration to select and utilize at each step, no sustained uphill progress toward nontrivial functionality occurs.

The choice *for potential* function at the decision-node programming level, prior to the realization of any phenotypic fitness, is always artificial rather than natural. No natural mechanism exists for selection of not-yet-existent function or not-yet-existent phenotypes. Natural selection does not even select for isolated existing function. It selects only for the fittest already-programmed, already-living phenotypic *organisms*.

It is incumbent upon "true believers" in spontaneous self-organization out of inanimate physical interactions to demonstrate the same. It remains to be seen whether any such observation has ever once been made in the history of ordinary human observation, let alone reported in scientific literature. Nontrivial organization is never observed to arise independent of purposeful steering, programming choices, and deliberate pursuits of potential function. Yes, the self-contradictory term "self-organization" is used extensively in scientific literature. But no observations of bona fide "self-organization" exist without behind the scenes purposeful steering of agents.

Theoretically, long RNA chains in an RNA sequence space (Ω) do have the potential to include a stochastic ensemble identical to a prescriptive informational RNA strand. No reason, exists, however, that an instructive polymer would be able to isolate itself out of Ω at the right place and time to instruct multiple formal functions or to cooperate with other random stands to organize metabolism.

A severe competition would have existed in any prebiotic environment for nucleoside resources. Ribonucleosides in a prebiotic environment are very difficult to activate. Even non-cyclic homopolymers of ribonucleosides in an aqueous solution are almost impossible to form. Only 3'5' bonds are acceptable. Only right-handed sugars can be used. Inanimate nature has no goal or straightforward means to distinguish between functional vs. non- functional bonds or the correct optical isomer of each ribose. The statistical prohibitiveness becomes staggering even with the simplest protometabolic scenario.

Eigen and Schuster, along with others, have pointed out that sequence space and hypercyclic advance would have been greatly limited by competition for resources [42-50]. This would have applied particularly to a theoretical RNA world where the number and length of RNA strands is greatly limited. In non-heated aqueous solution, a maximum of eight to ten RNA mers can polymerize [51, 52]. Up to 55 mers can polymerize on montmorillonite [51], but these chains are homopolymers. These chains are produced only at the expense of information content. Homopolymers like polyadenosines contain essentially no Shannon uncertainty. Such high order could not have contributed to any random algorithmic programming of genes. Even if all the right primary structures (digital messages) mysteriously emerged spontaneously at the same time from Ω, "a cell is not a bag of enzymes." And, as we have pointed out several times, there would be no operating system to read these messages [53].

David Deamer's group showed that *RNA-like* polymers can be synthesized non-enzymatically from mononucleotides in lipid environments [54]. Chemical activation of the mononucleotides was not required. "Synthesis of phosphodiester bonds is driven by the chemical potential of fluctuating anhydrous and hydrated conditions, with heat providing activation energy during dehydration. In the final hydration step, the RNA-like polymer is encapsulated within lipid vesicles."

Ernesto di Mauro's group recently has been able to produce linear chains of 120 mers of either cAMP or cGMP, without templates or enzymes, through slow heating of activated cyclic monomers in aqueous solution [55]. So far, they have not had much success with pyrimidines. As with clay adsorption, however, these 100-mer strands are sorely lacking in Shannon uncertainty. The RNA stands are so highly ordered that no information-rich instructional PI could be instantiated into them. Genetics could not have been born out of homopolymers. Without functional base sequencing, no biopolymer would be able to instruct the organization of a cooperative metabolic network. Even if a random string *resembled* an informational strand, without a processing system and nanocomputers to read meaning into such strings according to preformed rules of interpretation, they could not contribute to an organized holistic metabolism. No PI exists in random–sequence nucleic acid *or* highly ordered homopolymers.

4. What predictions of emergence have been fulfilled?

Another major component of the scientific method is prediction fulfillment. Have any prediction fulfillments of the self-organization been observed to date? Normally we would emphasize "so far" when asking this question, especially when predictions of a new model have only recently been published. But what about when a theory has been well-published and exercised in scientific literature for 160 years? Macroevolution presupposes and requires the notion of self-organization. Macroevolution is purported to be the only organizing theory that makes any sense of biology. Theodosius Dobzhansky, for example, argued, "Nothing in biology makes sense except in the light of evolution." But, tens of millions of species are presumed to have self-organized and self-programmed themselves into existence in the last 3.5 billion years. If every single species of living organism arose by duplication plus variation of DNA, wouldn't one expect to have at least one prediction fulfillment of a bona fide new self-programmed organism in the last 160 years? Most species have already become extinct. But we should have seen at least dozens of new species originate in the last 10 years. Since there is often some question as to exactly what qualifies as a species, it is perhaps best to think in terms of new genera. How many new genera have been observed to evolve in the last 160 years? In truth, not one new genus can be cited as having self-programmed itself via "duplication plus variation." Thus, not a single prediction fulfillment has been realized since the theory was proposed.

Worse yet, we are not just talking about the self-organization of a new kind of organism. We have not even seen a single prediction fulfillment of the self-organization of something as simple as a paper clip. A paper clip is nothing more than a long uniform cylinder of certain malleability bent back onto itself in such a way as to produce an efficient paper grasper. How many functional paper clips have spontaneously self-organized from the ground's iron ore in the history of human observation? Science is about *repeated* observation and prediction fulfillment. It is also about common sense. The notion of self-organization is not only rationally absurd, it is without both observation and prediction fulfillment.

In the absence of human thought and involvement, we simply have not seen any instances of spontaneous PI generation or formal organization of any kind. And we also have not seen any instances of chaos, probabilistic combinatorial complexity or catastrophe generating PI or formal organization either. Faith in the spontaneous emergence of true formal organization is blind belief.

No random number generator ever produced a nontrivial computational program. No reason or empirical justification exists to suppose that randomness could ever generate nontrivial organization. Random polyamino- acid strings do not fold into specifically needed functional proteins. Only one in 10^{77} stochastic ensembles fold into a functional protein fold of *any* kind [56, 57]. Even protometabolism requires folds of a certain kind at the right place and time. Nucleotide and codon sequencing must first be right in order to prescribe each needed protein fold. In the absence of sophisticated ribosomes (highly conceptually complex RNA and Protein complex machines), any protocell would have no access to proteins. Peptides, like ribozymes, are grossly inadequate to catalyze most of the needed integration functions necessary for even the most rudimentary metabolism and life.

It is not plausible to expect hundreds to thousands of random sequence polymers to all spontaneously and cooperatively self-organize into an amazingly efficient holistic metabolic network. Stochastic ensembles of ribonucleotides do not even generate ribozymes without extensive investigator involvement in experimental design [58]. Extensive artificial selection is required particularly in the choice of which iteration to pursue when starting from a random phase space.

Whereas plausibility used to be a purely qualitative and subjective impression, now, a quantitative cut-off of plausibility exists in science with which to evaluate extremely low probability notions. Plausibility can be measured weighing hypotheses of extremely low probability against highly relevant probabilistic resources [59]. But it is important to understand that The Universal Plausibility Metric (UPM) [60] is not a probability measure. It is a measure of the *plausibility* of scientific hypotheses. A numerical inequality is provided by the Universal Plausibility Principle (UPP) whereby any chance hypothesis can be definitively falsified when its UPM metric of ξ is < 1 [60]. Both UPM and UPP pre-exist and are independent of any experimental design and data set. Every spontaneous generation model thus far published in peer-reviewed literature is definitively falsified by the Universal Plausibility Metric calculation and Principle (See Chapter 11).

5.   Is the hypothesis of self-organized emergence falsifiable?

The notion of *emergence* can be traced back to Aristotle [61], but George H. Lewes was probably the first to define it in 1875: "The emergent is unlike its components insofar as these are incommensurable, and it cannot be reduced to their sum or their difference." [62. pg. 412]. The idea of emergence blossomed in the 1920's with contributions from C. Lloyd Morgan [63], Samuel Alexander [64], Roy Sellars [65], Henre Bergson [66], and Arthur O. Lovejoy [67]. Weak and strong versions of emergence exist [68], but life-origin models require convincing models of strong emergence. The whole is greater than the sum of its parts [69]. Novel functional qualities are believed to arise spontaneously from inanimate physical components [70-73]. First, second, third and now fourth order (Types I-IV) emergence are said to exist [74]. Heritable linear digital genetic prescription can produce three-dimensional protein molecular machines that bind, transport and catalyze metabolic integration. Strong and Type IV emergent theory together attempt to explain the source of these phenomena. Admits Mark Bedau, "Although strong emergence is logically possible, it is uncomfortably like magic." [75].

If Virchow's and Pasteur's First Law of Biology ("All life must come from previously existing life") is to be empirically falsified, direct observation of spontaneous generation is needed. In the absence of such empirical falsification, a plausible model of mechanism at the very least for both Strong and Type IV emergence (formal self-organization) is needed. Manfred Eigen [42-45, 76-83] and Tibor Ganti [84-88] have been leaders in the search for mechanisms of biologic emergence from abiotic environments. Shuster joined with Eigen to hypothesize hypercycles [48, 49, 89-94]. The Edge of Chaos [72, 73, 95-102] has been proposed as a possible source, though the description of all of the above models often seems more poetic or cartoon-like than real. Kauffman's and Dawkin's publications, for example, are often devoid of any consideration of the biochemical catastrophic realities that plague life-origin bench scientists [40, 71-73, 103-107].

Attempts to define complexity are on-going [95, 108-114]. Sequence complexity has been extensively studied, though far from exhaustively [8, 115-121].

Much debate has occurred over the relation of linear complexity to semantic information [122-134]. Some have attempted to reduce the information of linear digital prescription in genes to mere thermodynamics, combinatorial probabilism, and physicodynamic complexity [71, 135-149]. Other investigators tend to view genetic information as literal and real [2, 8, 16, 150-155]. The special case of semiotic linear digital complexity has fostered the whole new field of Biosemiotics [2, 156-176].

Wild complexity claims are frequently espoused in the literature [5, 177-184]. How complexity relates to life has attracted innumerable papers [16, 185-191]. Systems Biology emphasizes the growing genomic and epigenetic complexity [192-194]. Attempts to deal with Behe's "irreducible complexity" [195] are appearing more often in scientific literature [196-200]. von Neumann [201] and Pattee [202-204] attempted to deal with the issue of Complementarity between the formal and physical aspects of complexity. Hoffmeyer and Emmeche have addressed the same basic problem with Code Duality [205, 206]. Stein described the different sciences of complexity [207]. Norris has researched hypercomplexity [208]; Garzon dealt with bounded complexity [209]; and Levins the limits of complexity [210]. Bennett originated Logical Depth and its relation to physical complexity [211]. More recently, better quality attempts have been made to explain the cybernetic nature of life naturalistically, from a teleonomic rather than teleological approach [70, 149, 212-239].

The naturalistic scientific community, and complexity theorists in particular, should collectively pursue falsification of the following null hypothesis: "Spontaneous nontrivial algorithmic optimization is never observed in nature apart from either 1) already existing biological prescriptive information, or 2) investigator involvement in experimental design." Falsification of this null hypothesis could be achieved with a single exception. But great care must be taken to expose hidden artificial controls. Such artificial controls are frequently programmed into supposed "evolutionary software" (e.g., the thoroughly embarrassing "target phrase" naively incorporated into Richard Dawkin's "evolutionary" program [40]).

An algorithm is a step-by-step process or procedure for solving a computational problem. Algorithms are formal enterprises requiring optimization. To optimize requires goals and intentionality. By definition, evolution cannot pursue goal-oriented procedures. Evolution is not a programmer of linear digital instructions and code[5]. Natural selection provides no mechanism for the practice of formal representationalism at the genetic level using tokens in a material symbol system (MSS)[240]. Selection pressure cannot employ a Hamming "block code" of triplet codons to symbolically represent or signify each amino acid. Evolution is after-the-fact differential survival and reproduction of already-living phenotypic organisms. The fittest organisms survive and reproduce best. Less fit living organisms and populations tend to die out faster. Nothing in NeoDarwinism, punctuated equilibrium, or any recent modifications of evolutionary theory explains the initial programming of linear digital prescriptive information.

The latest and best discussion of emergence as it relates to life-origin is found in the October 2010 (No. 4-5) issue of Origins of Life and Evolution of the Biosphere (OLEB), Vol. 40, on contingency vs. determinism and emergence [241-251]. After studying all of these papers, one is still left with no clear sense of how the notion of spontaneous emergence of self-organization might be falsified. If the notion is not falsifiable, it is not scientific.

As pointed out in "The capabilities of chaos and complexity," [1, 18], stand-alone chaos, complexity and catastrophe should never be confused in our theories with what we intelligent humans do using abstract conceptual nonlinear dynamic models. ProtoBioCybernetics is not interested in:

a. Modern-day human applications of non-linear dynamical systems theory
b. Investigator involvement (artificial selection) in chaos, catastrophe, and complexity experimental designs.

c. Information defined in terms of the reduced uncertainty of subjective "observers" and "knowers" who did not exist 3.5 billion years ago.

Life origin science wants to know the capabilities of stand-alone chaos and complexity before any animal consciousness existed. If all known life depends upon genetic instructions, how was the first linear digital prescriptive genetic information generated by natural process? How were all of the additional layers of biological PI generated and organized? Can chance and/or necessity produce genomic control and regulation schemes?

Can we falsify this null hypothesis?

NH1: Prescriptive Information (PI) [2, 6, 8] cannot emerge spontaneously from physicodynamics alone.

Only one example would suffice to accomplish falsification. Not one example has ever been provided.

As explained in previous chapters, PI refers not just to intuitive or semantic information, but specifically to linear digital *instructions* using a symbol system. 0's and 1's could be used. Letter selections from an alphabet could be used, as could A, G, T, or C from a phase space of four nucleotides. But any symbol system requires the use of agreed-upon formal arbitrary rules by the sender and receiver. PI can also consist of purposefully programmed logic gates that provide cybernetic controls, and configurable switch-settings that integrate formal circuits.

Can we falsify this null hypothesis?

NH2: Formal Organization [9] cannot emerge spontaneously from physicodynamics alone.

By "formal" we mean abstract, nonphysical, mental, choice-contingent, arbitrary, cognitive behavior that is typically goal- and function-oriented. Formal behavior is typically linguistic and/or mathematical. It entails representationalism, generalizations, and groupings into larger classes or categories (forms) rather than specific physical characteristics. Formal behavior is often computationally successful, integrated-circuit producing, or algorithmically optimizing behavior arising from bona fide decision node choices (not just "bifurcation points" (forks in the road) [4, 7].

Providing falsification of the H2 null hypothesis should be easy if physicalism is an accurate total description of objective reality. Yet to date in scientific literature, neither of these two null hypotheses, H1 or H2, has been falsified despite various restatements and appeals having been published in many peer-reviewed papers, academic book chapters, and conference lectures for over a decade now.

Both PI and formal organization are abstract, conceptual, choice- contingent, nonphysical entities [2-6, 8, 9, 16-18, 20, 53, 252]. Scientific endeavors to better understand cybernetic reality in nature are confronted with the uneasy suggestion of its transcendence over the physicality it controls. The chance and necessity of physicodynamics cannot program. At the heart of all naturalistic life-origin models lies the metaphysical pre-assumption of self-organization of inanimate physicality into sophisticated formal utility.

Cellular automata can only be created using algorithms. Algorithms are formal stepwise procedures based on discrete choices. Physicodynamics cannot generate algorithms. Belief in the self-organization of formalisms from inanimate physicodynamics is a non-falsifiable notion with zero empirical, prediction-fulfillment, and rational support.

6. What is Life?

Defining life has remained quite elusive despite many papers [253-261] and books [262, 263]. The negentropy concept of life was started by Schrödinger in his *What Is Life?* [264]. Brillouin promoted a physical concept of information and organization [265-267]. Rizotti considered defining life to be the central problem of biology [262].

In 2000 an international conference was called in an attempt specifically to refine a scientific definition of life [263]. All participants at this conference were required to submit in writing their definition of life. This author participated in and lectured at that conference. No two definitions of "life" were the same.

Our best attempts to reduce life to mere combinatorial complexity have often resulted in a rather laughable naiveté [16, 268]. One of the questions raised at that conference by this author was, "To what degree can we reduce life without loss of life?" [252]. If anything is holistic, it is life. Vivisection tends to kill the very life being studied. The pursuit of protocell theory, while necessary, can rapidly lead to a fatal cellular dissection. The reduction of life to something amenable to naturalistic modeling most often seems to result in non-life that is only proclaimed to be living.

In one of the most recent attempts to define life, Bedeau [269] promotes The Program-Metabolism-Container (PMC) model. This model emphasizes that life is a *functionally integrated* triad of chemical systems. The PMC model illustrates the Aristotelian approach to life rather than a Cartesian one. But, as usual, no naturalistic explanation is provided for the phenomenon of "program." The problem is that "functional integration" is formally generated, not physicochemically generated.

Biophysicist Hubert P. Yockey makes the unique observation that "there is nothing in the physico-chemical world [apart from life] that remotely resembles reactions being determined by a sequence and codes between sequences. The existence of a genome and the genetic code divides living organisms from non-living matter." (*Computers and Chemistry*, 24 (2000) 105-123). This may well constitute the most concise and parsimonious dichotomization of animacy from inanimacy available in the literature. Yet every definition of life published thus far seems laughably naïve and incomplete [253, 261]. The only significant move towards clarity seems to have come from acknowledging that all known life is cybernetic [255, 256]. But confusion still reigns as to the nature of cybernetics and how nature could have produced steering controls. Howard Pattee sums up the problem quite nicely referring to the problem of symbolization needed to record any form of prescription: "The amazing property of symbols is their ability to control the lawful behavior of matter, while the laws, on the other hand, do not exert control over the symbols or their coded references."

We must remember, however, that the full complement of nucleic acid code, ribosomes, protein enzymes, regulatory peptides, polypeptides and microRNAs are still present immediately after cell death. Life, therefore, would appear not to be reducible to coded prescriptive information (instruction), nanocomputers and their operating systems alone. Formal algorithmic processes must be ongoing for life to be alive.

Decades ago we used to tell students, "Life is more than a bag of enzymes." "Life" is characterized by ongoing holistic, homeostatic, metabolic *processes,* optimized algorithmic function, and successful computation. Intracellular life is goal-oriented. This includes development, growth, and reproduction. Many of the key elements of life are related to *formal organization and control* rather than mere physical structure or chemical constraints and interactions. As we learned in earlier chapters, physicodynamics cannot possibly generate nonphysical formalisms. The role of formalism—purposeful decision-node choices needed to effect cybernetic controls—may well turn out to be the best single differentiating criteria of life from nonlife. Yockey's observation of the uniqueness of sequence and codes in life is just a subset of this formalism. *The ability to pursue and select for potential function is formal and unique to life. The use of representational symbol systems is also formal and unique to life.* This is true not only in terms of living organisms' actions. It is also true of the sub-cellular molecular biological programming and algorithmic processing that make life possible.

Neither the DNA molecule nor its instantiated instructions are themselves alive. We cannot underestimate the role that proteins, peptides, polypeptides and microRNAs play in their action *on* DNA, and all of the other processes (e.g., epigenetic) that make life alive. Life is a holistic, highly PI-controlled and regulated cybernetic metasystem of integrated processes and formal procedures. Life is an integrated cooperative concert.

A theoretical spontaneously self-replicating ribozyme, if one existed without extensive human engineering, would not be alive either. It might undergo self- or mutual-replication. It would likely accrue errors ("mutations") that we would label "evolution." But random self-replicative errors ("typographical errors") have never been shown to improve the PI of any instruction set or computational program. A self-replicative crude ribozyme with low fidelity would most likely quickly lose its happenstantially acquired self-replicative trait. There is no good reason other than wish-fulfillment to expect the noise pollution that degrades its self-replicative function to prescribe other needed and even more sophisticated protometabolic functions. No empirical evidence, prediction fulfillments, or sound reason provides plausibility to the contention that a crudely self-replicative ribozyme would spontaneously improve, acquire additional metabolic capabilities, or become alive. In addition, no explanation has ever been provided by theorists as to how the initial self-replicative ribozyme would have acquired its initial PI syntax.

Gerald Joyce is generally credited with being the source of the so-called "NASA definition of life" [254]. This definition attempts to reduce life to little more than self-replication and mutability. But many questions have been raised pointing to the inadequacy of this definition. The imaginary primordial life upon which most investigators wish to base a definition of life currently has no empirical accountability. We tend to "define down" life to make our models of life-origin "work for us." But at what point does our stripped-down definition of life cease to adequately describe life, let alone define life? The indivisible unit of life is the cell. No entity less than a cell has ever been found to be alive. We have not even observed non-living "chemotons"[88] spontaneously generate, let alone living ones.

A common misconception in life-origin literature is that being "far from equilibrium" is somehow synonymous with being alive. It is not. Both candle flames and tornadoes are "far from equilibrium" (FFE). But obviously neither is alive. A hurricane is a dissipative structure that is extremely far from equilibrium. A hurricane is not only not alive, it is not even organized! It is only self-ordered. *Organization requires purposeful choices in pursuit of formal utility.* Pure physicodynamics knows nothing of the kind. *Physicodynamics knows only cause-and-effect determinism that is oblivious to any goal or means of achieving pragmatism.* Thus, spontaneous inanimate self-ordering structure, just like order and pattern, has little to do with prescription of formal function.

Just because we *say* a minimal system is alive doesn't make it alive. Often our imaginings of what life is are in reality quite sterile. Nothing clarifies our understanding and appreciation of life better than death. We are able to smell death much better than we are able to define life. Absolutely no confusion exists about the difference between life and death when we smell death. We have no problem differentiating life from nonlife when we view a loved one in a casket.

What is far more important than the definition of life is the question of what *prescribes* life? How could inanimacy have become animate? This transition had to occur prior to natural selection [5, 41]. Natural selection depends upon life already existing. Natural selection cannot explain life origin. Say Koch and Silver, "The moment of origin of The First Cell is in a fundamental sense also the moment of the start of Darwinian organismic evolution." [270, pg. 5]. Notice that evolution cannot even begin until after life exists. Few evolutionary biologists, unfortunately, appreciate, let alone verbalize this insight. Even micro-evolution cannot begin until after living reproducing organisms already exist. Environmental selection is nothing more than differential survival and reproduction of already-programmed, already-living, fittest organism. Selection pressure does not explain how inanimate nature assembled all the needed components, instruction set, algorithmic processes, coherent integration, and successful computations leading to life-creating and life-sustaining metabolism. Say Kock and Silver,

> "The First Cell arose in the previously pre-biotic world with the coming together of several entities that gave a single vesicle the unique chance to carry out three essential and quite different life processes. These were: (a) to copy informational macromolecules, (b) to carry out specific catalytic functions, and (c) to couple energy from the environment into usable chemical forms. . . . but only when these three processes occurred together was life jump-started and Darwinian evolution of organisms began."[270, pg. 227]

Notice in this quote that "informational macromolecules" are just presupposed, not explained. The initial interest is merely in *copying* this information. Nobody seems to have a clue how this initial information that needs copying got written in the first place. For most of the last century life-origin science has centered on biochemistry and astrobiology. The problem of the source of initial formal Prescriptive Information has rarely been reluctantly acknowledged by metaphysical naturalism. The problem is in fact regularly swept under the rug. Evolution theory concerns itself only with the duplication and variation of already existing information.

The RNA World model provided hope of a catalytic biochemical system that could double as an information carrier. But to date no explanation has been provided as to how the sequencing of initial single positive strands of RNA could have acquired their functional sequence specificity needed for the strand to double back onto itself to form each secondary and tertiary catalytic structure.

The Gene Emergence Project and Origin of Life Prize have both sought to stimulate naturalistic models of how an initial linear digital symbol system and bijection (mapping) code could have been generated by physicodynamics alone, without any formal components. First we need to explain how arbitrary rules were set up to achieve such formally organized systems. We then can begin to try to explain how the specific instructions for each function were written into each biopolymeric informational strand. The point of focus of research needs to be how symbols were selected at the molecular/genetic level and how configurable switches were set and linked together with rigid covalent bonds prior to any folding, and prior to any phenotypic function. These are the questions that best define life and its uniqueness. But they are formal questions that mere physicodynamic interactions cannot answer.

Thus far, all attempts to define life have proved unsuccessful. The following is an editable attempt to provide an irreducible *description* of existing life. It is taken with permission from the discussion section of the Origin of Life Prize website (www.lifeorigin.org). Listed are essential characteristics and criteria exhibited by *all known free-living organisms.* Minimal empirical life could be described as any system which from its own inherent set of biological instructions, however crude, can perform all ten of the following functions:

1) Delineate itself from its environment through the production and maintenance of membrane equivalent, most probably a rudimentary or quasi-active-transport membrane necessary for selective absorption of nutrients, excretion of wastes, and overcoming osmotic and toxic gradients,

2) Write, store, and pass along into progeny Prescriptive Information (PI; linear digital cybernetic programming) needed for organization; provide steering, control, regulation, and management for usable energy derivation and for needed metabolite production and function; symbolically encode and communicate functional messages through a transmission channel to a receiver/decoder/destination/effecter; establish and operate a semiotic material symbol system (MSS [240, 271, 272]) using "messenger molecules;" integrate past, present and future time into its biological prescriptive information content,

3) Bring to pass through algorithmic processing the above recipe instructions into the production or acquisition of actual catalysts, coenzymes, cofactors, small RNAs, etc.; physically orchestrate the biochemical processes/pathways of metabolic reality; manufacture and maintain physical cellular architecture. The algorithmic processing of PI must also be inherited.

4) Capture, transduce, store, call up when needed, and carefully utilize energy for formal, useful work,

5) Actively self-replicate and eventually reproduce, not just passively polymerize or crystallize; pass along the apparatus and "know-how" for homeostatic metabolism and reproduction into progeny,

6) Self-monitor and repair its constantly deteriorating physical matrix of bioinstruction retention/transmission, and its architecture,

7) Develop and grow from immaturity to reproductive maturity,

8) Productively react to environmental stimuli. Respond in an efficacious manner that is supportive of survival, development, growth, and reproduction,

9) Possess relative phenotypic stability, yet sufficient genetic variability to allow for adaptation and potential evolution.

10) Be capable of dying

Differences of opinion still seem to prevail as to whether *Mycoplasma genitalium* is a free-living organism. Certainly the even simpler organism *Carsonella ruddii*, the endosymbiont of psyllids, is not free-living. But even *Mycoplasma genitalium* manifests nearly all (if not all) ten of the above characteristics. All classes of archaea, bacteria, and every other known free-living organism, meet *all* ten of the above criteria. Eliminate any one of the above ten requirements, and it remains to be demonstrated whether that system is or could be considered truly "alive." Simpler descriptions and definitions of life arising from abiogenic imaginings suffer from fictional departures from known responsible parameters of life. Purely metaphysical imperatives then elevate such imaginings to the level of scientific necessity: "The spontaneous generation of life HAD to have happened because here we are." What an absurd, embarrassing contention for any academic to seriously state! We just presuppositionally pre-assume what we purport to have proven. We have not proven the spontaneous generation of life. The original First Law of Biology, "all life must come from previously existing life," is still alive and well.

Ribozyme, polypeptide, protein, prion, riboprotein, aptamer and ligand conglomerations do not meet many of these minimal criteria of free-living life. Neither do viroids and viruses. Even in historical science, there must be some degree of empirical accountability to our theories. *Proposing a plausible mechanism that explains the origin of life must not consist of "defining down" the meaning and essence of the observable phenomenon of "life" to include "nonlife" in order to make our theories "work for us."* Any scientific life-origin theory must connect with "life" as we observe it (the "continuity principle"). Science will never be able to abandon its empirical roots in favor of purely theoretical conjecture. Science must also constantly guard itself against Kuhnian paradigm ruts. The fact that most scientists currently believe a certain model does not establish its veracity. But we must also be open-minded to the possibility that life has not always existed in the form that we currently observe. And we must take into consideration the limitations of any historical science where the observation of past realities is impossible.

7. Can a computer analogy be applied to life?

Recently the exaggerated claim has been made of the creation of synthetic life [272]. Anyone doubting the role of and necessity for PI in life should listen to Craig Venter's discussion of his own claims of having synthesized life:

> http://www.guardian.co.uk/science/video/2010/may/20/craig-venter-new-life-form

The discussion only affirms the contention that "All known life is cybernetic." Venter's methodologies always start with what he calls "software." Says Donald E. Johnson (who holds Ph.Ds in both chemistry and Information/Computer Science), "Since all of the components used to manufacture Craig Venter's synthetic organism were produced by living organisms, Craig Venter's accomplishment was definitely not life from non-life." But Johnson agrees with Venter's cybernetic paradigm of life:

> The DNA is equivalent to physical memory (RAM, ROM, disk) – the memory hardware. The genome is the memory content: the implemented prescriptive algorithm with its functional data. DNA is hardware. Genome is formal software instantiated into the material symbol system of DNA. Both hardware and software must be designed to have a working system. Any functional hardware is an implementation of a prescriptive algorithm. The control unit of a CPU, for example, can have the control algorithm implemented in hardware (electronics), firmware (microcode ROM), or software (writable control store). There are many aspects of the computing systems of life that are not yet known, such as how much of the operating system is "firmware" (designed into the hardware as an integral part of the instruction set) and how much is software (interpreted by the hardware)." [personal communication].

Johnson's insights are spelled out in great detail in his chapter in this anthology, in an excellent book entitled *Programming of Life* [273] and also in a book entitled *Probability's Nature and Nature's Probability (A call to scientific integrity)* [274].

Perhaps the DNA hardware is the concert of collective interaction of the DNA, body of proteins, peptides, polypeptides, ribosomes, and regulatory microRNAs of the cell. These players constitute the primary algorithmic processors in millions of nanocomputers in each cell. The linear digital prescription instantiated into nucleotide sequencing is a little like a Turing Tape, except that the multi-dimensional nature of genomics renders the Turing tape analogy far too simplistic. The rapidly unfolding added dimensions of biological PI include:

- Transcriptional editing
- reading DNA in both directions
- the non-protein-coding prescription of sRNAs by the anti-sense strand
- gene overlapping
- the assembling of gene fragments from multiple chromosomes
- the spatial grouping of related genes in the mass of chromosomes
- proof reading and error repair mechanisms
- the editing of post translational polyamino acid strings

The rapid growth of recognized multiple layers of PI only compound the sophistication of life's control mechanisms. To try to attribute all of these ingenious cybernetic innovations to mere "duplication plus variation (noise)" is nothing less than laughable.

The heuristic/operational value of using linguistic and computational analogies to describe genetic programming is widely accepted by naturalistic science. Some try to dismiss parallels with cybernetics as being merely metaphorical. The limits of the metaphor have been explored [141, 275-281].

Some investigators have questioned whether semantic information about phenotypic traits exists at all [143, 144, 154, 282-286]. Lwoff felt that we often take the genetic information and linguistic metaphors too far [287]. Others assert that the metaphor is misleading [71, 142-145, 288, 289]. Rocha [290, 291] seeks to explain formal self-organization and sign systems physcodynamically despite acknowledging the reality of Pattee's epistemic cut [292] and the need for semantic closure [204]. Others view genetic information as quite real [2, 4-6, 8, 20, 150-155], though not the sole key to understanding life.

The contention that biological programming is merely metaphorical and nothing more than a heuristic tool of molecular biology professors is simply not tenable. Linear digital prescription and the codon table both predate humans, their consciousness, and the very existence of metaphors. In addition, the entire field of computer science was inspired by molecular biology, not the other way around. Turing [293], von Neumann [294], and Wiener [295, 296] all got most of their ideas, inspiration and understanding of cybernetic principles from observing the growing knowledge of Mendelian genetics, Watson and Crick's discovery, and various cellular control mechanisms. If what is known today about molecular biology had been known 40 years ago, computer science would have advanced far faster.

8. Astrobiological and multiverse considerations of life-origin

Few seem to think through the value, or lack thereof, of appealing to panspermia to overcome the scientific implausibility of spontaneous abiogenesis on earth. The age of the cosmos is estimated to be only three times that of the age of the earth. Of what value is a mere time factor of 3 in solving the statistical prohibitiveness of spontaneous generation on earth?

It is for good reason that many theorists have found it necessary to appeal to the purely metaphysical notion of "multiverse" to salvage any naturalistic hope of spontaneous life-origin. Multiverse models imagine that our universe is only one of perhaps countless parallel universes [297-299]. It could be argued that multiverse notions arose only in response to the severe time and space constraints arising out of Hawking, Ellis and Penrose's singularity theorems [300-302]. Solutions in general relativity involve singularities wherein matter is compressed to a point in space and light rays originate from a curvature. These theorems place severe limits on time and space since the Big Bang. Many of the prior assumptions of limitless time and sample space in naturalistic models were eliminated by the demonstration that time and space in the cosmos are quite finite, not infinite. For instance, we only have 10^{17}-10^{18} seconds at most to work with in any responsible cosmological universe model since the Big Bang.

The notion of multiverse is literally "beyond physics and astronomy," the very meaning of the word "metaphysical." Appeals to the Multiverse worldview are becoming more popular in life-origin research as the statistical prohibitiveness of spontaneous generation becomes more incontrovertible in a finite Universe [303-305]. The problem is that belief in multiverse is no more scientifically responsible than appealing to superstition. If the only way we can prop up a supposedly scientific model is to appeal to the equivalent of superstition, the plausibility and worth of such a notion as a scientific theory are virtually non-existent. It has no place in science. Such notions belong only in science fiction novels.

The notion of multiverse has no observational support, let alone repeated observations. Empirical justification is completely lacking. It has no testability: no falsification potential exists. Multiverse imagination provides no prediction fulfillments. The non-parsimonious construct of multiverse grossly violates the principle of Ockham's (Occam's) Razor [306]. No logical inference seems apparent to support the strained belief other than a perceived need to rationalize what we know is statistically prohibitive in the only universe that we *do* experience. Multiverse fantasies tend to constitute a back-door fire escape for when our models hit insurmountable roadblocks in the observable cosmos. When none of the facts fit our favorite model, we conveniently create imaginary extra universes that are more accommodating. This is not science. Science is interested in falsification within the only universe that science can address. Science cannot operate within mysticism, blind belief, or superstition. A multiverse may be fine for theoretical metaphysical models. But no justification exists for inclusion of this "dream world" in the observational science of astrophysics.

Even if multiple physical cosmoses existed, it is still a logically sound deduction that linear digital genetic instructions using a representational material symbol system (MSS) [291] cannot be programmed by the chance and/or fixed laws of physicodynamics [1-6, 8, 9, 17, 20, 41]. This fact is not only true of the physical universe, but would be just as true in any imagined physical multiverse. Physicality cannot generate nonphysical PI [6]. Physicodynamics cannot practice formalisms (The Cybernetic Cut) [4, 307]. Constraints cannot exercise formal control unless those constraints are themselves chosen to achieve formal function [1]. Environmental selection cannot select at the genetic level of arbitrary [308] symbol sequencing (e.g., the polymerization of nucleotides and codons) (The GS Principle [Genetic Selection Principle] [5]). Polymeric syntax (sequencing; primary structure) prescribes future (potential; not-yet-existent) folding and formal function of small RNAs and DNA. Symbol systems and configurable switch-settings can only be programmed with choice contingency, not chance contingency or fixed law, if nontrivial coordination and formal organization are expected [6, 9]. The all-important determinative sequencing of monomers is completed with rigid covalent bonds before any transcription, translation, or three-dimensional folding begins. Any editing of the initial sequence is highly refined and purposeful, not haphazard. It is only made possible by extremely sophisticated molecular machines and highly tailored helper molecules. Very specific microRNAs regulate real time transcription of newly edited PI in order to meet metabolic goals and needs. Thus, imagining multiple physical universes or infinite time does not solve the problem of the origin of *formal* (nonphysical) biocybernetics and biosemiosis using a linear digital representational symbol system. The source of PI [6, 309] in a metaphysically presupposed material-only world is closely related to the problem of gene emergence from physicodynamics alone. The latter hurdles remain the number-one enigmas of life-origin research [310] when begun from purely physicalistic metaphysical presuppositions.

The main subconscious motivation behind multiverse conjecture seems to be, "Multiverse models can do anything we want them to do to make our models work for us." We can argue Multiverse models ad infinitum because their potential is limitless. The notion of Multiverse has great appeal because it can explain everything (and therefore nothing). Multiverse models are beyond scientific critique, falsification, and prediction fulfillment verification. They are purely metaphysical.

Even if panspermia or the notion of multiverse were accurate descriptions of a presumed objective reality, the origin of the extraordinary array of nanohardware, firmware, wetware, operating systems, languages, software applications, and specific prescriptive genetic information (source code and ap) would remain unexplained. No fixed laws or formulae can program metabolic programming and computation. Instruction is abstract and conceptual. Yet instruction and control are exactly what genomes and epigenetics do. In addition to instructing, they actually perform and regulate through algorithmic processing the entire metabolic symphony. They achieve such integrated and holistic function through the same formal hardware and firmware implementation as our engineered computers. The only difference is that our finest computers seem archaic compared to the cybernetics found within any prokaryote, let alone eukaryotic and metazoan cell systems.

9. Mere replication is not the primary issue of life origin

Base-pairing is easy to explain. It is purely physicodynamic. It has nothing to do with the generation of the initial informational sequence. Base pairing cannot possibly program the algorithmic instructions instantiated into the positive informational DNA strand. Yet the particular sequencing of nucleotides and codons in the positive strand is not the only controller of life. As we shall see, the negative strand is filled with PI too. Multiple layers of PI exist that prescribe the integration and computation of cellular metabolism. How did the inanimate environment program linear digital instructions (PI) using a material symbol system (MSS) [271]? How did nature know how to write noise-correcting Hamming block codes (a fixed number of nucleotides *representing* each amino acid letter of the protein word) in order to reduce noise pollution in the Shannon channel? How did a purely physical nature encrypt and decrypt arbitrary (choice-contingent) coding? Coding and translation are formal functions, not physicodynamic interactions or phase changes. There is no direct physicochemical reaction between mRNA and amino acids. The mRNA is dynamically inert in its instructive role. The codon table is arbitrary [5] and formal [4], not physical. It is also conceptually ideal [311-313]. The prescription is all in the physicodynamically indeterminate sequencing of nucleotides and codons, not in chemical reactions [308]. Says Stegmann, "Aboutness, misrepresentation and storage are semantic properties, but such properties are not posited by ordinary biochemistry." [154]

Inanimate mass/energy interactions cannot generate computational solutions. Physicality cannot program integrated circuits. Only already-existing algorithms from a pool of "potential solutions" can be optimized. The inanimate physical environment cannot generate or optimize algorithms. Prior to an algorithm having computational function, no basis exists in nature for selection. So the question becomes, "How did *any* computational program arise in nature? Computation is formal, not physical. Natural selection cannot generate formalisms. It can only prefer *the results* of formal computations, and only then after those computations have generated living organisms [5]. What would be the basis of natural selection favoring a half-written program that does not yet compute? Even if a formal computational program were to somehow spontaneously arise, why would an inanimate environment value and preserve it? What would process it? No basis for recognition of computational success exists in a prebiotic environment.

The only basis for natural selection from Darwin to this day has been survival of the fittest already-living organisms. But *no* organism exists without hundreds of cooperating formal algorithms all organized into one holistic scheme. The more computational steps that are required to achieve integrative success and computational halting, the harder it becomes for an inanimate environment to explain optimization of any purposeful multi-step procedure. Nature doesn't pursue formal function. And the more algorithms that must be simultaneously optimized and integrated to achieve overall organization, the harder it is to explain homeostatic metabolism.

Natural selection resembles public consumption of the best available software. The programming details and methodology of production are of no interest to retail purchasers of software. Pre-programmed, bug-free, superior utility is the only criterion of public selection. The consumer plays no role whatever in the writing or refinement of the program's computational efficiency. The finished product with the best reputation, availability, and lowest cost becomes "the fittest species." Just as consumers are oblivious to how the best software was produced, natural selection is oblivious to how the fittest species was produced. Natural selection offers no explanation whatever for programming at the genetic level. Similarly, natural selection does not explain the derivation of the many cooperative computational processes leading up to the origin of metabolism or life.

10. The generation of initial Prescriptive Information is the real issue of life origin.

When it comes to life-origin studies, we have to address how symbol selection in the genetic material symbol system came about objectively in nature [2]. Life origin science must address the derivation of objective organization and control in the first protocells. How did prescriptive information and control arise spontaneously out of the chaos of a Big Bang explosion, primordial slime, vent interfaces in the ocean floor, or mere tide pools?

Self-ordering phenomena arise spontaneously out of phase space, but we have no evidence whatsoever of formal organization arising spontaneously out of physical chaos or self-ordering phenomena [9]. Chance and necessity has not been shown to generate the choice contingency required to program computational success, algorithmic optimization, or sophisticated function [53].

If chance and necessity, order and complexity cannot produce formal function, what does? *Selection for potential* utility is what optimizes algorithms, not randomness (maximum complexity), and not fixed law (highly patterned, unimaginative, redundant order with no information retaining potential). Utility lies in a third dimension imperceptible to chance and necessity (See Chapter 4, Figure 3). What provides this third dimension is when each token in a linear digital programming string is arbitrarily (non-physicodynamically, but formally) selected for potential function. The string becomes a cybernetic program capable of computation only when signs/symbols/tokens are purposefully *chosen* from an alphabet to *represent* utilitarian logic-gate and configurable-switch settings. The choice represented by that symbol can then be instantiated into physicality using a dynamically inert (physicodynamically decoupled or incoherent) [290, 291, 314] configurable switch setting. At the moment the switch knob seen in Chapter 2, Figure 1a is pushed, nonphysical formalism is instantiated into physicality. Then and only then does algorithmic programming become a physical reality. Once instantiated, we easily forget the requirement of instantiation of *formal instructions and controls* into the physical system to achieve engineering function. It was the formal voluntary pushing of the configurable switch knob in a certain direction that alone *organized* physicality [3, 4, 8, 9, 17, 20, 315].

Degrees of integration are achieved through *a combination* of configurable switch-settings which we can reduce to binary representation. The selection of any combination of multiple switch settings to achieve degrees of organization is called programming. But purposefully flipping the very first binary configurable switch is the foundation and first step of any form of programming. Programming requires purposeful choice contingency. The measure of algorithmic compressibility requires a second dimension to visualize from the bidirectional vector graph of order vs. complexity. Only this 2nd dimension shows us where to place a sequence on the uni-dimensional vector graph showing varying degrees of order and complexity (see Figs 1 and 2 in Chapter 4, section 1).

Just as it takes an additional dimension to measure the algorithmic compressibility of a sequence, it takes a third dimension to measure the formal utility of any sequence. Formalisms are abstract, conceptual, representational, algorithmic, choice-contingent, nonphysical activities of mind. Formalisms typically involve steering toward utility. Formalisms employ controls rather than mere physicodynamic constraints. Formalisms require obedience to arbitrarily prescribed rules rather than forced laws. Physicodynamics cannot visualize, let alone quantify formal utility. No known natural process spontaneously writes an informational message string. As Howard Pattee has repeatedly pointed out, any type of measurement is a formal function that cannot be reduced to physicodynamics [204, 292, 316, 317]. We do not plug initial conditions into the formal equations known as "the laws of physics." We plug *symbolic representations* of those initial conditions into the laws of physics. Then we do formal mathematical manipulations of these equations to reliably predict physicodynamic interactions and outcomes. In this sense formalism governs physicality. The role that mathematics plays in physics is alone sufficient to argue for formalism's transcendence over physicality.

11. Mutations do not produce new Prescriptive Information

Stunningly, information has been shown *not* to increase in the coding regions of DNA with evolution. Mutations do not produce increased information. Mira et al [318] showed that the amount of coding in DNA actually decreases with evolution of bacterial genomes, not increases. This paper parallels Petrov's papers starting with [319] showing a net DNA loss with Drosophila evolution [319, 320]. Konopka [128] found strong evidence against the contention of Subba Rao et al [321, 322] that information increases with mutations. The information content of the coding regions in DNA does not tend to increase with evolution as hypothesized. Konopka also found Shannon complexity not to be a suitable indicator of evolutionary progress over a wide range of evolving genes. Konopka's work applies Shannon theory to known functional text.

Kok et al. [323] also found that information does not increase in DNA with evolution. As with Konopka, this finding is in the context of the change in mere Shannon uncertainty. The latter is a far more forgiving definition of information than that required for *Prescriptive* Information (PI) [6, 8, 9, 121]. It is all the more significant that mutations do not program increased PI. PI either instructs or directly produces formal function in an appropriately designed operating system and hardware. No increase in Shannon or PI occurs in duplication. What the previous chapters in this anthology show is that not even *variation* of the duplication produces new information, not even Shannon "information," and certainly not PI. Variation can reduce sequence order, moving the sequence toward randomness and thereby increasing its bit content of Shannon uncertainty. But it cannot generate a nontrivial increase in Functional Information (FI), of which PI is a subset along with merely Descriptive Information (DI).

All of the above work correlates well with Weiss et al [324] finding only 1% deviation from randomness in coding regions. One cannot increase "information" (really "uncertainty") very much when starting from only 1% deviation from randomness in the coding regions. Only 1% deviation from randomness is already nearly maxed out in uncertainty. How did a text that deviates only slightly from seeming randomness get *so* instructional and biofunctional? Clearly, mere combinatorial uncertainty is not going to explain the phenomenon of cybernetic genetic prescription.

No empirical evidence exists of mere variation ever having generated sophisticated PI, computational halting, or cybernetic integration of large numbers of pathways and cycles, or the achievement of metabolic goals.

12. Evolution requires a mutable genetic MSS separate from its phenotype

In all known current life, a Material Symbol System (MSS) using nucleotide and codon tokens is used to "represent" genetic instruction not only of the genes themselves, but of microRNAs that regulate those genes. Regulatory peptides and polypeptides much shorter than proteins are also prescribed by the DNA MSS. Many of the microRNAs are transcribed in reverse direction from the antisense strand unwound from the sense strand that prescribes the gene that instructs polyamino acid sequencing. Says Rocha, "Representations are used to, literally, materialize dynamical systems." [291, pg. 15] "Syntax is required for communication in reproduction and for variation, both essential for natural selection and [open-ended evolution] OEE." [291, pg. 14]

Physicist Howard Pattee explains that the matter symbol-problem is referred to as Philosophy's "problem of reference." Since all known life depends upon a MSS, one of the most fundamental questions of life-origin science is

"How do symbols come to stand for material structures [325-327]." [204, pg. 11]   Pattee also points out that,

> Self-reference that has open-ended evolutionary potential is an auton-
> omous closure between the dynamics (physical laws) of the material
> aspects and the constraints (syntactic rules) of the symbolic aspects of a
> physical organization. I have called this self-referent relation semantic
> closure [328] because only by virtue of the freely selected symbolic as-
> pects of matter do the law-determined physical aspects of matter be-
> come functional (i.e., have survival value, goals, significance, meaning,
> self-awareness, etc). Semantic closure requires complementary models
> of the material and symbolic aspects of the organism. [204, pg. 9-10]

Pattee and Rocha have demonstrated in many publications [329-336] that open-ended evolution (OEE) is impossible without a linear digital genetic symbol system *that can mutate independent of the real-time living of the phenotypic organisms that harbor them.* Outwardly, the same relatively stable phenotypes exist and mate while tremendous modifications can be occurring in their genomes.

Ruiz-Mirazo, et al. agree with the necessity of "phenotype-genotype decoupling" for open-ended evolution to be possible [222, 260]. Open-ended evolution (OEE) requires a mutable genetic Material Symbol System (MSS) separate from its phenotype. The linear digital genome must be able to undergo substantive changes in its instructive PI sequencing without disrupting phenotypic viability. Says Howard Pattee, "Separate description and construction components are necessary for complex systems that can adapt and evolve." [337, pg 261]

In addition, Pattee points out that, "A necessary condition for hereditary transmission is a classification process or a many-to-one mapping." [338, pg. 410]. Three nucleotide selections are mapped to one amino acid prescription. This many-to-one bijection, along with codon redundancy with multiple codons all prescribing the same amino acid, affords degrees of freedom for the genome to vary during maintenance of phenotypic form and function. The many non-critical regions of nucleotide sequence also permits random drift without affecting genetic prescription of proteins. Nucleotide sequencing in DNA is now known to prescribe significantly more critical function than gene coding [339, 340]. This will greatly reduce the number and size of sections considered to be inconsequential.

Most mutations are silent. Genetic drift would be impossible without a genetic material symbol system (MSS) that can experience abundant variation within the same basic "phenotype" [240, 272, 291]. The phase space of potential new instructional sequences would be severely limited if genetic drift via successive point mutations, duplications, inversions, and transpositions could not progress at the genetic level independent of initial phenotype realization.

How were metabolic unity and coherence established in any living organism?   The lone answer that withstands careful scrutiny is, "Only algorithmically; only cybernetically; only computationally."   All of these enterprises are nonphysical, choice-based, and formal. They depend upon sign/symbol/token use. The programming choices are represented by each physical nucleoside token selection. But the *selection* of each token itself is nonphysical. It is decoupled from physicodynamic determinism. It is a programming feature that requires freedom of selection at bona fide decision nodes. Only secondarily does each selection become instantiated into the physical medium of nucleotide syntax.

Prebiotic metabolic unity and coherence could only have been established through genetic algorithms, formal optimizations for utility, and cybernetic programming. Holistic protometabolism would have needed successful formal computation. The self-ordering processes of chaos theory can generate none of these formal interventions [2, 16, 53]. Despite abundant confusion in the literature, self-ordering is not self-organization. Physicodynamic constraint cannot steer toward formal function. Life cannot arise from order and monotonous patterning. It arises from cybernetic management mediated through linear digital programming. Life requires controls and constant regulation, not mere constraints. Four-way configurable switch settings (in the form of each nucleotide selection from among four options) must be set a certain way to prescribe integrated circuits and pragmatic computational success.

Howard Pattee argues that living matter is distinguished from nonliving matter by its ability to select particular initial conditions. But what aspect of physicality would enable it to "choose" its own initial conditions? The ability *to select constraints* prior to the unfolding of cause-and effect necessity amounts to *formal control*. The exercise of formal control over physicality traverses The Cybernetic Cut [4, 307] (Section 3.4) via the one-way Configurable Switch (CS) Bridge. The CS Bridge permits formalisms to be instantiated into physicality either through the selection of physical tokens or through the setting of physicodynamically indeterminate configurable switches and logic gates [4, 307]. No return traffic across the one-way CS Bridge occurs. Physicodynamics is never observed arbitrarily controlling formalisms. Formalisms require freedom from physical constraints.

The formal aspects of programming and Prescriptive Information (PI) using a Material Symbol System (MSS) must experience semantic closure with the physicodynamics into which the instructions and control mechanisms are instantiated. Says Pattee:

"I have called this self-referent relation semantic closure [328] because only by virtue of the freely selected symbolic aspects of matter do the law-determined physical aspects of matter become functional (i.e., have survival value, goals, significance, meaning, self-awareness, etc). Semantic closure requires complementary models of the material and symbolic aspects of the organism." [204, pg. 9-10]

13. Evolution cannot *pursue* organization and *potential* protometabolic schemes

No natural basis exists for optimization of a ribozyme's primary structure leading to folds that will only later enable self-replication, specific catalyses, or participation in *potential* protometabolic schemes.

As explained in Chapter 7 (The GS Principle), evolution cannot work at the molecular/genetic level of nucleic acid sequence prescription. And clearly chance and necessity cannot program functional nucleic acid sequence.

The genetic-like function of ribozymes is quite different from DNA's prescription of function. Much of DNA's prescription is indirect via codon sequencing, transcription, transcription-editing, micro RNA regulation, and translation into a completely different language. Ribozymal prescription, however, is direct. The sequencing of ribonucleotides directly determines secondary and tertiary folding and catalytic function. In addition, this form of linear digital prescription of folds and catalysis is only half of ribozymal capabilities. The other half consists of their direct genetic potential through self-replication.

The first problem, however, with evolution of both ribozymal functions is that neither function is selectable until after it exists. Catalytic function exists only after the ribozyme sequence polymerizes and folds. Any genetic-like function of ribozymes can only be realized through self-replication of the particular sequence optimized for self-replication. This is not the same sequence as one that would contribute best to some protometabolic function. Selection must take place at each decision node or logic gate of ribonucleotide selection. At that point in time (polymerization of the primary structure), no naturalistic (purely physicodynamic) basis for selection for function exists. Programming is finished before selection begins. Selection, therefore, is all or none.

Function must be optimized prior to natural selection by the prebiotic environment. But this leaves no basis for selection at each decision node—

each nucleotide polymerization—where the sequence is established that determines folding and function. No evolutionary mechanism exists. It is just imagined. What was supposed to be the scientific explanation of progress is found to be a fairy tale.

14. Conclusions

By what supposedly "natural" process did inanimate nature generate phenomena like

1) A genetic representational sign/symbol/token system?
2) Bona fide decision nodes and logic gates (as opposed to just random "bifurcation points")?
3) Physicodynamically-indeterminate (dynamically inert, incoherent) [291] configurable switch-settings that instantiate functional "choices" into physicality?
4) formal operating system and the hardware on which to run such software?
5) an abstract encoding/decoding system jointly intelligible to both source and destination?
6) many-to-one Hamming "block codes" (triplet-nucleotide codons prescribing each single amino acid) used to reduce the noise pollution in the Shannon channel of genetic messages?
7) the ability to achieve functional computational success in the form of homeostatic metabolism?

All of these attributes of life are nonphysical and formal, not physical and natural. They cannot have a materialistic, naturalistic explanation.

References

1. Abel, D.L. 2009, The capabilities of chaos and complexity, Int. J. Mol. Sci., 10, (Special Issue on Life Origin) 247-291 Open access at http://mdpi.com/1422-0067/10/1/247

2. Abel, D.L.; Trevors, J.T. 2006, More than metaphor: Genomes are objective sign systems, Journal of BioSemiotics, 1, (2) 253-267.

3. Abel, D.L. 2007, Complexity, self-organization, and emergence at the edge of chaos in life-origin models, Journal of the Washington Academy of Sciences, 93, (4) 1-20.

4. Abel, D.L. 2008, 'The Cybernetic Cut': Progressing from description to prescription in systems theory, The Open Cybernetics and Systemics Journal, 2, 234-244 Open access at www.bentham.org/open/tocsj/articles/V002/252TOCSJ.pdf

5. Abel, D.L. 2009, The GS (Genetic Selection) Principle, Frontiers in Bioscience, 14, (January 1) 2959-2969 Open access at http://www.bioscience.org/2009/v14/af/3426/fulltext.htm.

6. Abel, D.L. 2009, The biosemiosis of prescriptive information, Semiotica, 2009, (174) 1-19.

7. Abel, D.L. 2010, Constraints vs. Controls, Open Cybernetics and Systemics Journal, 4, 14-27 Open Access at http://www.bentham.org/open/tocsj/articles/V004/14TOCSJ.pdf.

8. Abel, D.L.; Trevors, J.T. 2005, Three subsets of sequence complexity and their relevance to biopolymeric information., Theoretical Biology and Medical Modeling, 2, 29 Open access at http://www.tbiomed.com/content/2/1/29.

9. Abel, D.L.; Trevors, J.T. 2006, Self-Organization vs. Self-Ordering events in life-origin models, Physics of Life Reviews, 3, 211-228.

10. Lagos-Quintana, M.; al., e. 2001, Identification of novel genes coding for small expressed RNA's, Science, 294, 853-858.

11. Lau, N.C.; Lim le, E.P.; Weinstein, E.G.; Bartel da, V.P. 2001, An Abundant Class of Tiny RNAs with Probable Regulatory Roles in Caenorhabditis elegans, Science, 294, (5543) 858-62.

12. Misteli, T. 2002, A new continent in the RNA world., Trends Cell Biol, Feb 12, (2) 61-2.

13. Riddihough, G. 2002, The other RNA world., Science, May 17;296, (5571) 1259.

14. Dinger, M.E.; Pang, K.C.; Mercer, T.R.; Mattick, J.S. 2008, Differentiating Protein-Coding and Noncoding RNA: Challenges and Ambiguities, PLoS Computational Biology, 4, (11) e1000176.

15. Tsokolov, S. 2010, A Theory of Circular Organization and Negative Feedback: Defining Life in a Cybernetic Context, Astrobiology, 10, (10) 1031-1042.

16. Abel, D.L. 2002, Is Life Reducible to Complexity? In *Fundamentals of Life*, Palyi, G.; Zucchi, C.Caglioti, L., Eds. Elsevier: Paris, pp 57-72.

17. Abel, D.L. 2006 Life origin: The role of complexity at the edge of chaos, Washington Science 2006, Headquarters of the National Science Foundation, Arlington, VA

18. Abel, D.L. 2008 The capabilities of chaos and complexity, Society for Chaos Theory: Society for Complexity in Psychology and the Life Sciences, International Conference at Virginia Commonwealth University, Richmond, VA., Aug 8-10.

19. Abel, D.L. 2011, Moving 'far from equilibrium' in a prebitoic environment: The role of Maxwell's Demon in life origin. In *Genesis - In the Beginning: Precursors of Life, Chemical Models and Early Biological Evolution* Seckbach, J.Gordon, R., Eds. Springer: Dordrecht.

20. Abel, D.L.; Trevors, J.T. 2007, More than Metaphor: Genomes are Objective Sign Systems. In *BioSemiotic Research Trends*, Barbieri, M., Ed. Nova Science Publishers: New York, pp 1-15

21. Overman, D.L. 1997, *A Case Against Accident and Self-Organization*. Rowman and Littlefield Publishers, Inc.: New York.

22. Chomsky, N. 1957, *Syntactic Structures*. Mouton: The Hague/Paris.

23. Sullivan, A. 2000, The problem of naturalizing semantics", Language & Communication 20, (2 April) 179-196.

24. Cheng, L.K.L.; Unrau, P.J. 2010, Closing the Circle: Replicating RNA with RNA, Cold Spring Harb Perspect Biol.

25. Ma, W.; Yu, C.; Zhang, W.; Zhou, P.; Hu, J. 2009, The emergence of ribozymes synthesizing membrane components in RNA-based protocells, Biosystems.

26. Fedor, M.J. 2009, Comparative Enzymology and Structural Biology of RNA Self-Cleavage, Annual Review of Biophysics, 38, (1) 271-299.

27. Bagby, S.C.; Bergman, N.H.; Shechner, D.M.; Yen, C.; Bartel, D.P. 2009, A class I ligase ribozyme with reduced Mg2+ dependence: Selection, sequence analysis, and identification of functional tertiary interactions, Rna, 15, (12) 2129-46.

28. Szathmary, E. 2007, Coevolution of metabolic networks and membranes: the scenario of progressive sequestration, Philos Trans R Soc Lond B Biol Sci, 362, (1486) 1781-7.

29. Vorobjeva, M.; Zenkova, M.; Venyaminova, A.; Vlassov, V. 2006, Binary Hammerhead Ribozymes with Improved Catalytic Activity, Oligonucleotides, 16, (3) 239-252.
30. Szathmary, E. 2006, The origin of replicators and reproducers, Philos Trans R Soc Lond B Biol Sci, 361, (1474) 1761-76.
31. Shechner, D. 2009, Revealing the RNA World?, Science, 326, (5957) 1159-a-.
32. Link, K.H.; Breaker, R.R. 2009, Engineering ligand-responsive gene-control elements: lessons learned from natural riboswitches, Gene Ther, 16, (10) 1189-201.
33. Jonikas, M.A.; Radmer, R.J.; Laederach, A.; Das, R.; Pearlman, S.; Herschlag, D.; Altman, R.B. 2009, Coarse-grained modeling of large RNA molecules with knowledge-based potentials and structural filters, RNA, 15, (2) 189-99.
34. Paul, N.; Joyce, G.F. 2002, Inaugural Article: A self-replicating ligase ribozyme, PNAS, 99, (20) 12733-12740.
35. Hutton, T.J. 2002, Evolvable self-replicating molecules in an artificial chemistry, Artif Life, 8, (4) 341-56.
36. Ellington, A.D.; Szostak, J.W. 1990, In vitro selection of RNA molecules that bind specific ligands, Nature, 346, (6287) 818-822.
37. Tuerk, C.; Gold, L. 1990, Systematic evolution of ligands by exponential enrichment -- RNA ligands to bacteriophage - T4 DNA-polymerase, Science, 249, 505-510.
38. Robertson, D.L.; Joyce, G.F. 1990, Selection in vitro of an RNA enzyme that specifically cleaves single-stranded DNA, Nature, 344, 467-468.
39. Tuerk, C.; Gold, L. 1990, Systematic evolution of ligands by exponential enrichment: RNA ligands to bacteriophage T4 DNA polymerase, Science, 249, (4968) 505-10.
40. Dawkins, R. 1986, *The Blind Watchmaker*. W. W. Norton and Co.: New York.
41. Abel, D.L. The GS (Genetic Selection) Principle [Scirus Topic Page]. http://www.scitopics.com/The_GS_Principle_The_Genetic_Selection_Principle.html (Last accessed September, 2011).
42. Eigen, M. 1971, Self-organization of matter and the evolution of biological macromolecules, Naturwissenschaften, 58, (In German) 465-523.
43. Eigen, M. 1971, Molecular self-organization and the early stages of evolution, Experientia, 27, (11) 149-212.
44. Eigen, M. 1983, Life from the test tube?, MMW Munch Med Wochenschr, Suppl 1, S125-135.
45. Eigen, M. 1987, New concepts for dealing with the evolution of nucleic acids, Cold Spring Harb Symp Quant Biol, 52, 307-320.
46. Eigen, M. 1992, *(with Winkler-Oswatitsch, R.), Steps Toward Life: A Perspective on Evolution*. Oxford University Press: Oxford, UK.
47. Smith, J.M. 1979, Hypercycles and the origin of life, Nature, 280, (5722) 445-446.
48. Eigen, M ; Gardiner, W.C., Jr.; Schuster, P. 1980, Hypercycles and compartments. Compartments assists--but do not replace--hypercyclic organization of early genetic information, J Theor Biol, 85, (3) 407-411.
49. Eigen, M.; Schuster, P.; Sigmund, K.; Wolff, R. 1980, Elementary step dynamics of catalytic hypercycles, Biosystems, 13, (1-2) 1-22.
50. Schuster, P. 1984, Polynucleotide evolution, hypercycles and the origin of the genetic code, Adv Space Res, 4, (12) 143-151.
51. Ferris, J.P.; Hill, A.R., Jr.; Liu, R.; Orgel, L.E. 1996, Synthesis of long prebiotic oligomers on mineral surfaces, Nature, 381, (6577) 59-61.
52. Joyce, G.F.; Orgel, L.E. 1999, Prospects for understanding the origin of the RNA World. In *The RNA World*, Second ed.; Gesteland, R. F.; Cech, T. R.Atkins, J. F., Eds. Cold Spring Harbor Laboratory Press: Cold Spring Harbor, NY, pp 49-78.
53. Trevors, J.T.; Abel, D.L. 2004, Chance and necessity do not explain the origin of life, Cell Biology International, 28, 729-739.
54. Rajamani, S.; Vlassov, A.; Benner, S.; Coombs, A.; Olasagasti, F.; Deamer, D. 2008, Lipid-assisted synthesis of RNA-like polymers from mononucleotides, Orig Life Evol Biosph, 38, (1) 57-74.
55. Costanzo, G.; Pino, S.; Ciciriello, F.; Di Mauro, E. 2009, RNA: Processing and Catalysis: Generation of Long RNA Chains in Water, J. Biol. Chem., 284, 33206-33216.
56. Axe, D.D. 2000, Extreme functional sensitivity to conservative amino acid changes on enzyme exteriors, J Mol Biol, 301, (3) 585-95.
57. Axe, D.D. 2004, Estimating the prevalence of protein sequences adopting functional enzyme folds, J Mol Biol, 341, (5) 1295-315.
58. Lohse, P.A.; Szostak, J.W. 1996, Ribozyme-catalysed amino-acid transfer reactions, Nature, 381, (6581) 442-4.
59. Dembski, W. 1998, *The Design Inference: Eliminating Chance Through Small Probabilities*. Cambridge University Press: Cambridge.
60. Abel, D.L. 2009, The Universal Plausibility Metric (UPM) & Principle (UPP), Theor Biol Med Model, 6, (1) 27 Open access at http://www.tbiomed.com/content/6/1/27.
61. Aristotle Metaphysics, Book 8.6.1045a:8-10 In.

62. Lewes, G.H. 1875, *Problems of Life and Mind (First Series)*. Trübner London, Vol. 2.

63. Morgan, C.L. 1925, *Emergent Evolution*. Henry Holt and Co.

64. Alexander, S. 1920, *Space, Time, and Deity* Kessinger Publishing reprint.

65. Sellars, R.W. 1922, *Evolutionary Naturalism*. Open Court, pdf file: Chicago.

66. Bergson, H. 1911, *Creative Evolution*. Henry Holt and Company.

67. Lovejoy, A.O. 2008, The meanings of 'emergence' and its modes, with an introduction by Alicia Juarrero and Carl A. Rubino, E:CO, 10, (1) 62-78.

68. Chalmers, D.J. 2006, Strong and Weak Emergence. In *The Re-Emergence of Emergence*, Clayton, P.Davies, P., Eds. Oxford Univeristy Press: Oxford.

69. Steels, L. 1991, Towards a Theory of Emergent Functionality In *Animals to Animats 1*, Meyer, J.-A.Wilson, S., Eds. MIT Press: Cambridge, Mass.

70. Corning, P.A. 2002, The Re-Emergence of "Emergence": A Venerable Concept in Search of a Theory, Complexity, 7, (6) 18-30.

71. Kauffman, S.A. 1993, *The Origins of Order: Self-Organization and Selection in Evolution*. Oxford University Press: Oxford.

72. Kauffman, S. 1995, *At Home in the Universe: The Search for the Laws of Self-Organization and Complexity*. Oxford University Press: New York.

73. Kauffman, S.A. 2000, *Investigations*. Oxford University Press: New York.

74. Fromm, J. 2005, Types and Forms of Emergence, arXiv:nlin, 0506028v1 [nlin.AO].

75. Bedau , M.A. 1997, Weak emergence. In *Philosophical Perspectives: Mind, Causation, and World*, Tomberlin, J., Ed. Blackwell Publishers: pp 375-399.

76. Eigen, M. 1993, The origin of genetic information: viruses as models, Gene, 135, (1-2) 37-47.

77. Eigen, M. 1994, Selection and the origin of information, Int Rev Neurobiol, 37, 35-46; discussion 47-50.

78. Eigen, M.; Biebricher, C.K.; Gebinoga, M.; Gardiner, W.C. 1991, The hypercycle. Coupling of RNA and protein biosynthesis in the infection cycle of an RNA bacteriophage, Biochemistry, 30, (46) 11005-18.

79. Eigen, M.; de Maeyer, L. 1966, Chemical means of information storage and readout in biological systems, Naturwissenschaften, 53, (3) 50-7.

80. Eigen, M.; Winkler-Oswatitsch, R. 1981, Transfer-RNA: the early adaptor, Naturwissenschaften, 68, (5) 217-28.

81. Eigen, M.; Winkler-Oswatitsch, R. 1981, Transfer-RNA, an early gene?, Naturwissenschaften, 68, (6) 282-92.

82. Eigen, M.; Winkler-Oswatitsch, R. 1990, Statistical geometry on sequence space, Methods Enzymol, 183, 505-30.

83. Eigen, M.; Winkler-Oswatitsch, R.; Dress, A. 1988, Statistical geometry in sequence space: a method of quantitative comparative sequence analysis, Proc Natl Acad Sci U S A, 85, (16) 5913-7.

84. Gánti, T. 1975, Organization of chemical reactions into dividing and metabolizing units: the chemotons, Biosystems, 7, (1) 15-21.

85. Gánti, T. 1980, On the organizational basis of the evolution, Acta Biol, 31, (4) 449-59.

86. Gánti, T. 1997, Biogenesis itself, J Theor Biol, 187, (4) 583-93.

87. Gánti, T. 2002, On the early evolutionary origin of biological periodicity, Cell Biol Int, 26, (8) 729-35.

88. Gánti, T. 2003, *The Principles of Life*. Oxford University Press: Oxford, UK.

89. Eigen, M.; Gardiner, W.; Schuster, P.; Winkler-Oswatitsch, R. 1981, The origin of genetic information, Sci Am, 244, (4) 88-92, 96, et passim.

90. Eigen, M.; Gardiner, W.; Schuster, P.; Winkler-Oswatitsch, R. 1981, The origin of genetic information, laws governing natural selection of prebiotic molecules have been inferred and tested, making it possible to discover how early RA genes interacted with proteins and how the genetic code developed, Scientific American, 244, 88-118.

91. Eigen, M.; Schuster, P. 1977, The hypercycle. A principle of natural self-organization. Part A: Emergence of the hypercycle, Naturwissenschaften, 64, (11) 541-65.

92. Eigen, M.; Schuster, P. 1979, *The Hypercycle: A Principle of Natural Self Organization*. Springer Verlag: Berlin.

93. Eigen, M.; Schuster, P. 1981, Comments on "growth of a hypercycle" by King (1981), Biosystems, 13, (4) 235.

94. Eigen, M.; Schuster, P. 1982, Stages of emerging life--five principles of early organization, J Mol Evol, 19, (1) 47-61.

95. Waldrop, M.M. 1992, *Complexity*. Simon and Schuster: New York.

96. Kauffman, S.A.; Johnsen, S. 1991, Coevolution to the edge of chaos: coupled fitness landscapes, poised states, and coevolutionary avalanches, J Theor Biol, 149, (4) 467-505.

97. Bratman, R.L. 2002, Edge of chaos, J R Soc Med, 95, (3) 165.

98. Ito, K.; Gunji, Y.P. 1994, Self-organisation of living systems towards criticality at the edge of chaos, Biosystems, 33, (1) 17-24.

99. Munday, D. 2002, Edge of chaos, J R Soc Med, 95, (3) 165.

100. Forrest, S. 1999, Creativity on the edge of chaos, Semin Nurse Manag, 7, (3) 136-40.

101. Innes, A.D.; Campion, P.D.; Griffiths, F.E. 2005, Complex consultations and the 'edge of chaos', Br J Gen Pract, 55, (510) 47-52.

102. Mitchell, M.; Hraber, P.T.; Crutchfield, J.T. 1994, Dynamics, computation, and "the edge of chaos:" a re-examination. In *Complexity: Metaphors, Models, and Reality*, Cowan, G. P., D. and Melzner, D. , Ed. Addison-Wesley: Reading, MA pp 1-16.

103. Kauffman, S. 1970, Behavior of randomly constructed genetic nets. In *Towards a Theoretical Biology Vol. 3*, Waddington, C. H., Ed. Aldine Publishing Co.: Chicago, Vol. 3, p 18.

104. Kauffman, S. 2007, Beyond Reductionism: Reinventing the Sacred, Zygon, 42, (4) 903-914.

105. Kauffman, S.A. 2001, Prolegomenon to a general biology, Ann N Y Acad Sci, 935, 18-36; discussion 37-8.

106. Dawkins, R. 1976, *The Selfish Gene*. 2 ed., Oxford Univerisy Press: Oxford.

107. Dawkins, R. 1996, *Climbing Mount Improbable*. W.W. Norton & Co: New York.

108. Gell-Mann, M. 1995, What is complexity?, Complexity, 1, (1) 16-19.

109. Ricard, J. 2003, What do we mean by biological complexity?, C R Biol., Feb;326, (2) 133-40.

110. van de Vijver, G.; van Speybroeck, L.; Vandevyvere, W. 2003, Reflecting on complexity of biological systems: Kant and beyond?, Acta Biotheoretica, 51, (2) 101-109.

111. Edelman, G.M.; Gally, J.A. 2001, Degeneracy and complexity in biological systems, PNAS, 98, 13763-13768.

112. Simon, H.A. 1962, The architecture of complexity, Proc.Am. Philos. Soc., 106, 467-482.

113. Nicolis, G.; Prigogine, I. 1989, *Exploring Complexity*. Freeman: New York.

114. Badii, R.; Politi, A. 1997, *Complexity : hierarchical structures and scaling in physics*. Cambridge University Press: Cambridge ; New York.

115. Yockey, H.P. 1992, *Information Theory and Molecular Biology*. Cambridge University Press: Cambridge.

116. Yockey, H.P. 2005, *Information Theory, Evolution, and the Origin of Life*. Second ed., Cambridge University Press: Cambridge.

117. Lenski, R.E.; Ofria, C.; Collier, T.C.; Adami, C. 1999, Genome complexity, robustness and genetic interactions in digital organisms, Nature, 400, (6745) 661-4.

118. Lempel, A.; Ziv, J. 1976, On the complexity of finite sequences, IEEE Trans Inform. Theory, 22, 75.

119. Konopka, A.K.; Owens, J. 1990, Complexity charts can be used to map functional domains in DNA, Genet Anal Tech Appl, 7, (2) 35-8.

120. Adami, C.; Cerf, N.J. 2000, Physical complexity of symbolic sequences, Physica D, 137, 62-69.

121. Durston, K.K.; Chiu, D.K.; Abel, D.L.; Trevors, J.T. 2007, Measuring the functional sequence complexity of proteins, Theor Biol Med Model, 4, 47 Free on-line access at http://www.tbiomed.com/content/4/1/47.

122. Ebeling, W.; Jimenez-Montano, M.A. 1980, On grammars, complexity, and information measures of biological macromolecules, Math. Biosciences, 52, 53-71.

123. Gell-Mann, M.; Lloyd, S. 1996, Information measures, effective complexity, and total information, Complexity, 2, (1) 44-52.

124. Zurek, W.H. 1990, *Complexity, Entropy, and the Physics of Information*. Addison-Wesley: Redwood City, CA.

125. Farre, G.L.; Oksala, T. 1998, Emergence, Complexity, Hierarchy, Organization; Selected and Edited Papers from ECHO III. Acta Polytechnia Scandinavica;Espoo: Helsinki.

126. Rosen, R. 1985, On information and complexity. In *Complexity, Language, and Life: Mathematical Approaches*, Casti, J. L.Karlqvist, A., Eds. Springer: Berlin.

127. Zvonkin, A.K.; Levin, L.A. 1970, The complexity of finite objects and the development of the concepts of information and randomness by means of the theory of algorithms, Russ. Math Serv, 256, 83-124.

128. Konopka, A.K. 1984, Is the information content of DNA evolutionarily significant?, J Theor Biol, 107, (4) 697-704.

129. Konopka, A.K. 1985, Theory of degenerate coding and informational parameters of protein coding genes, Biochimie, 67, (5) 455-468.

130. Konopka, A.K. 1994, Sequences and Codes: Fundamentals of Biomolecular Cryptology. In *Biocomputing: Informatics and Genome Projects*, Smith, D., Ed. Academic Press: San Diego, pp 119-174.

131. Konopka, A.K. 2003, Systems biology: aspects related to genomics. In *Nature Encyclopidia of the Human Genome*, Cooper, D. N., Ed. Nature Publishing Group Reference: London, Vol. 5, pp 459-465.

132. Konopka, A.K. 2003, Information theories in molecular biology and genomics. In *Nature Encyclopedia of teh Human Genome*, Cooper, D. N., Ed. Nature Publishing Group Reference: London, Vol. 3, pp 464-469.

133. Konopka, A.K. 2003, Sequence complexity and composition. In *Nature Encyclopedia of the Human Genome. Vol. 5.*, Cooper, D. N., Ed. Nature Publishing Group Reference: London, pp 217-224.

134. Koonin, E.V.; Dolja, V.V. 2006, Evolution of complexity in the viral world: the dawn of a new vision, Virus research, 117, (1) 1-4.

135. Toussaint, O.; Schneider, E.D. 1998, The thermodynamics and evolution of complexity in biological systems, Comp Biochem Physiol A Mol Integr Physiol, 120, (1) 3-9.

136. Barham, J. 1996, A dynamical model of the meaning of information, Biosystems, 38, (2-3) 235-41.

137. Stonier, T. 1996, Information as a basic property of the universe, Biosystems, 38, (2-3) 135-40.

138. Boniolo, G. 2003, Biology without information, History and Philosophy of the Life Sciences, 25, 255-73.

139. Sarkar, S. 1996, Biological information: a skeptical look at some central dogmas of molecular biology. In *The Philosophy and History of Molecular Biology: New Perspectives*, Sarkar, S., Ed. Kluwer Academic Publishers: Dordrecht, pp 187-231.

140. Sarkar, S. 2000, Information in genetics and developmental biology: Comments on Maynard Smith, Philosophy of Science, 67, 208-213.

141. Sarkar, S. 2003, Genes encode information for phenotypic traits. In *Comtemporary debates in Philosophy of Science*, Hitchcock, C., Ed. Blackwell: London, pp 259-274.

142. Stent, G.S. 1981, Strength and weakness of the genetic approach to the development of the nervous system, Annu Rev Neurosci, 4, 163-94.

143. Griffiths, P.E. 2001, Genetic information: A metaphor in search of a theory, Philosophy of Science, 68, 394-412.

144. Godfrey-Smith, P. 2003, Genes do not encode information for phenotypic traits. In *Contemporary Debates in Philosophy of Science* Hitchcock, C., Ed. Blackwell: London, pp 275-289.

145. Noble, D. 2002, Modeling the heart--from genes to cells to the whole organ, Science, 295, (5560) 1678-82.

146. Mahner, M.; Bunge, M.A. 1997, *Foundations of Biophilosophy*. Springer Verlag: Berlin.

147. Kitcher, P. 2001, Battling the undead; how (and how not) to resist genetic determinism. In *Thinking About Evolution: Historical Philosophical and Political Perspectives*, Singh, R. S.; Krimbas, C. B.; Paul, D. B.Beattie, J., Eds. Cambridge University Press: Cambridge, pp 396-414.

148. Chargaff, E. 1963, *Essays on Nucleic Acids*. Elsevier: Amsterdam.

149. Kurakin, A. 2010, Order without design, Theoretical Biology and Medical Modelling, 7, (1) 12.

150. Jacob, Francois 1974, *The Logic of Living Systems---a History of Heredity*. Allen Lane: London.

151. Alberts, B.; Bray, D.; Lewis, J.; Raff, M.; Roberts, K.; Watson, J.D. 2002, *Molecular Biology of the Cell*. Garland Science: New York.

152. Davidson, E.H.; Rast, J.P.; Oliveri, P.; Ransick, A.; Calestani, C.; Yuh, C.H.; Minokawa, T.; Amore, G.; Hinman, V.; Arenas-Mena, C.; Otim, O.; Brown, C.T.; Livi, C.B.; Lee, P.Y.; Revilla, R.; Rust, A.G.; Pan, Z.; Schilstra, M.J.; Clarke, P.J.; Arnone, M.I.; Rowen, L.; Cameron, R.A.; McClay, D.R.; Hood, L.; Bolouri, H. 2002, A genomic regulatory network for development, Science, 295, (5560) 1669-78.

153. Wolpert, L.; Smith, J.; Jessell, T.; Lawrence, P. 2002, *Principles of Development*. Oxford University Press: Oxford.

154. Stegmann, U.E. 2005, Genetic Information as Instructional Content, Phil of Sci, 72, 425-443.

155. Barbieri, M. 2004, Biology with information and meaning, History & Philosophy of the Life Sciences, 25, (2 (June)) 243-254.

156. Deely, J. 1992, Semiotics and biosemiotics: are sign-science and life-science coextensive? In *Biosemiotics: The Semiotic Web 1991*, Sebeok, T. A.Umiker-Sebeok, J., Eds. Mouton de Gruyter: Berlin/N.Y., pp 46-75.

157. Sebeok, T.A.; Umiker-Sebeok, J. 1992, Biosemiotics: The Semiotic Web 1991. Mouton de Gruyter: Berlin.

158. Hoffmeyer, J. 1997, Biosemiotics: Towards a new synthesis in biology, European Journal for Semiotic Studies, 9, 355-376.

159. Sharov, A. 1992, Biosemiotics. A functional-evolutionary approach to the analysis of the sense of evolution. In *Biosemiotics: The Semiotic Web 1991*, Sebeok, T. A.Umiker-Sebeok, J., Eds. Mouton de Gruyter: Berlin, pp 345-373.

160. Kull, K. 1999, Biosemiotics in the twentieth century: A view from biology., Semiotica, 127, 385-414.

161. Kawade, Y. 1996, Molecular biosemiotics: molecules carry out semiosis in living systmes, Semiotica, 111, 195-215.

162. Barbieri, M. 2005, Life is 'artifact-making', Journal of Biosemiotics, 1, 113-142.

163. Pattee, H.H. 2005, The physics and metaphysics of Biosemiotics, Journal of Biosemiotics, 1, 303-324.

164. Salthe, S.N. 2005, Meaning in nature: Placing biosemitotics within pansemiotics, Journal of Biosemiotics, 1, 287-301.

165. Kull, K. 2005, A brief history of biosemiotics, Journal of Biosemiotics, 1, 1-36.

166. Nöth, W. 2005, Semiotics for biologists, Journal of Biosemiotics, 1, 195-211.

167. Artmann, S. 2005, Biosemiotics as a structural science, Journal of Biosemiotics, 1, 247-285.

168. Barbieri, M. 2006, Is the Cell a Semiotic System? In *Introduction to Biosemiotics: The New Biological Synthesis*, Barbieri, M., Ed. Springer-Verlag New York, Inc. : Secaucus, NJ, USA

169. Barbieri, M. 2006, Introduction to Biosemiotics: The New Biological Synthesis. Springer-Verlag Dordrecht, The Netherlands.

170. Barbieri, M. 2007, Has biosemiotics come of age? In *Introduction to Biosemiotics: The New Biological Synthesis*, Barbieri, M., Ed. Springer: Dorcrecht, The Netherlands, pp 101-114.

171. Jämsä, T. 2006, Semiosis in evolution. In *Introduction to Biosemiotics: The New Biological Synthesis*, Barbieri, M., Ed. Springer-Verlag New York, Inc. : Dordrecht, The Netherlands; Secaucus, NJ, USA

172. Hoffmeyer, J. 2006, Semiotic scaffolding of living systems. In *Introduction to Biosemiotics: The New Biological Synthesis*, Barbieri, M., Ed. Springer-Verlag New York, Inc. : Dordrecht, The Netherlands; Secaucus, NJ, USA pp 149-166.

173. Kull, K. 2006, Biosemiotics and biophysics--The fundamental approaches to the study of life. In *Introduction to Biosemiotics: The New Biological Synthesis*, Barbieri, M., Ed. Springer-Verlag New York, Inc. : Dordrecht, The Netherlands; Secaucus, NJ, USA

174. Barbieri, M. 2007, The Codes of Life: The Rules of Macroevolution (Biosemiotics). Springer: Dordrecht, The Netherlands.

175. Barbieri, M. 2008, Biosemiotics: a new understanding of life, Naturwissenschaften, 95, 577-599.

176. Hodge, B.; Caballero, L. 2005, Biology, semiotics, complexity: An experiment in interdisciplinarity Semiotica, 2005, Proc. Natl. Acad. Sci. USA.

177. Adami, C.; Ofria, C.; Collier, T.C. 2000, Evolution of biological complexity, Proc Natl Acad Sci U S A., 97, (9) 4463-8.

178. Goodwin, B. 1994, *How the Leopard Changed Its Spots: The Evolution of Complexity*. Simon and Schuster; Charles Scribner & Sons: New York.

179. Mao, C. 2004, The emergence of complexity: lessons from DNA, PLoS Biol, 2, (12) e431.

180. Holland, J.H. 1995, *Hidden Order: How Adaptation Builds Complexity*. Addison-Wesley: Redwood City, CA.

181. Mikulecky, D.C. 2001, The emergence of complexity: science coming of age or science growing old?, Computers Chem., 25, 341-348.

182. Salthe, S.N. 1993, *Development and Evolution: Complexity and Change in Biology*. MIT Press: Cambridge, MA.

183. Pattee, H.H. 2000, Causation, Control, and the Evolution of Complexity. In *Downward Causation: Minds, Bodies, and Matter*, Andersen, P. B.; Emmeche, C.; Finnemann, N. O.Christiansen, P. V., Eds. Aarhus University Press: Aarhus, DK, pp 63-77.

184. Szathmary, E.; Smith, J.M. 1995, The major evolutionary transitions, Nature, 374, (6519) 227-32.

185. Sole, R.; Goodwin, B. 2000, *Signs of Life: How Complexity Pervades Biology*. Basic Books: New York.

186. Stano, P.; Luisi, P.L. 2007, Basic questions about the origins of life: proceedings of the Erice international school of complexity (fourth course), Orig Life Evol Biosph, 37, (4-5) 303-7.

187. Homberger, D.G. 2005, Ernst Mayr and the complexity of life, J Biosci, 30, (4) 427-33.

188. Pross, A. 2005, On the emergence of biological complexity: life as a kinetic state of matter, Origins of life and evolution of the biosphere, 35, (2) 151-66.

189. Bedau, M.A. 2003, Artificial life: organization, adaptation and complexity from the bottom up, Trends in cognitive sciences, 7, (11) 505-12.

190. Umerez, J. 2001, Howard Pattee's theoretical biology--a radical epistemological stance to approach life, evolution and complexity, Biosystems, 60, (1-3) 159-77.

191. Branca, C.; Faraone, A.; Magazu, S.; Maisano, G.; Migliardo, P.; Villari, V. 1999, Suspended life in biological systems. Fragility and complexity, Ann N Y Acad Sci, 879, 224-7.

192. Oltvai, Z.N.; Barabasi, A.L. 2002, Systems biology. Life's complexity pyramid, Science, 298, (5594) 763-4.

193. Rosen, R. 1977, Complexity and system description. In *Systems, Approaches, Theories, Applications*, Harnett, W. E., Ed. D. Reidel: Boston, MA.

194. Rosen, R. 1987, On Complex Systems, Euro. J. of Operational Rsrch., 30, 129-134.

195. Behe, M.J. 1996, *Darwin's Black Box*. Simon & Shuster: The Free Press: New York.

196. Anderson, E. 2004, Irreducible complexity reduced: An integrated Approach to the complexity space, PCID 3.1.5 November, 1-29.

197. Thompson, C. 2004, Fortuitous phenomena: on complexity, pragmatic randomised controlled trials, and knowledge for evidence-based practice, Worldviews Evid Based Nurs, 1, (1) 9-17; discussion 18-9.

198. Pennock, R.T. 2003, Creationism and intelligent design, Annu Rev Genomics Hum Genet, 4, 143-63.

199. Aird, W.C. 2003, Hemostasis and irreducible complexity, J Thromb Haemost, 1, (2) 227-30.

200. Keller, E.F. 2002, Developmental robustness, Ann N Y Acad Sci, 981, 189-201.

201. von Neumann, J.; Burks, A.W. 1966, *Theory of Self-Reproducing Automata*. University of Illinois Press: Urbana,.

202. Pattee, H.H. 1978, The complementarity principle in biological and social structures, Journal of Social and Biological Structure, 1, 191-200.

203. Pattee, H.H. 1979, Complementarity vs. reduction as explanation of biological complexity, Am J Physiol, 236, (5) R241-6.

204. Pattee, H.H. 1995, Evolving Self-Reference: Matter, Symbols, and Semantic Closure, Communication and Cognition-Artificial Intelligence, 12, 9-28.

205. Hoffmeyer, J. 2000, Code-duality and the epistemic cut, Ann N Y Acad Sci, 901, 175-86.

206. Hoffmeyer, J. 2002, Code duality revisited, SEED, 2, 1-19.

207. Stein, D.L., Ed 1988, Lectures in the Sciences of Complexity. Addison-Wesley: Redwood City, CA.

208. Norris, V.; Cabin, A.; Zemirline, A. 2005, Hypercomplexity, Acta Biotheor, 53, (4) 313-30.

209. Garzon, M.H.; Jonoska, N.; Karl, S.A. 1999, The bounded complexity of DNA computing, Bio Systems, 52, (1-3) 63-72.

210. Levins, R. 1971, The limits of complexity. In *Biological Hierarchies: Their Origin and Dynamics*, Pattee, H., Ed. Gordon and Breach: New York.

211. Bennett, D.H. 1988, Logical depth and physical complexity. In *The Universal Turing Machine: a Half-Century Survey*, Herken, R., Ed. Oxford University Pres: Oxford.

212. Yan, K.-K.; Fang, G.; Bhardwaj, N.; Alexander, R.P.; Gerstein, M. 2010, Comparing genomes to computer operating systems in terms of the topology and evolution of their regulatory control networks, Proceedings of the National Academy of Sciences, 107, (20) 9186-9191.

213. Corning, P.A.; Kline, S.J. 2000, Thermodynamics, information and life revisited, Part II: Thermoeconomics and Control information, Systems Research and Behavioral Science, 16, 453-482.

214. Corning, P.A.; Kline, S.J. 2000, Thermodynamics, information and life revisited, Part I: To be or entropy, Systems Research and Behavioral Science, 15, 273-295.

215. Deacon, T.W. 2010, 8. What's missing from theories of information? In *Information and the Nature of Reality: From Physics to Metaphysics*, Davies, P.Gregersen, N. H., Eds. Cambridge University Press: Cambridge, p In Press.

216. Wills, P.R. 2009, Informed generation: Physical origin and biological evolution of genetic codescript interpreters, Journal of Theoretical Biology.

217. Takeuchi, N.; Hogeweg, P. 2009, Multilevel Selection in Models of Prebiotic Evolution II: A Direct Comparison of Compartmentalization and Spatial Self-Organization, PLoS Comput Biol, 5, (10) e1000542.

218. Moreno, A.; Ruiz-Mirazo, K. 2009, The problem of the emergence of functional diversity in prebiotic evolution, Biology and Philosophy, 24, (5) 585-605.

219. Vesterby, V. 2008, *Origins of Self-organization, Emergence and Cause*. ISCE Publishing: Goodyear, Arizona.

220. Turner, S. 2008, Homeostasis, Complexity, and the Problem of Biological Design, Emergence: Complexity and Organization, 10.2.

221. Spinelli, G.; Mayer-Foulkes, D. 2008, New Method to Study DNA Sequences: The Languages of Evolution, Nonlinear Dynamics, Psychology, and Life Sciences, 12, (2, April) 133-151.

222. Ruiz-Mirazo, K.; Umerez, J.; Moreno, A. 2008, Enabling conditions for 'open-ended evolution', Biol Phil, 23, (1) 67-85.

223. Raginsky, M.; Anastasio, T. 2008, Cooperation in self-organizing map networks enhances information transmission in the presence of input background activity, Biological Cybernetics, 98, (3) 195-211.

224. Juarrero, A.; Rubino, C.A. 2008, Emergence, Complexity, and Self-Organization: Precursors and Prototypes. ISCE Publishing: 17947 W. Porter Lane, Goodyear, AZ 85338.

225. Deacon, T.; Sherman, J. 2008, The Pattern Which Connects Pleroma to Creatura: The Autocell Bridge from Physics to Life. In *A Legacy for Living Systems: Gregory Bateson as Precursor to Biosemiotics* Hoffmeyer, J., Ed. Springer Netherlands: Netherlands.

226. Schiffmann, Y. 2007, Self-organization in and on biological spheres, Prog Biophys Mol Biol.

227. Lozneanu, E.; Sanduloviciu, M. 2007, Self-organization scenario grounded on new experimental results, Chaos, Solitons and Fractals.

228. Gershenson, C. 2007, *Design and Control of Self-Organizing Systems*. Vrije Universiteit Brussel, Brussels.

229. Fishkis, M. 2007, Steps towards the formation of a protocell: the possible role of short peptides, Orig Life Evol Biosph, 37, (6) 537-53.

230. Ruiz-Mirazo, K.; Moreno, A. 2006, On the origins of information and its relevance for biological complexity, Biological Theory, 1, (3) 227-229.

231. Moreno, A.; Ruiz-Mirazo, K. 2006, The maintenance and open-ended growth of complexity in nature: information as a decoupling mechanism in the origins of life. . In *Rethinking Complexity* Capra, F.; Sotolongo, P.; Juarrero, A.van Uden, J., Eds. ISCE Publisher: pp 55-72.

232. Kurakin, A. 2006, Self-organization versus Watchmaker: molecular motors and protein translocation., Biosystems 84, (1) 15-23.

233. Jacob, E.B.; Shapira, Y.T., Alfred I. 2006, Seeking the foundations of cognition in bacteria: From Schrödinger's negative entropy to latent information, Physica A, 359, 495-524.

234. Humphreys, P. 2006, Self-Assembling Systems Philosophy of Science, 73, 595-604.

235. Gabora, L. 2006, Self-Other Organization: Why early life did not evolve through natural selection, J Theor Biol, 241, (3) 443-450.

236. Feltz, B.; Crommelinck, M.; Goujon, P. 2006, *Self-organization and emergence in life sciences*. Springer: Dordrecht.

237. Deacon, T.W.; Sherman, J. 2006, How teleology emerged: Bridging the gap from physics to life.

238. Deacon, T.W.; Cashman, T.; Sherman, J. 2006, Disembodiment and Intentionality. Disembodiment: Absence as the root of intentionality. In *Embodiment*.

239. Deacon, T.W. 2006, Reciprocal linkage between self-organizing processes is sufficient for self-reproduction and evolvability, Biological Theory, 1, (2) 136-149.

240. Rocha, L.M. 2000, Syntactic autonomy: or why there is no autonomy without symbols and how self-organizing systems might evolve them, Annals of the New York Academy of Sciences, 207-223.

241. Luisi, P.L. 2010, Contingency and Determinism in the origin of life, and elsewhere, OLEB, 40, (4-5 October) 356-361.

242. Mitchell, S.D. 2010, Determinism vs. Contingency: A false dichotomy, OLEB, 40, (4-5 October) 361-362.

243. Pohorille, A. 2010, Was the emergence of life on earth a likely outcome of chemical evolution?, OLEB, 40, (4-5) 362-365.

244. Norris, V.; Delaune, A. 2010, Contingency vs. Determinism, OLEB, 40, (4-5) 365-370.

245. Bich, L.; Bocchi, G.; Damiano, L. 2010, An epistemology of Contingency: Chance and Determinism at the orgin of life, OLEB, 40, (4-5) 370-375.

246. Moya, A. 2010, Godel, biology and emergent properties, OLEB, 40, (4-5) 375-377.

247. Pizzarello, S. 2010, The chemistry that preceded life's origin: When is an evolutionary story an emergent story?, OLEB, 40, (4-5) 378-380.

248. Kauffman, S.A. 2010, On emergence, OLEB, 40, (4-5) 381-383.

249. Hanczyc, M.; Ikegami, T. 2010, Emergence of self-movement as a precursor to Darwinian evolution, OLEB, 40, (4-5) 383-384.

250. Pohorille, A. 2010, Emerging Properties in the origins of life and Darwinian evolution, OLEB, 40, (4-5) 384-386.

251. Norris, V.; Grondin, Y. 2010, Emergence, OLEB, 40, (4-5) 386-391.

252. Abel, D.L. 2000 To what degree can we reduce "life" without "loss of life"?, Workshop on Life: a satellite meeting before the Millenial World Meeting of University Professors, Modena, Italy,

253. Cleland, C.E.; Chyba, C.F. 2002, Defining 'life', Origins of life and evolution of the biosphere, 32, (4) 387-93.

254. Joyce, G.F. 1994, *Origins of Life: The Central Concepts   see Forward.* Jones and Bartlett: Boston, MA.

255. Korzeniewski, B. 2001, Cybernetic Formulation of the Definition of Life, Journal of Theoretical Biology, 209, (3) 275-286.

256. Korzeniewski, B. 2005, Confrontation of the cybernetic definition of a living individual with the real world, Acta Biotheor, 53, (1) 1-28.

257. Lahav, N. 1985, The synthesis of primitive 'living' forms: definitions, goals, strategies and evolution synthesizers, Orig Life Evol Biosph, 16, (2) 129-49.

258. Nealson, K.H.; Tsapin, A.; Storrie-Lombardi, M. 2002, Searching for Life in the Universe: unconventional methods for an unconventional problem, Int Microbiol, 5, 223-230.

259. Pereto, J. 2005, Controversies on the origin of life, Int Microbiol, 8, (1) 23-31.

260. Ruiz-Mirazo, K.; Pereto, J.; Moreno, A. 2004, A universal definition of life: autonomy and open-ended evolution, Orig Life Evol Biosph, 34, (3) 323-46.

261. Tsokolov, S.A. 2009, Why Is the Definition of Life So Elusive? Epistemological Considerations, Astrobiology, 9, (4) 401-412.

262. Rizzotti, M. 1996, Defining Life: The Central Problem in Theoretical Biology. University of Padova Press: Padova, p 208.

263. Palyi, G.; Zucchi, C.; Caglioti, L. 2002, *Fundamentals of Life.* Elsevier: Paris.

264. Schrödinger, E. 1944, *What is Life: The Physical Aspect of the Living Cell.* Cambridge Univ. Press: Cambridge.

265. Brillouin, L. 1953, The negentropy principle of information, Journal of Applied Physics, 24, 1153.

266. Brillouin, L. 1962, *Science and Information Theory.* 2nd ed., Academic Press: New York.

267. Brillouin, L. 1990, Life, thermodynamics, and cybernetics. In *Maxwell's Demon, Entropy, Information, and Computing,* Leff, H. S.Rex, A. F., Eds. Princeton University Press: Princeton.

268. Abel, D.L. 2000 Is Life Reducible to Complexity?, Workshop on Life: a satellite meeting before the Millennial World Meeting of University Professors, Modena, Italy,

269. Bedau, M.A. 2010, An Aristotelian Account of Minimal Chemical Life, Astrobiology, 10, (10) 1011-1020.

270. Koch, A.L.; Silver, S. 2005, The first cell, Adv Microb Physiol, 50, 227-59.

271. Rocha, L.M. 1997, *Evidence Sets and Contextual Genetic Algorithms: Exploring uncertainty, context, and embodiment in cognitive and biological systems.* . State University of New York, Binghamton.

272. Rocha, L.M. 1998, Selected self-organization and the semiotics of evolutionary systems. In *Evolutionary Systems: Biological and Epistemological Perspectives on Selection and Self-Organization,* Salthe, S.; van de Vijver, G.Delpos, M., Eds. Kluwer: The Netherlands, pp 341-358.

273. Johnson, D.E. 2010, *Programming of Life.* Big Mac Publishers: Sylacauga, Alabama.

274. Johnson, D.E. 2010, *Probability's Nature and Nature's Probability (A call to scientific integrity).* Booksurge Publishing: Charleston, S.C.

275. Konopka, A.K. 2002, Grand metaphors of biology in the genome era, Computers & Chemistry, 26, 397-401.

276. Lackoff, G.; Johnson, M. 1980, *Metaphors We Live By.* University of Chicago Press: Chicago, IL.

277. Lackoff, G. 1993, The contemporary theory of metaphor. In *Mataphor and Thought, 2nd Edition,* Ortony, A., Ed. Cambridge University Press: Cambridge, pp 11-52.

278. Fiumara, G.C. 1995, *The Metaphoric Process: Connections Between Language and Life*. Rutledge: London.
279. Torgny, O. 1997, *Metaphor--A Working Concept. KTH, Royal Institute of Technology*. CID: Stockholm, Sweden.
280. Rosen, R. 1993, Bionics revisited. In *The Machine as a Metaphor and Tool*, Haken, H.; Karlqvist, A.Svedin, U., Eds. Springer-Verlag: Berlin, pp 87-100.
281. Emmeche, C.; Hoffmeyer, J. 1991, From language to nature: The semiotic metaphor in biology, Semiotica, 84, 1-42.
282. Atlan, H.; Koppel, M. 1990, The cellular computer DNA: program or data, Bulletin of Mathematical Biology, 52, (3) 335-348.
283. Maynard Smith, J. 2000, The concept of information in biology, Philosophy of Science, 67, (June) 177-194 (entire issue is an excellent discussion).
284. Moss, L. 2003, *What Genes Can't Do*. MIT Press: Cambridge, MA.
285. Sterelny, K.; Smith, K.; Dickison, M. 1996, The extended replicator, Biology and Philosophy, 11, 377-403.
286. Wheeler, M. 2003, Do genes code for traits? In *Philosophic Dimensions of Logic and Science: Selected Contributed Papers from the 11th International Congress of Logic, Methodology, and Philosophy of Science*, Rojszczak, A.; Cachro, J.Kurczewski, G., Eds. Kluwer: Dordrecht, pp 151-164.
287. Lwoff, A. 1962, *Biological Order*. MIT Press: Cambridge, MA.
288. Kay, L. 2000, *Who Wrote the Book of Life? A History of the Genetic Code*. Stanford University Press: Stanford, CA.
289. Keller, E.F. 2000, Decoding the genetic program. In *The Concept of the Gene in Development and Evolution*, Beurton, P.; Falk, R.Rheinberger, H.-J., Eds. Cambridge University Press: Cambridge, pp 159-177.
290. Rocha, L.M.; Hordijk, W. 2005, Material representations: from the genetic code to the evolution of cellular automata, Artif Life, 11, (1-2) 189-214.
291. Rocha, L.M. 2001, Evolution with material symbol systems, Biosystems, 60, 95-121.
292. Pattee, H.H. 1995, Artificial Life Needs a Real Epistemology. In *Advances in Artificial Life* Moran, F., Ed. Springer: Berlin, pp 23-38.
293. Turing, A.M. 1936, On computable numbers, with an application to the *entscheidungs problem*, Proc. Roy. Soc. London Mathematical Society, 42, (Ser 2) 230-265 [correction in 43, 544-546].
294. von Neumann, J. 1950, Letter to physicist George Gamow (first scientist to elucidate *triplet* codons) on July 25, 1950. Cited by Steve J. Heims in "John von Neumann and Norbert Wiener: *From Mathematics to the Technologies of Life and Death*," Canbridge, MA, MIT Press, 1980. In.
295. Wiener, N. 1948, *Cybernetics*. J. Wiley: New York.
296. Wiener, N. 1961, *Cybernetics, its Control and Communication in the Animal and the Machine*. 2 ed., MIT Press: Cambridge.
297. Barrau, A. 2007, Physics in the multiverse. In CERN Courier, Vol. December.
298. Carr, B. 2007, Universe or Multiverse? Cambridge University Press: Cambridge.
299. Garriga, J.; Vilenkin, A. 2008, Prediction and explanation in the multiverse. In Phys.Rev.D 77:043526,2008.
300. Hawking, S.; Ellis, G.F.R. 1973, *The Large Scale Structure of Space-Time*. Cambridge University Press. : Cambridge.
301. Hawking, S. 1988, *A Brief History of Time*. Bantam Books: New York.
302. Hawking, S.; Penrose, R. 1996, *The Nature of Space and Time*. Princeton U. Press: Princeton, N.J.
303. Axelsson, S. 2003, Perspectives on handedness, life and physics, Med Hypotheses, 61, (2) 267-74.
304. Koonin, E.V. 2007, The Biological Big Bang model for the major transitions in evolution, Biol Direct, 2, 21.
305. Koonin, E.V. 2007, The cosmological model of eternal inflation and the transition from chance to biological evolution in the history of life, Biol Direct, 2, 15.
306. Vitányi, P.M.B.; Li, M. 2000, Minimum Description Length Induction, Bayesianism and Kolmogorov Complexity, IEEE Transactions on Information Theory, 46, (2) 446 - 464.
307. Abel, D.L. The Cybernetic Cut [Scirus Topic Page]. http://www.scitopics.com/The_Cybernetic_Cut.html (Last accessed Sept, 2011).
308. Stegmann, U.E. 2004, The arbitrariness of the genetic code, Biology and Philosophy, 19, (2) 205-222.
309. Abel, D.L. Prescriptive Information (PI) [Scirus Topic Page]. http://www.scitopics.com/Prescriptive_Information_PI.html (Last accessed September, 2011).
310. Origin of Life Science Foundation, I. Origin of Life Prize. http://www.lifeorigin.org
311. Bradley, D. 2002, Informatics. The genome chose its alphabet with care, Science, 297, (5588) 1789-1791.
312. Freeland, S.J.; Hurst, L.D. 1998, The genetic code is one in a million, Journal of Molecular Evolution, 47, 238-248.
313. Itzkovitz, S.; Alon, U. 2007, The genetic code is nearly optimal for allowing additional information within protein-coding sequences, Genome Res, 17, (4) 405-12.
314. Rocha, L.M. 2001, The physics and evolution of symbols and codes: reflections on the work of Howard Pattee, Biosystems, 60, 1-4.

315. Allweis, C. 1988, Proposal for APS-IUPS convention for diagraming physiological mechanisms, Am J Physiol, 254, (5 Pt 2) R717-26.
316. Pattee, H.H. 1989, The measurement problem in artificial world models, Biosystems, 23, (2-3) 281-9; discussion 290.
317. Pattee, H.H. 2007, Laws, constraints, and the modeling relation--History and interpretations, Chemistry & Biodiversity, 4, 2272-2295.
318. Mira, A.; Ochman, H.; Moran, N.A. 2001, Deletional bias and the evolution of bacterial genomes, Trends Genet, 17, (10) 589-96.
319. Petrov, D.A.; Chao, Y.C.; Stephenson, E.C.; Hartl, D.L. 1998, Pseudogene evolution in Drosophila suggests a high rate of DNA loss, Mol Biol Evol, 15, (11) 1562-7.
320. Petrov, D.A.; Hartl, D.L. 1998, High rate of DNA loss in the Drosophila melanogaster and Drosophila virilis species groups, Mol Biol Evol, 15, (3) 293-302.
321. Subba Rao, G.; Hamid, Z.; Subba Rao, J. 1979, The information content of DNA and evolution J. Theoretical Biology, 81, 803.
322. Subba Rao, J.; Geevan, C.P.; Subba Rao, G. 1982, Significance of the information content of DNA in mutations and evolution, J Theor Biol, 96, (4) 571-7.
323. Kok, R.A.; Taylor, J.A.; Bradley, W.L. 1988, A statistical examination of self-ordering of amino acids in proteins, Origins of life and evolution of the biosphere, 18, (1-2) 135-42.
324. Weiss, O.; Jimenez-Montano, M.A.; Herzel, H. 2000, Information content of protein sequences, J Theor Biol, 206, (3) 379-86.
325. Whitehead, A.N. 1927, *Symbolism: Its meaning and effect.* Macmillan: New York.
326. Cassirer, E. 1957, *The Philosophy of Symbolic Forms, Vol 3: The Phenomena of Knowledge.* Yale Univ. Press: New Haven, CT.
327. Harnad, S. 1990, The symbol grounding problem, Physica D, 42, 335-346.
328. Pattee, H.H. 1982, Cell psychology: an evolutionary approach to the symbol-matter problem, Cognition and Brain Theory, 5, 325-341.
329. Pattee, H.H. 1968, The physical basis of coding and reliabiity in biological evolution. In *Prolegomena to Theoretical Biology*, Waddington, C. H., Ed. University of Edinburgh: Edinburgh.
330. Pattee, H.H. 1986, Universal principles of measurement and language functions in evolving systems. In *Complexity, Language, and Life: Mathematical Approaches*, Casti, J. L.Karlqvist, A., Eds. Springer-Verlag: Berlin, pp 579-581.
331. Pattee, H.H. 1997, The physics of symbols and the evolution of semiotic controls. In *Proc. Workshop on Control Mechanisms for Complex Systems*, Coombs, M. e. a., Ed. Addison-Wesley: p http://www.ssie.binghamton.edu/pattee/semiotic.html.
332. Pattee, H.H. 1961, On the origin of macromolecular sequences, Biophys J, 1, 683-710.
333. Pattee, H.H. 1969, How does a molecule become a message? In *Communication in Development; Twenty-eighth Symposium of the Society of Developmental Biology.*, Lang, A., Ed. Academic Press: New York, pp 1-16.
334. Pattee, H.H. 1971, The nature of hierarchichal controls in living matter. In *Foundations of Mathematical Biology*, Rosen, R., Ed. Academic Press: New York, Vol. 1, pp 1-22.
335. Pattee, H.H. 1972, Laws and constraints, symbols and languages. In *Towards a Theoretical Biology*, Waddington, C. H., Ed. University of Edinburgh Press: Edinburgh, Vol. 4, pp 248-258.
336. Pattee, H.H. 1973, Physical problems of the origin of natural controls. In *Biogenesis, Evolution, and Homeostasis*, Locker, A., Ed. Springer-Verlag: Heidelberg, pp 41-49.
337. Pattee, H.H. 1977, Dynamic and linguistic modes of complex systems, Int. J. General Systems, 3, 259-266.
338. Pattee, H.H. 1967, Quantum mechanics, heredity and the origin of life, J Theor Biol, 17, (3) 410-20.
339. Ledford, H. 2010, Mystery RNA spawns gene-activating peptides: Short peptides that regulate fruitfly development are produced from 'junk' RNA. In NATURE, Vol. Published online 15 July.
340. Robertson, M. 2010, The evolution of gene regulation, the RNA universe, and the vexed questions of artefact and noise, BMC Biology, 8, (1) 97.

The First Gene, David L. Abel, Editor 2011,   pp 231-286   ISBN: 978-0-9657988-9-1

9. Examining specific life-origin models for plausibility

David L. Abel

Department of ProtoBioCybernetics/ProtoBioSemiotics
Director, The Gene Emergence Project
The Origin-of-Life Science Foundation, Inc.
113 Hedgewood Dr. Greenbelt, MD 20770-1610 USA

Abstract: All models of life-origin, whether Protometabolism-First or pre-RNA / RNA World early informational self-replicative models, encounter the same dead-end: no naturalistic mechanism exists to steer objects and events toward eventual functionality. No insight, motive, foresight or impetus exists to integrate physicochemical reactions into a cooperative, organized, pragmatic effort.   Inanimate nature cannot pursue the goal of homeostasis; it cannot scheme to locally and temporarily circumvent the 2nd Law. This deadlock affects all naturalistic models involving hypercycles, composomes and chemotons. It precludes all spontaneous geochemical, hydrothermal, eutectic, and photochemical scenarios. It affects the Lipid, Peptide and Zinc World models. It pertains to Co-evolution and all other code-origin models. No plausible hypothetical scenario exists that can convert chance and/or necessity into an organized protometabolic scheme. In this paper the general principles of previous chapters are applied to the best specific models of life origin in the literature. Tibor Ganti's chemoton model and the pre-RNA and RNA World models receive more attention, as they are the most well-developed and preferred scenarios.

Correspondence/Reprint request: Dr. David L. Abel, Department of ProtoBioCybernetics/ProtoBioSemiotics, The Origin-of-Life Science Foundation, Inc., 113 Hedgewood Dr. Greenbelt, MD 20770-1610 USA E-mail: life@us.net

Introduction: Every life-origin model encounters the same great impasse

Every naturalistic hypothetical path seems to lead to the same dead-end. More accurately stated, no "paths" exist in the first place within inanimate physicodynamic reaction space that would lead to any spontaneous protometabolic schema. "Paths" presuppose directionality and goal. Not even evolutionary selection manifests directionality or pursues a goal. Paths to pragmatism exist only in the minds of theorists, not in inanimate nature. The inanimate physical world is constrained, not controlled. Cross-reactive, resource-consuming, spontaneous biochemical reactions prevail, not metabolic "success." Only agency pursues paths to function. Life and its attributes must be presupposed in order to actually generate any hypothetical scheme of life-origin.

To avoid redundancy, we shall avoid reviewing here the general principles established in all of the other chapters of this anthology. Instead, we shall simply examine all of the leading abiogenic scenarios that exist in the literature. These begin with Miller-Urey amino acid syntheses and synthetic routes from inorganic gases to organic molecules that would hopefully polymerize. Miller's publication in 1953 [1] was originally thought to solve the life-origin mystery. Life-origin investigators learned very quickly that the formation of amino acids from electrical sparks was a long way from explaining abiogenesis [2].

Jeffrey Bada's group recently found in some of Stanley Miller's old spark-discharge vials from 1958 amino acids that had never been reported. The addition of H_2S gas apparently increased the yield of amino acids. "A total of 23 amino acids and 4 amines, including 7 organosulfur compounds, were detected in these samples." [3].   Not all of these amino acids were relevant to life, and they were not homochiral.

Life-origin models proceed from Dyson's "Garbage Can" model all the way up through self-replicative, auto-catalytic informational modes in sophisticated vesicles. It is beyond the scope of this one short chapter to critique in detail each model. But we can legitimately classify them into groups requiring common pre-assumptions. Those pre-assumptions often have no basis in observational fact. They also strain rational credibility at every turn. Not one prediction fulfillment has ever occurred independent of artificial selection that steers events, controls and regulates outcomes. The leading hypotheses are rarely, if ever, falsifiable, disqualifying them as scientific theories.

1. Cairns-Smith's clay life

In a class all its own was Cairns-Smith's "clay life" [4]. This was far from the first model of early abiogenesis. But we begin with it because it was clearly unique, intriguing and utterly inanimate. Cairns-Smith examined clay matrix crystals and considered them to provide a possible self-replicative pro-life scaffolding upon which living systems could be built [4-8]. Cairns-Smith, Ingram and Walker [7] addressed formose production by minerals. But Cairns-Smith explored more generally primitive "genes," "genetic takeover" from those primitive genes, and finally an imagined primitive metabolism.

Cairns-Smith deserves much credit for realizing that the key to any info-genesis scenario, whether in clay crystals or any other matrix, would not be found in their highly self-ordered crystalline state, but rather in their crystal *irregularities*. To this day, many otherwise brilliant investigators are still thoroughly confused as to the relation between information and order. Order and information are antithetical [9]. Without uncertainty (almost nonexistent in highly ordered states), no hope of information generation and recordation is possible in any physical medium. Cairns-Smith was wise enough to realize that any potential genetic information that could be "taken over" from clay crystals would have to be found in the irregularities of those crystals, not in their high order or regular structure.

What Cairns-Smith could not explain was how happenstantial uncertainty (random crystal irregularities) could write meaningful, functional Prescriptive Information (PI) (Chapter 1, sections 3 and 4). To this day the question remains, "Take-over of what?!" What would be the basis for assuming that randomly distributed clay crystal irregularities could generate Functional Information (FI) [10-14], whether mere Descriptive Information (DI) or the much more difficult to explain Prescriptive Information (PI) [15-18]? The notion makes no more sense than expecting a random number generator to write a sophisticated computational program.

Many other problems existed with the clay life model. The crystal irregularities were buried in inaccessible layers. The crystal irregularities, even if they had meaning, could not be "read." The number of such irregularities would be insufficient to retain the amount of information required by life. In addition, no satisfactory mechanism of genetic takeover was ever proposed. The information would have to be translated from inaccessible clay crystal irregularities into nucleotide sequences. No theoretical means of code bijection (one-to-one correspondence between "languages") exists. The automaticity of crystal formation was thought to mimic genetic replication. But crystallization provided no Turing tape readability with which to covey instructions to future generations. No mechanism of heritability of irregularities was provided by the model. No replicative genetic system existed to maintain and propagate any PI that might arise. The clay remained as dead and as when the model was first imagined. And the clay contained no meaningful or functional PI to replicate, let alone be "taken over."

An important lesson from Cairns-Smith's work at the time was how plausible and convincing a case could be made for a totally hypothetical scenario that had no real basis in fact. After reading the thoroughly entertaining and popular book *Seven Clues to the Origin of Life* [4], the transition from mere clay to life seemed not only possible, but likely. Some had such vivid imaginations and were operating under such a naturalistic metaphysical imperative as to brazenly call the transition "inevitable." The same history has been repeated a hundred times. Fads move through the life-origin scientific community almost as though it were a seasonal fashion show. Science fiction quickly gains scientific respectability when materialistic metaphysical presuppositions trump sound scientific principles. We become "true believers," all in the name of science, in scenarios with little more than superstition to support them.

2. Silicone and boron-based life

Early on astrobiologists toyed with the idea of life forms that might not use carbon as their main "backbone." Both non-layered silicone and boron were considered.

Interest in silicon [19-22] did not last long because silicon polymers cannot gain sufficient length for adequate information retention. Silicon also forms bonds with other elements that would interfere with silicon-silicon chain formation. Silicon also lacks the relatively easily-broken-and-rejoined covalent-like bonds enjoyed by carbon, hydrogen, and oxygen in organic compounds [23]. Says Trevors and Abel,

> Silicon bonds are too rigid and irreversible for cellular metabolic recycling of structural, enzymatic, regulatory, and informational biopolymers. Silicon is too insoluble in an aqueous environment. Sand, a typical silicon compound, is a good example. No organisms could have been produced except in an aqueous environment [24-26]. Carbon, unlike silicon, is amenable with the help of catalysts to dehydration synthesis even in an aqueous environment. Yet carbon-based organisms do not dissolve in ponds, rivers, and oceans. Carbon chains are unique. Finally, silicon chains lack the ability of carbon chains to establish a lipoprotein-like connection between different kinds of biomolecules.

Lipids have a different solubility and serve different functions from proteins. Both are needed for life as we know it. Carbon-carbon bonds provide both kinds of branching using the same basic building blocks. Lipoprotein molecules can cooperate to contribute to cellular survival through such functions as membrane formation. Silicone oxide can form layers, but lacks the unique properties of lipoprotein needed for semi-permeable membranes, active transport, secretion, and excretion. [23, pg..]

Silicon was also of interest because it can serve as a surface adsorbent and catalyst for proper alignment and polymerization of polyadenosines and poly-uradine [20, 27-41]. But polyadenosines and polyuradines, like the monotonous clay crystals to which they adsorb, contain almost no Shannon uncertainty. Without Shannon uncertainty, no opportunity exists for instructions to be instantiated into any medium. Clay-adsorbed homopolymers could not possibly be the source of highly informational genetic instructions [23].

A side issue related to silicon dioxide and silica carbonate is that such minerals display filamentous-looking features that have called supposedly ancient microfossils into question [42, 43]. Patterns suggesting fossilized filamentous bacteria (e.g. cyanobacterial) can form independent of life. Scientists have trouble telling what is or was alive on Earth. Microscopic structures uncovered in the roughly 3.5-billion-year-old Apex Chert formation in western Australia were originally described as the oldest microbial fossils. These structures were thought to be blue-green algae embedded in a silica-loaded rock. The branching structures were always suspect. University of Kansas geospectroscopist Craig Marshall and his colleagues determined that these structures might not be carbon-based after all [44], but rather a series of fractures filled with quartz and iron-rich hematite. The signature for hematite is very similar to that of carbon. The case has grown stronger that these microstructures are not ancient microfossils of cyanobacteria they were originally thought to be.

The notion of boron life did not receive serious attention for very long. Insufficient boron seems to exist in the cosmos to support life on any planet. Large quantities of boron compounds would be needed to provide enough diversity from which the environment could have "selected" accidental algorithmic metabolic function. The only problem is that no one has ever observed a nontrivial algorithm arise accidently. Borate minerals, however, can stabilize ribose [45]. The instability of ribose constitutes a major problem for the RNA World model.

Neither silicon nor boron provides the replicative potential of carbon bi-opolymers. In addition, both silicon and boron fail to provide the peculiar secondary and tertiary folding versatility needed to catalyze and support life. The unique lock-and-key binding fits that are so important to carbon biochemistry are not provided by boron or silicon molecules. In short, silicon and boron life-origin models face way too many serious hurdles to provide plausibility. No empirical evidence exists for any form of life other than carbon-chemistry life.

3. Geochemical self-organization models

Many geochemical models have been published suggesting how a very early protometabolism could have "self-organized." The main problem with all of them is that no basis for formal organization exists except in the mind of the theorist. Every investigator mentally constructs the needed and desired biochemical pathways to metabolic success with no basis within inanimate nature for such cooperative integration of tasks Relentless progress up multiple foothills towards the mountain peak of pragmatic success is imagined. Little possibility is allowed for the effect of gravity back down any of those foothills, let alone the mountain top.

Wächtershäuser explored the "Iron-Sulfur-World" [46-48] in which complex organic molecules were formed from the catalysis and energy release of the redox system $FeS + H2SFeS2 + H2$. Alpermann, et al [49] have adapted Wächtershäuser's model using vesicles to achieve the needed compartmentalization. They call their initial prebiotic unit "Polymersomes." In April of 2011, Wächtershäuser's model of a precursor to the RNA World hypothesis was updated by Frederick Kundell [50]. Kundell suggested a cubic pyrite crystal edge serving as a catalytic surface for the production of a condensed ribose, and potentially a proto-nucleic acid. Martin, Russell and Hall have worked in the same area for over two decades, concentrating on serpentinization of ultramafic crust, iron monosulphide bubbles and hydrothermal vents [51-57].

Many others have investigated hydrothermal vents [51, 58, 59]. In one of the most recent [60], Lane argues that the first donor was hydrogen and the first acceptor $CO2$. Martin, Russell, et al [51] had previously noted striking parallels between the chemistry of the $H(2)-CO(2)$ redox couple present in hydrothermal systems and the core energy metabolic reactions of some modern prokaryotic autotrophs. Lane points out that "synthesis of ATP by chemiosmosis today involves generation of an ion gradient by means of vectorial electron transfer from a donor to an acceptor." [60].

Recently a Zinc World has been suggested [61]. "ZnS surfaces (1) used the solar radiation to drive carbon dioxide reduction, yielding the building blocks for the first biopolymers, (2) served as templates for the synthesis of longer biopolymers from simpler building blocks, and (3) prevented the first biopolymers from photo-dissociation, by absorbing from them the excess radiation."[61] This was believed to be powered by UV-rich solar radiation at photosynthetically active porous edifices made of precipitated zinc sulfide (ZnS). Similar conditions are found around deep-sea hydrothermal vents.

Among the many problems with hydrothermal vent models is that homochirality and the polymers themselves break down at high temperatures. An aqueous environment is no friend of "dehydration synthesis." The removal of a molecule of H_2O to form each peptide bond is rather difficult in an environment of water molecules! RNA bases and even some of the essential amino acid monomers are degraded in hot aqueous environments. Other problems include dilution factors and lack of containment in compartments. Thus along with heat's advantages (e.g., speeding up reaction rates) comes a slew of problems. This has led some to postulate a cold origin of life where denaturization and other breakdowns would not occur.

Various eutectic ice approaches have been tried [62-68]. The main problem with eutectic ice environments is that the reaction rates are slowed to a snail's pace. No sophisticated enzyme catalysts exist yet in any type of envisioned protocell, whether peptide first or pre-RNA world models. In order for any macroevolutionary scenario to even seem plausible at cold temperatures, huge phase spaces of varying polymers and efficient enzymatic catalysis would have been required. Neither were available. And time was limited since earth's cooling.

Eshenmoser pursued the chemical etiology of nucleic acid structure [69-71]. Rode suggested a salt-induced peptide formation reaction in connection with adsorption processes on clay minerals as the source of a possible Peptide World [72]. Carney and Gazit have further explored peptide assemblies that "possess the ability to bind and stabilize ribonucleotides in a sequence dependent manner." [73]

Many other geochemical models explore hypothetical chemolithoautotrophic pathways [74-81] [82]. Many geochemical models arise out of astrobiological research programs sponsored by NASA grants. In all these geochemical models, no naturalistic basis is ever presented for inanimate nature's *pursuit* of any potential functional pathway. Teleology is of course disallowed by any naturalistic model. Yet without pursuit of formal function, not even the naturalistic version of teleology, teleonomy, is ever observed to arise spontaneously.

4. Protometabolism First models

Most prebiotic molecular evolution models involve organic components and are many and varied [83-140]. As Fry points out [141], autotrophy versus heterotrophy and "soup" versus hydrothermal vent environments remain the major dividing theoretical camps [2, 47].

Freeman Dyson proposed an initial "Garbage Can model" in 1982 [142, 143]. The Dyson model in 1982 was not specific about the nature of the proposed inorganic reactions and catalysts [144]. No experimental basis is provided for high discrimination factors between similar molecules (e.g., amino acids). As Anet points out, the high discrimination factors are far more difficult to achieve than mere catalysis [144, pg. 655]. Dyson's oligopeptides contained around a five-monomer active site in twenty-monomer strings. But the non-active-site monomer sequences are critical to folding and function.

von Kiedrowski [136, 137, 145] demonstrated a self-replicating hexadeoxy nucleotide [145]. Short double helix templates can self-replicate easily. The two strands separate with a certain probability to form templates. Szathmary points out that longer strands can serve as templates too, but do not replicate "because the two strands do not separate spontaneously; intervention is required."[146, pg. 32].

Lifson refutes theories of self-organization that fail to incorporate natural selection [90]. But what exactly is the prebiotic mechanism of selection for function at the level of monomer sequence formation? Biopolymers form with rigid covalent bonds prior to any folding. Selection occurs at the phenotypic level, not at the genetic level of polymer formation [147]. Even in Peptide and Protein Worlds, sequencing would have determined minimum-free-energy folding space. Some have argued that sequences of at least 80-100 mers are necessary for any substantial selective catalysis [148].

Differential autocatalytic doubling is seen as a substitute for informational genetic replication and the basis for natural selection in protometabolism first models. Constrained geochemical conditions are actually seen as advantageous by some in limiting the possibilities of early protolife development [48, 149, 150]. This would reduce the likelihood of cross reactions and metabolic dead-ends. This more deterministic hypothesis resembles Dean Kenyan's "biochemical predestination" model of life origin [151] which Kenyan himself eventually disowned.

Protometabolism First models such as Robert Shapiro's [152] are seen as more probable by metabolists than Information First genetic models. The extreme improbability of RNA world scenarios is the reason. But others such as

Leslie Orgel have challenged this [144, 153, 154]. Plausibility is perhaps just as lacking for protometabolism first models as replication-first models.

4.1 Composomes

The naturalistic dream upon which all macroevolutionary theory is founded is a "self-reproducing and evolving proto-metabolic network." Given all of the many biochemical problems with RNA World models, many investigators have returned to previously-abandoned Metabolism First models [53, 56, 143, 146, 152, 155-159]. The self-replication and ligation of peptides has received some attention [160, 161]. But the catalytic properties of peptides is well known to be little more than those of ribozymes—extremely poor compared to proteins.

Lancet, Segre and Shenhav have championed "composomes" [162], Graded Autocatalytic Replication Domains (GARD) [163], and Lipid World models [164]. Progress was limited with these models. Composomes are theoretical Metabolism First protocellular entities. Early on they are believed to have found themselves contained within lipid vesicles or micelles. They are not seen as full-fledged living systems. But they example one of the simplest conceivable models of protolife. Compositional "genomes" (assemblies of varying molecular species) are thought by some to be able to propagate evolvable chemical and structural information.

Catalytic organic amphiphiles within vesicles provide what Lancet's group calls "compositional information." [163, 165-167]. Lancet's protocells are able to propagate without polynucleotides. Irregularities that arise in the envisioned catalytic networks are considered to be mutations [167].

It has never been clear exactly how "compositional information" is read, or exactly how this three-dimensional composition organizes and instructs a protometabolism. The model resembles the old imaginative protoplasm gel theory of life. The devil is in the details of any model. Detail is sorely lacking in the composome model. The problem with a non-digital means of analog replication is that the fidelity of information quickly deteriorates.[168-171]. The units of replication and selection need to be discrete. Nevertheless, Sterelny and Griffiths [172, 173], Lancet and Shenhav [167], and Pohorille [174] argue for analog instructions playing the major role.

Autocatalytic amphiphilic assemblies are hypothesized to co-evolve along with their surroundings. Auto- or perhaps mutually-catalytic "metabolic networks" are envisioned. They are "devoid of sequence-based biopolymers, yet could exhibit transfer of chemical information" with the hope that they might also undergo selection and evolution without a genetic apparatus [162]. They are viewed as rudimentary "compositional genomes." The chemical composition of the environment would have theoretically governed the chemical repertoire generated within molecular assemblies in these compositional protocells. Segre et al's "lipid world" includes evolutionary genetic membranes that do not require protein or RNA enzyme catalysis (Segre, Ben-Eli, Deamer, et al., 2001).

The compounds generated within the composomes would have then altered the chemical composition of the environment. This is called the environment exchange polymer graded autocatalysis replication domain (EE-GARD) model [162]. In the computer models of composome evolution, early stage composomes disappeared, while others emerged.

EE-GARD is of course not the only non-genetic Metabolism-First life-origin model. Many models throughout recent decades have been proposed. Freeman Dyson's two-step model attempted to bridge the gap and provide transition from garbage-can Metabolism First to more template–directed informational genetic models [143]. Yarus posits a more contemporary two-step model [175]. Many of these models emphasized a three-dimensional physico-chemical and structural concept of information as opposed to the linear digital PI messaging found in all known current life. Eshenmoser [71], Russell, Martin and Hall [51, 55, 176] Shapiro [177, 178], Kauffman [179, 180], de Duve [150], Wachtershauser [48], Morowitz [181], Deamer [182], Pohorille [174] and Lindhal [183] models provide specific chemical detail. Wong [184, 185], Guamaeres [159, 186] and Di Giulio [187, 188] offer models that attempt to link early protometabolism to code development. But, as we shall see below, virtually all Metabolism First models lack organizational motive, ability and naturalistic explanation.

4.2 Compartmentalization

Compartmentalization becomes a major issue early on in life-origin research. Most life-origin models today incorporate some form of an early micelle, vesicle, or pseudo membrane.

Pier Luigi Luisi [189-194] and David Deamer [182, 195-204] have been leaders in primordial membrane research. The problem is that most primordial pseudo membranes that would keep needed protometabolites in the compartment also keep toxic metabolites in with them. In addition, needed nutrients are kept out without highly selective active transport mechanisms. The simple bilipid layers of vesicle models lack this sophistication. Simple but serious osmotic problems even exist with bilipid vesicular pseudomembranes. Active transport and highly specific differentiations are necessary in channels to exclude cross-reacting factors, yet provide needed metabolites. Even Wächtershäuser's "surface metabolist" model requires a water-resistant pseudomembrane on pyrite crystal surfaces [47]. Mansy et al have done some of the most recent and impressive work on protomembranes [205-208].

The latest RNA World models of life origin all require a membrane of sorts that can expand and divide. This pseudomembrane encircles an RNA replicase ribozyme [209]. An additional ribozyme is thought to have formed within this vesicle that catalyzed the synthesis of essential membrane-genome relationship factors [193].

4.3 The problem of sequencing

It is only recently that the origin-of-life scientific community has been candid in acknowledging the problem of co-polymer sequencing. Perhaps the Origin of Life Prize was instrumental in stimulating previously neglected discussion and research into gene emergence (www.lifeorigin.org). A co-polymer is a polymer consisting of two or more different monomers. This is distinguished from homopolymers where all monomers are identical. It is much easier to form homopolymers under prebiotic conditions than co-polymers. As Pier Luigi Luisi point out, "the kinetics and thermodynamics attending the synthesis of copolymers poses stringent constraints for the biogenesis and growth of specific sequences." [210] Luisi goes on to say that although co-oligopeptide chains can be produced by prebiotic reactions, "It is not possible by the bottom up approach to find the conditions for the synthesis of our actual proteins—lysozyme, chymotrypsin or the like, . . ." [210] Sequencing is critical in proteins and nucleic acid. But of course the sequencing of amino acids to form proteins stems back to the prescriptive codon sequencing and editing of DNA and mRNA. The chemical bonds between nucleotides are all identical. The chemical bonds between amino acids are also identical. The problem of sequencing, especially the prescriptive sequencing of nucleotides will not be solved by any physical law [9, 15, 18, 211]. The same is true of the formal Hamming block-coding and codon table used to reduce noise pollution in the Shannon channel.

4.4 Hypercyles

Few concepts are more important to any protometabolic scheme than Eigen and Schuster's original notion of hypercycles [212-226]. Hypercycles have been proposed as a source of spontaneous naturalistic self-organization [215, 218, 219, 225]. Biochemical cycles act as catalysts. Hypercycle theory is based on positive and negative feedback constraints [48, 162, 174, 227-230]. These mere circular constraints are typically confused with formal controls. These deterministic cycles of interaction are envisioned to give rise spontaneously to pragmatic pathways, networks, and finally integrated metabolic processes. The key word is "envisioned." An imaginative human mind can envision extraordinary creativity through natural process without any empirical or prediction fulfilling confirmation. Actual experimental observation of physicodynamic interactions say otherwise. Eigen was able to *engineer* a hypercycle of cooperating RNAs that catalyzed each others' replication [231]. But engineering always involves formal controls. Wächtershäuser attributes what he considers to be non-genetic memory to the branching products feeding back into the cycle.

The reason Eigen's hypercycles are so appealing is that the simpler chemical cycles obviate the need for protein enzymes and even ribozymes. The cycle itself provides catalysis. Hypercycles are envisioned to provide fuel for a bottom-up theory of self-organization and life origin, all from purely chemical "systems." A cyclic process sign is used to replace equality in mass-balance equations. Once again, mere cyclical constraints are confused with formally programmed feedback controls needed to organize any bona fide pragmatic system (e.g., metabolism).

Blomberg, the head of NASA's Astrobiology program at the time, reminded us of the Eigen-Schuster "error theshold,"

> There is a limitation of size due to the accuracy, which can be called the Eigen Schuster error threshold [217, 232]. If the size of a replicating macromolecule becomes too large, then there will be too many errors, and no systematic reproduction. This leads to a dilemma, sometimes called the Eigen dilemma: it was necessary to have a high accuracy to obtain long, functionally active macromolecules. But to achieve high accuracy, long active controlling macromolecules were needed. There are many chicken and egg problems of that kind for this stage that obscures a direct step by step development. [233]

4.5 Tibor Ganti's well-developed chemoton model

Perhaps the most well-developed, comprehensive Protometabolism-First model published by any one author is Tibor Ganti's "chemoton theory" [86, 87, 146, 234-237]. Tibor Ganti's chemoton model is a hypercycle take-off. Ganti envisioned interconnected autocatalytic, or at least mutually catalytic [223], Eigen-Schuster type hypercycles [213-219] in well-organized elementary units of life called "chemotons." [87, 234, 236, 237]

Ganti's basic idea is that stoichiometric cycles act as catalysts. No proteins exist yet, and therefore no linear digital prescription is needed. He uses a cyclic process sign to replace equality in mass-balance equations [238]. The cycles do the work of enzymes and ribozymes. It's a bottom-up theory that is very appealing to those faithfully committed to purely materialistic metaphysical presuppositions. The model depicts a transition directly from chemicals to living "systems" without any formal controls.

Chemotons have three self-producing (autocatalytic) stoichiometric subsystems that are coupled to one another: autocatalytic metabolism, a genetic polymer, and a membrane. The key to life is "fluid automata," complex systems of chemical reactions in fluid phase that function like machines. They have no solid parts. But they can be regulated.

Ganti's models are quite different from Lancet, Segre and Shenhav's composomes discussed earlier. Ganti recognizes and even emphasizes the need for cybernetic controls. "Chemical reactions as building blocks can be assembled into regulated and program-controlled chemical automata without including any solid components." Ganti realizes that true organization is essential for life to exist. The chemoton model's "organizational principles must be present in every living being." Ganti calls his chemical cycles and networks "cycle stoichiometry." Ganti sees a direct link between genetic cybernetics and computer science. "Program control must control a functional system and enzymatic regulation must also regulate a functional system."[146, pg xii] "Chemoton theory is concerned primarily with this machinery aspect."

Says Ganti, "A chemoton is the simplest chemical machine which shows the generally accepted characteristics of life." [146] The first problem with this is that no such machine exists. The simplest living organism is a cell containing multiple operating systems, hundreds if not thousands of programs, multiple layers and dimensions of Prescriptive Information (PI), and huge numbers of molecular nanocomputers all cooperating in one concerted integrative effort [239, 240]. The second problem is that "the generally accepted characteristics of life" depend upon who is defining life. No two scientists' definitions of life seem to be the same. Life-origin science has a long history of defining-down life to something far less than life in order to make our models "work for us." [241]

Ganti repeatedly refers to his chemotons as not only "interconnected systems of chemical reactions," but as "organized regulated processes." "Organizational principles must be present in every living being," [146, pg 1]. The question of exactly how chance and or necessity could sense, obey or pursue a formal "organizational principle" is never addressed. Until this missing essential piece of the puzzle is supplied, the model falls apart as a supposedly naturalistic explanation.

Ganti envisioned his fluid automata chemotons to have two parts: an operating part—the automaton; and a controlling part—the genetic programs

[146, pg. 13]. Regarding the controlling part, Ganti makes a stunning admission: "Of course, *the sequence of signs is not material* [italics mine]. But neither is it independent of material, since the sign is carried by some material substance." [146, pg 13] In other words, Ganti regards the information found in the sequence of signs as nonmaterial even though that sequence is instantiated into a physical medium of monomeric sequence. It is surprising to this author that much of the supposedly controversial material presented in *The First Gene* Ganti himself would probably have had to agree with.

What in nature for Ganti could possibly be "not material" (not mass/energy)? Naturalistic science is physicalistic. Reality tends to be defined solely in terms of mass/energy. When Ganti talks about non-material representational sign syntax, does he believe in some sort of non-material "super nature"? Ganti would probably assure us of his commitment to a naturalistic worldview. The whole point of naturalistic life-origin science is to avoid nonphysical explanations of what is claimed to be a purely physical reality. But is his cybernetic model tenable under the naturalistic metaphysical imperative? Representationalism is a little hard to explain from a physicalist perspective. So are the mathematics and reason upon which the scientific method relies. To argue logically for physicalism is to deny physicalism. There is nothing physical about the exercise of logic theory.

What does Ganti's word "program" mean? Programming requires purposeful logic-gate settings. Exactly how are programs instantiated into this liquid chemoton's physicality? What is the basic unit of selection and instruction in this liquid? By what chemical mechanism did this programming arise? Can we cite any examples of chemical reactions in liquid phase, especially, programming and optimizing algorithms? Says Ganti, "The living system is a program-controlled cybernetic system." [146, pg 12] Ganti continues, "Cybernetics itself originated from the study of the regulated and controlled operation of living systems, and program control is already familiar from the genetic program." [146 pg 12]

Says Ganti, "Chemical reactions as building blocks can be assembled into regulated and program-controlled chemical automata without including any solid components." [146, pg xiv]. But what is the nature of this mysterious liquid program? Is it a liquid crystal? How could programmed information be instantiated into a liquid OR highly ordered solid structure? And what exactly *does* the "regulating" of these chemical reactions? What sort of magic cybernetic liquid crystal is this?

There is no basis for logic theory, quantification, decision theory, scientific debate, computation, computer science, controlled experimentation, or engineering within the physicalist world view. But we ought not be too hard on Ganti for acknowledging the reality of representational signs in the cybernetics of life. The sequence of physically instantiated "signs" ("physical symbol vehicles;" tokens) is everything when it comes to the message of messenger molecules.

The most serious problem with Ganti's model is that he is unable to generate a basis for "programming" OR "organization" from mere physicodynamics. Interconnected hypercycles are even more constrained than individual cycles. Constraints are not controls. Constraints cannot steer events toward pragmatic goals.

Ganti's chemotons are assumed to be already evolved prior to the appearance of catalytic RNA's [146, pg vii]. So RNA linear digital prescription cannot explain Ganti's acknowledged need for programming and organization in the forming of chemotons. RNA was supposedly assembled only later by substrates already present in the chemoton [146, pg vii]. This notion alone does not explain how the particular functional sequencing of "signs" [actually tokens—"physical symbol vehicles" in a Material Symbol System (MSS) [242, 243]] was achieved.

Exons, the protein-coding regions of eukaryotic DNA, were found to contain a seemingly random statistical distribution of the four nucleotides. Weiss et al. reported that "Protein sequences can be regarded as slightly edited random strings." [244]. These facts led this author to originally surmise that nucleotide sequencing in highly informational strands must be physicodynamically indeterminate not only in exons, but probably in the very first informational single-stranded, non-templated RNAs. All of the 3'5' phosphodiester bonds are the same between coding nucleotides. With a seemingly random nucleotide frequency of distribution, one would expect no physicodynamic "preference" in sequencing. But the recent finds of extensive highly-functional regulatory microRNAs in non-coding introns raises new questions. Introns contain a great deal of redundant order. They were once thought to represent non-informational junk DNA. For configurable switch-settings to have cybernetic function, the switch must exhibit near equal physicodynamic opportunity to be flipped either way. "Off" must be just as feasible as "On" from the standpoint of natural law's influence on the physical switch. If the laws of physics in any way militate against "Off" (e.g., the force of gravity favored the down position of the switch knob on a vertical switch board), the switch setting becomes less significant algorithmically. To whatever degree the switch-setting is forced by physicochemical necessity, its Shannon uncertainty drops. Shannon uncertainty is nothing more than a probability function. Its ability to instantiate *prescriptive information* nose-dives with increasing physicodynamic determinism. Maximum information retention at that switch is realized only if "On" and "Off" are equally possible from the standpoint of physics and chemistry.

Constraints and physical laws are normally poisonous to control unless configurable switches are specifically designed to be physicodynamically inert. This is exactly the case with the coding regions of DNA. Protein coding sequences are non-ordered by "necessity" (law). Weiss et al found that proteins are "slightly edited random strings." [244] The 1% nonrandom factors were thought to arise out of secondary structure requirements and low complexity regions.

Very recent work has also shown that up to 120 mer cyclic homopolymers of RNA can form spontaneously in heated aqueous environments. No catalysts or templates are required. Oligomerization of 3', 5'-cGMP to ~25-nucleotide-long RNA molecules, and of 3',5'-cAMP to 4- to 8-nucleotide-long molecules were achieved [245]. The authors hoped that this research might explain the first genetic polymers. But they failed to address how such high redundancy could instantiate much Shannon uncertainty, let alone PI. Highly ordered strings should have a very low information-retaining ability. Programming with highly redundant, intron-like sequences would have to be very simple and limited. Perhaps the information requirements for regulation are minimal, with only short segments needing specificity, and the bulk of the redundancy going into basic noncritical carrier structure of the active sites. But with the discovery of so many highly functional, yet highly-ordered microRNAs, it does raise the question of whether ordered strings, not just seemingly random strings, can be used to prescribe function. Functional sequencing can include repeated selection of the same nucleotides in monomeric sequencing (syntax). Order in a sentence or programming string is not necessarily from physicodynamic causation. In a linguistic string, for example, some letters have a high frequency of arbitrary re-use. Morse code assigns the shortest symbols to represent the most frequently used letters ("." represents and "e"). "Freedom" from physicochemical law exists in the coding regions of DNA to program biopolymeric messenger molecules. This programming freedom includes frequent re-use of the same nucleotide in microRNA prescriptions. The situation is a little like a programmer re-using redundant modules, or an engineer using many highly-ordered components to manufacture a sophisticated product.

Also surprising, if not shocking, is the discovery that the negative complementary strand of coding DNA that is physicodynamically base-paired can simultaneously prescribe regulatory function independent of the positive strand's prescriptive function. Both PI sets are sequence dependent. Clearly some formal factors are transcending what seems to be the purely physicodynamic base-pairing to prescribe multiple unrelated formal functions with each complementary strand. It is a false conclusion that these formal functions were prescribed by base-pairing itself. Purely physicodynamic base-pairing is incapable of programming any formal function. The programming is so formally sophisticated that it prescribes and organizes even physicodynamics to accomplish computational cybernetic goals. The evidence only continues to mount for the Formalism > Physicality (F > P) Principle presented in Chapter 12.

The existence of Ganti's initial PI is just presupposed rather than explained by his model. Ganti repeatedly acknowledges the need for programming and organization, but provides no naturalistic model for either formalism.

Ganti is right that, "A living being is a controlled system." [146, Pg. 13] But Ganti thoroughly confuses circular physicodynamic constraints with formal organizational *controls* needed to generate "usefulness." [246] Constraints and Laws cannot possibly generate cybernetics. We should never confuse mere order with algorithmic programming. Ganti says, "A strictly defined order exists in our television set, radio, or computer." [146, pg. 19] Physicodynamic order (e.g., redundant crystalline structure or oscillations) doesn't make television sets. Says a peer reviewer of this paper, "Actors collecting the necessary and defined resources using processes to act on them through formal cybernetic and algorithm organization make televisions. The televisions, in turn, run electronics which are taking incoming analog or digital inputs and converting them to audio-visual displays through algorithm organization."

Ganti pursues order in his liquid automatons in an effort to explain programming, organization and the beginnings of cellular cybernetics. He looks for "forced coupling between the forced trajectories so that the solution is capable of operating as an automaton with a given function." [146, pg 19] Neither cause-and-effect ordering (necessity) nor oscillating (redundant) chemical reactions can explain phenomena such as chemotaxis. Chemotaxis requires freedom from forced constraint and law. Genetic programming is similar to computer programming wherein degrees of user freedom are programmed into the software. The user gets to choose from among real options. No trajectory is forced. No highly ordered oscillation allows a bacterium to avoid noxious stimuli. Logic gates require programming freedom, not fixed order.

Could circular constraints ever generate formal regulatory control and organization of protometabolism? Many publications have argued that this is possible. But they are all purely theoretical with no observational or prediction fulfilling support. The circular intertwined diagrams of Tibor Ganti certainly look plausible at first glance [146, 236, 237]. Upon more careful, critical analysis however, they are merely *self-constraining* feedbacks, not formally self-controlling feedbacks. No choice contingency is involved. No steering or selective programming in pursuit of formal utility is achieved by purely chemical positive or negative circular cause-and-effect constraints. Minimal uncertainty is involved in circular constraint, and therefore almost no Prescriptive Information (PI) is required. Indeed, PI control is nearly impossible. Any feedback is forced in a circular cause-and-effect deterministic chain. Increased production of reactant A will always lead to increased feedback production of reactant A_1 or B in cases of positive self-constrained feedback. The opposite is true in cases of negatively self-constrained feedback   No opportunity exists for formal *"regulation"* of anything. The product ratios are forced by the relative constraints irrespective of any formal need or desired utility. No fine-tuning in pursuit of function optimization is possible or even desired by a self-constrained cycle. The term "feedback *mechanism*" is an illusion. Only unimaginative redundant, cyclical, physicodynamic determinism prevails without regard to mechanistic (machine like) pragmatic benefit. Sophisticated machines don't spontaneously generate any more than life spontaneously generates.

Ganti's model of life-origin was doomed from the start for two major reasons: 1) his failure to understand the difference between "constraints" and "controls," and 2) his confusion of physicodynamic laws with the arbitrary cybernetic rules of life's programming.

Controlled paths can be paths with purposefully preset switches. *Preset switches are formal decision-node choice commitments.* They are logic gates. And they must be freely set to either open or closed positions with programming intent. Neither coin flips at each decision node (fair or weighted) nor physicodynamic laws will achieve sophisticated algorithmic function. Life is algorithmic and dependent upon linear digital prescription using a material symbol system (MSS) [243].

Ganti's notion of cybernetics is nothing more than physicodynamic coupling—cause-and-effect chemical reactions interconnected by forced stoichiometric connections between three auto-catalytic cycles. The three cycles are unable to function without each other. Their cooperation together forms an interdependent super system, a model which, if it actually existed, Michael Behe might claim exemplifies "irreducible complexity" [247].

In addition, "If A, then B" reactions are constrained, not formally controlled to achieve desired function. Such cycles of constraint, even if they lead to positive or negative feedback, are not true examples of formal control, regulation, pragmatic organization, or true systems. Cycle stoichiometry would have to be steered by controls, not mere reaction constraints, to achieve formal integration and final metabolic function. Only by highly selective catalysis and formal regulation of the cyclic pathway can stoichiometry produce any desired useful work.

Ganti considers an oscillatory chemical reaction a chemical automaton (p 19) because of his sometimes loose definition of "cybernetic." At other times, he more accurately defines "cybernetic." A major problem with Ganti's model is that he shifts back and forth between the two definitions of cybernetic within his model. Mere oscillation is not cybernetic, at least not to the degree we could draw any analogies with sophisticated machine generation or function. A pendulum swing doesn't do very much "useful work." A reviewer counter argued that crystal oscillators can be used in computers to provide input function for circuitry timing. "Crystal oscillation provides useful work—timing input function—which allows all dependent circuits to be coordinated." But the key to this contention is "can be used" (by agents). A hill does not become a simple machine (an inclined plane) until an agent decides to use that hill as the means to achieve a desired function. The hill itself does not perform useful work. The hill just exists in a formally neutral sense. A hill is not "an inclined plane" until that hill is used by human agents to do useful work—to achieve a desired purpose—to accomplish formal utility. A hill is not a simple machine merely because the hill exists. The hill can be used by us to increase the efficiency of raising a heavy object to a higher altitude against the force of gravity. The choice and act of an agent rolling a heavy object up the hill, rather than lifting it, alone makes the hill a simple machine. The wind can blow a tumble weed up the hill, but no *functional work* is accomplished. The inanimate environment does not value the tumble weed winding up at a higher altitude. No formal goal is pursued, and no utility accomplished, in seeming opposition to nature's relentless trend towards disorganization and loss of sophisticated utility.

Ganti talks about "successive chemical transformations involving organic acids in biological oxidation." [146, pg xi]. Ganti did not seem to concern himself with the prevailing serious problem of how interminably-long biochemical reactions take without sophisticated enzymes with remarkable rate constants (10^7 to 10^8 times the acceleration rate). Szathmary, in critiquing Ganti's model, seems to believe that a non-enzymatic chemoton would be impossible [146, pg 41]. Over a hundred reaction steps would be required for the simplest chemoton. Two thirds of these reactions have been shown in the literature to be infeasible in a prebiotic environment [146, pg 41]. Even if sophisticated protein enzymes had been available from the start, how would a hundred different chemoton reactions all be integrated into a unified and coherent pragmatic system in a prebiotic environment?

More problems with Ganti's model are listed by Stegmann [248]. The Krebs Cycle, like a biopolymer, can be viewed as an "ordered [sequenced] set." Stegmann [248] asks which factors determine the identity and the sequence of the elements of this 8-tuple? "In the Krebs cycle, each reaction product becomes the substrate of the subsequent reaction." Oxalacetate (O) accepts an acetyl residue, producing citrate (C). The C next becomes the substrate for the reaction that produces isocitrate (I). The sequence of the 8-tuple $\langle O, C, I, G, ... \rangle$ is a direct result of "If A, then B" reactions. Thus says Stegmann, "the order [the sequence] is not determined by the properties of a molecule present before the reactions occur. It is not even necessary that the set of substrates is present before the cycle starts, because the substrates are produced as the cycle unfolds."

Where is this chemoton *unit of life* in observational nature? Chemotons as elementary units of life have never been observed. No prediction fulfillment has ever been realized of spontaneously forming chemotons coming to life. For all of the tens of millions of life forms that exist on earth, one would expect to be able to identify innumerable examples of spontaneously generated chemoton units from which all these organisms supposedly arose originally. And we would also expect to be able to vivisect organisms down to their simplest chemoton units of life for study. But any attempt to reduce life down to its imagined sub cellular chemoton level invariably kills that life.

Glass et al. [249] identified 382 of the 482 *Mycoplasma genitalium* protein-coding genes as essential, plus five sets of disrupted genes that encode proteins with potentially redundant essential functions, such as phosphate transport. Genes encoding proteins of unknown function constituted 28% of the essential protein-coding genes set. It remains to be seen how many peptides, polypeptides, and small RNAs are essential for regulation in this Mycoplasma in order for it to be alive. This is the simplest form of life known. It cannot be reduced to a smaller living chemoton unit. Just because we *say* life is reducible to only five criteria does not make it so in objective reality.

Says Ganti, "Thinking in a suitable abstract chemical state space" (a cycle stoichiometry state space), "It is possible to *design* fluid chemical automata in a similar way to that used by mechanical *engineers* when *planning* mechanical automata or by electrical *engineers* when *designing* a radio or a computer. *The same method is used* and *the work can be done at the desk.*" [Italics mine] [146, pg 23] Ganti goes on to point out that, "The author has *designed* several 'chemical machines'" [146, pg 23]. We might ask in response to this argument, "Are these contentions supposed to provide support for a prebiotic naturalistic spontaneous generation of a chemoton?!

The primary problem with the chemoton model is that the reaction chain always branches into innumerable "wild goose chases." Controls, not mere constraints, are needed at every fork in the road to achieve computational success. Ganti's main figure of the minimum chemoton (Fig 1.1) in *Principles of Life* is telling. It depicts a non-existent abstract scenario featuring careful, deliberate exclusions of the innumerable dead-end branches, cross-reactions, and negative feedbacks that shatter hypercyclic dreams. In addition, Fig. 1.1 in his book does not show the rapid consumption of valuable resources by both the "right" and the "wrong" paths that could be taken. There is no right or wrong path in a purely physicalistic reality. Whatever cause-and-effect determinism militates is "right," whether it produces any functional benefit or not. No basis exists for "correct" (computationally successful) switch-settings within Monod's chance or necessity (Monod, 1972). Freedom of selecting each fork in the road of the potential reaction network is essential to generate formal function. In the absence of choice-with-intent, we simply do not observe three conceptually complex autocatalytic cycles simultaneously integrating themselves into a protometabolism. Algorithmic programming is required to couple them appropriately to achieve unified and coherent holistic function. Ganti just presupposes—blindly believes—all this spontaneous cooperation. No observational support exists in the history of human experience for such spontaneous pragmatic self-organization.

Says Ganti, "It is also a splendid recognition that living systems are complicated fluid machines consisting of invisible wheels in an imaginary field. But if this is true, then those claiming that such a machine could not be developed by itself are probably correct." [146, pg 24] One wonders how "machines consisting of invisible wheels in an imaginary field" would be any more respectable a scientific hypothesis than vitalism.

Ganti continues, "Obviously, a constructor was needed who designed the controlling program first, but also planned the controlled machine. Who was the constructor who designed these congenial machines and who was the chemist who realized these plans?" Somehow Ganti's model does not seem to be measuring up to the plausible naturalistic model it is claimed to be.

Biochemical products of cycle stoichiometry can occur spontaneously. But these products are not directed toward or integrated into any desired or needed task. The reactions just happen without regard for any pragmatism. They are not programmed or organized into any process. "A cell is not a bag of enzymes."

Contrary to the current preferred scenario, the environment doesn't "co-evolve" with physicodynamics towards ever increasing spontaneous functionality. The inanimate environment could care less whether anything "functions" or whether "useful" products are produced. No basis exists for the environment preferring them. The only way hundreds, if not thousands, of co-evolution steps could progress towards an organized protometabolism is via Freudian "wish-fulfillment" in our minds, not in a physical inanimate environment. Not only would all available resources be consumed in the militated automaticity of Ganti's first few coupled reactions, but the products would cross-react and lead to biochemical dead-ends. The number of cycles and cycle couplings would have to far exceed by orders of magnitude the number of current-life enzymes needed to accomplish just the reductive citric acid cycle alone.

Like Stuart Kauffman, Ganti asks many of "the right questions" which others run from: "Living systems are program-controlled chemical machines. So where is the program?" [146 pg 30]. Ganti answers by saying that in self-regulating chemical processes, reaction networks are regulated by feedback. He views feedback as being the precursor or primordial program itself. But what exactly is feedback in the case of Ganti's chemoton model? The forced or "necessary" products of chemical reactions become substrates and/or catalysts for more automatic cycles of the same constrained process. Autocatalysis and self-replication are self-ordering phenomena of physicochemical necessity, not formal programming freedom. They are like the redundant oscillation of a pendulum, or one full revolution of a tornado or hurricane. Ganti senses something is wrong with his attempted equating of controls with constraints: "However, for program control, external intervention is necessary which regulates, via external information, the operation of the machine." What external information might that be? Could it be Prescriptive Information (PI)? What would be the source of PI a prebiotic naturalistic environment? Mere stoichiometric cycles cannot generate formal PI *or* organization.

Programming is something much more than mere auto-catalysis. No programming whatsoever is required for trivial autocatalysis. "Automaticity" is not synonymous with "program." The automaticity of trivial autocatalysis relies solely on cause-and-effect forces, self-ordering, and the laws of nature. Programming, on the other hand, relies upon *freedom from* that order and necessity. Algorithmic programs are possible only because the programmer can make real choices at each successive decision node. Switch-settings must not be physicodynamically determined by interlocking chemical reactions! Programming decisions require freedom from physicochemical determinism.

Ganti keeps presupposing "process" in his model. He pre-assumes that natural events are progressing relentlessly toward ever-improving functionality. No natural mechanism is provided for such a preference. Then he repeatedly argues that he has logically proven his thesis. According to Ganti, a clear unambiguous answer has been given to the question of the genesis of life (Ganti, 2003, pg. 41):

> Is it possible to *prepare a plan*—a metabolic *map*—of a chemoton in which, compounds present in the primordial atmosphere can be substituted for the letters in the *abstract* reactions? If so, then this will represent a credible *pathway* for the genesis of life which is *not based on chance*. [146 pg 39, italics mine].

The problem is that it would not be based on physical determinism either. Preparing a plan is formal, not physical [15, 211, 250, 251]. Ganti is right that chance cannot explain abstract plans and maps, as he calls them. But neither can the fixed invariant laws of nature. Ganti continues: "If such a *design* can be realized, this will show that the spontaneous genesis of life cannot be regarded as an accidental improbable miracle, but as a process directed by the laws of nature . . ." Ganti simply does not understand, or refuses to acknowledge, that fixed law precludes the very programming freedom and organization which his model requires. *Physicodynamic determinism locks logic gates into a fixed position and destroys all hope of formal programming of utility [250].*

The fixed laws of nature cannot "prepare a plan." The laws of nature cannot "map" out a journey through a maze, "represent" in "abstract" terms, or "design" a single engineering function. All of the latter accomplishments are formal and algorithmic. They are phenomena arising from a different category entirely from redundant natural law. Natural law cannot program anything. Ganti continues to reason from within a certain internal inconsistency:

> The chemoton model can be used to design program-controlled proliferating fluid automata which form spontaneously [!] and show properties characteristic of living systems in an exact and concrete manner if the necessary data are known, just as an engineer designs his machines and instruments [!]. ([146, pg 41], exclamation marks mine)

Ganti is first saying that his abstract, human-crafted model can be used to design program-controlled fluid automata. He is describing obvious engineering functions here. Both his model and the program-controlled fluid automata designed by his model are algorithmic. Neither chance nor necessity could produce any of these objects. Under no circumstances would they be expected to form spontaneously in nature. Even if they did, they would display none of the properties or characteristics of living systems, which are invariably algorithmic to an extraordinary degree. Spontaneous events have almost nothing in common with how "an engineer designs his machines and instruments." An engineer designs through an integrated succession of purposeful decision-node choice commitments. Each and every choice is made with intent to achieve a desired function. The function comes into existence only upon implementation and halting of the finished program. The initial program may not be ideal. But it must produce the basic computational function to even be considered a program.

But Ganti is right to emphasize "the property of specially organized systems."

Ganti is correct when he says, "Life itself is the continuous organized functioning of the system, which can only be maintained at the price of continuous performance of work" [146, pg. 72]. But his correctness is realized only within the context of proper definitions of "organized" and "work." "Organized" is always algorithmic, never self-ordered. An algorithm is a stepwise procedure governed by purposeful programming choices. Organization cannot arise from law-like self-ordering phenomena. "Work" is always defined by utility relevant to some need, desire, or goal. Life is cybernetic. All known living organisms manifest the ability to harness, transduce, store, and call up when needed chemical energy for the work of staying alive. As Ganti points out, organisms must maintain themselves permanently far from equilibrium. This requires the constant expenditure of captured and algorithmically-transduced energy.

Where is the empirical support for this chemoton pipe dream? Ganti repeatedly argues in effect, "See, no miracle is needed." But in the absence of empirical support for spontaneous programming, faith in miracles is exactly what is being presented as though it were a scientific model. Even spontaneous multiple couplings of organic autocatalytic cycles has not been empirically supported, especially in prebiotic environments.

Can bona fide "systems" exist without control and regulation? Undeniable life is empirically unknown without the essential ingredient of highly fine-tuned metabolism made possible only by highly organized systems of sophisticated enzymes. The fine-tuning cannot be so easily dismissed from the "necessary and sufficient" definition of empirical life. But it is all-too-easily dismissed from Ganti's abstract model of life.   The efficiency of regulation may well be the most significant component of life. It is the key to homeostasis in a constantly changing, hostile environment. The primordial environment was far more hostile than it is today. Abstract theoretical models of life all-too-easily escape the checks and balances of empirical accountability and predictability which are so crucial to science. Ganti's model dangerously departs from scientific empirical accountability. We have abandoned observable biological categories in favor of purely metaphysical mental constructions. Life and its criteria have now become philosophic and psychological abstractions. As in theoretical physics, all sorts of "realities" can be deduced in such a fantasy world that have no connection with "the real world."

Antirealists no doubt recoil at the notion "objective reality." We tend to be obsessed with our "epistemological problem." This often leads to committing a non-sequitur. We fallaciously conclude that because we cannot know inside our minds that objective reality exists outside of our minds, that objective reality doesn't exist, or cannot exist, or doesn't matter. Objective reality will have the last laugh over our dying anthropocentric and solipsistic brains and minds.

Near the end of his long career, the esteemed Leslie Orgel saw little hope of any kind of Metabolism First model being successful: "In my opinion, there is no basis in known chemistry for the belief that long sequences of reactions can organize spontaneously---and every reason to believe that they cannot."

Metabolism First models have no purely physicodynamically steered pathway from self-ordered or random molecular assemblies toward the Aristotelian "final" state of true organization and metabolic pragmatic benefit. In fact, "efficient" causation of formal function and organization cannot even be explained from physicodynamics alone. Molecular assemblies can spontaneously self-order. But they do not self-organize into bona fide function-valuing and function-pursuing systems. Thus it is not surprising that only computer models exist of composomal evolution with no empirical realizations of spontaneous metabolism arising within a truly "natural" nature. Computer models almost always have smuggled-in hidden experimenter teleology to make the computer model work for the investigator. The supposed evolution model invariably is found to exhibit "directed evolution," which is NO evolution at all. "Directed evolution" is a self-contradiction. If the supposed evolution is directed, it is not evolutionary. If it is evolutionary, it cannot be directed. The whole point of evolutionary theory is to obviate the necessity of any artificial steering. So-called "natural process" must be free of any hint of teleology. Simply renaming "teleology" to the more naturalistic-sounding "teleonomy" doesn't help. If the process is directed or engineered in any way, it is not "natural." No more "natural process" mechanism is provided in the literature for "teleonomy" than for "teleology."

Non-genetic Metabolism First models require constantly re-inventing the wheel from scratch with each new generation. No means exist of preserving and propagating any already hard-won organizational and pragmatic advances. But, no basis was ever provided for any *initial* metabolic organizational success, either. It was just blindly believed because such faith was required for any materialistic model to get off the ground. We conveniently imagine spontaneous self-organization of formal metabolic integration and success in the first place. From there it is an easy matter to further imagine a relentless uphill refinement of metabolism. Optimization is firmly believed to occur in the complete absence of any goals or purposeful steering. Finally, the process is of course proclaimed to be "inevitable," all in the name of science! Anyone who might raise an eyebrow of educated skepticism about this scenario is immediately labeled a "vitalist" or "religionist." No possibility of practicing quality skeptical science is entertained.

The inability of mere "chance and necessity" to optimize, and the lack of evolvability of self-sustaining autocatalytic networks presents serious, if not fatal, problems for all protometabolism-first models [252].

5. Self-replicative, auto-catalytic, informational models.

Protometabolism First models depend upon physicodynamic coupling of circular constraints. The presumed constancy of these integrated cycles is envisioned to eliminate the need for both enzymes and the replication of genetic instructions. From Protometabolism First models we move to self-replicative, auto-catalytic, informational models. As we consider these models, it is wise to keep in mind what Kovac et al warned: "A system of self-replication has to consist of both replicators and replicants." [253]

5.1 RNA World

RNA World theories initially envisioned free-standing RNA oligomers folding into ribozymes that spontaneously acquire both catalytic and information-retaining functions. Primordial ribozymes are best thought of as single strands of RNA folded back onto themselves with a considerable degree of base-pairing stems and loops—bulge, internal, hairpin, multi-loops with branches, etc. This folding is determined by the sequencing of ribonucleotides. Ribozyme sequencing does not provide DNA-like, coded, codonic prescription of amino acid sequencing. For an RNA World to get off the ground, ribonucleotide sequencing must be such that when the single strand folds back onto itself, base-pairing causes a secondary structure to form that will fold into a tertiary structure that is self-replicative.

RNA chemists quickly encountered many serious road bocks to any natural RNA World model. Ribonucleotide oligermization did not occur spontaneously, especially in aqueous environments. Ribose was difficult to form and was unstable.   An RNA World model "assumes a large prebiotic source of D-ribose. The problem of obtaining a homochiral population of pure D-ribose is daunting. But even prior to that dilemma are problems with any ribose. The generally accepted prebiotic synthesis of ribose, the formose reaction, yields numerous sugars without any selectivity. Even if there were a selective synthesis specifically of ribose, there is still the problem of stability. Sugars are known to be unstable in strong acid or base, . . ." [254]. "These results suggest that the backbone of the first genetic material could not have contained ribose or other sugars because of their instability." [254].

Pyrimidine ribonucleotides have been synthetized in the presence of phosphate using a precursor of both ribose and nucleobases [255, 256]. But a plausible prebiotic synthesis of purines is still lacking. Ribonucleotides are difficult to activate. No basis for functional sequencing existed. Cytosine was extremely difficult to make, even by highly intelligent chemists [257, 258]. Even Gerald Joyce and Leslie Orgel became skeptical of the RNA World [259, pg. 213]. Says Orgel,

> I believe that it is very unlikely that RNA did arise prebiotically on the primitive earth. Ribonucleotides are such complicated molecules that they are not likely to have formed in sufficient amounts and with sufficient purity on the primitive earth to permit the formation of even the simplest self-replicating RNA molecule. [260 pg. 213]

As Fry points out [141], even as recently as in the 3rd Edition of *The RNA World* [148], Joyce and Orgel still refer to the RNA World model as "the prebiotic chemist's nightmare."

Nevertheless, adherence to the RNA World [261-264] and early information models still prevails in many circles [265-279]. It has become impossible to deny the essential role of informational biopolymers. All Protometabolism First models suffer primarily from a lack of heritable mutability and retained progress. Open-ended evolution (OEE) requires separation of genotypic and phenotypic functions [242, 280, 281]. Even early Information First models are all plagued with a need for molecular evolution to improve the genetic material symbol system's functional token sequencing. Yet no basis for selection for phenotypic function exists at the point of genetic polymerization [147, 282].

A self-replicative ribozyme has been a lot harder to design and engineer than RNA chemists originally supposed. Some advancements in ribozyme engineering have occurred, however [283]. But artificially engineered self-replicative ribozymes usually catalyze only self-replication of that one ribozyme [284]. Paul and Joyce successfully engineered a self-replicating ligase ribozyme [285, 286]. The authors explain, "A self-replicating molecule directs the covalent assembly of component molecules to form a product that is of identical composition to the parent." Note that the "component molecules" are just presupposed rather than explained. "When the newly formed product also is able to direct the assembly of product molecules, the self-replicating system can be termed autocatalytic." Paul and Joyce designed "a self-replicating system based on a ribozyme that catalyzes the assembly of additional copies of itself through an RNA-catalyzed RNA ligation reaction. The R3C ligase ribozyme was redesigned so that it would ligate two substrates to generate an exact copy of itself, which then would behave in a similar manner."[285] The problem with this ribozyme for life-origin specialists is its lack of prebiotic plausibility. Would it have formed spontaneously under harsh conditions with limited component resources, with no steering and regulatory controls, and no enzyme catalysis? As admitted by the authors, a "rational design approach" was used to create this sophisticated molecule. This hardly provides evidence for a stochastic ensemble arising in a prebiotic world with all of the attributes of this ribozyme. In addition, the authors admit, "Exponential growth was limited, however, because newly formed ribozyme molecules had greater difficulty forming a productive complex with the two substrates. Further optimization of the system may lead to the sustained exponential growth of ribozymes that undergo self-replication." [285]. Notice that only "further optimization" "may" lead to sustained exponential growth of this ribozyme. What would be the naturalistic impetus for pursuing functional optimization? The environment doesn't care whether anything functions. And if further optimization did somehow continue, most all resources in the environment would be consumed in the massive self-replication of this one ribozyme. What resources would be left for all of the hundreds, if not thousands, of other needed ribozymes to form? Ribozymes are very poor catalysts compared to proteins. Any protometabolism would require a large number of ribozymes in the same place and time. What would organize all these reactions into a productive protometabolic effort? What would prevent all of the many cross-reactions? [287].

Ma and Yu suggested that two RNA synthesis ribozymes may be integrated into one RNA molecule, as two functional domains which could catalyze the copy of each other. Thus the RNA molecule could self-replicate and be referred to as "intra-molecular replicase" [288]. Ribozymes that function as RNA polymerases have been humanly engineered [283, 285, 287]. But no catalytic polymerases have been found among natural ribozymes, and no protometabolism is organized by self-catalyzed, self-replicative ribozymes.

Anet [144] differs with Shapiro, Dyson, Kauffman, de Duve, Wachtershauser, Morowitz, Deamer, Lancet, Lindhal, Guamaeres, and Russell, all of whom promote models of early spontaneous protometabolism. Anet likes instead Nicholas Hud's molecular midwives, intercalations and base-stacking as a source of functional nucleic acid molecules [289-293]. But it is not at all clear how intercalations and base-stacking arranges varying monomers into prescriptive strings using a linear digital symbol system. The model purports to provide a source of new untemplated information, but mere Shannon uncertainty is erroneously equated with "information." Certainly no basis is provided for the generation of Prescriptive Information (PI). Even Shannon uncertainty would be compromised by a tendency toward self-ordering.

More recent RNA World models all include vesicle containment [205, 294-296]. Yarus and Janas found that membrane-binding RNAs coat artificial phospholipid membranes relatively uniformly, except for a frequent tendency to concentrate at bends and membrane junctions. Yarus calls RNA protocells "ribocytes" [297, 298].

The biggest problem with bilipid vesicle pseudo-membranes is their lack of highly specific active transport tunnels and mechanisms to control the "ingestion" of needed metabolites, the rejection of deleterious molecules from entry into the protocell, and the excretion of toxic waste products from the vesicle. Some slight progress has been made with osmotic factors which can actually promote vesicle division [253, 299, 300]. But Kovac et al remind us that "logical possibility does not equal thermodynamic feasibility." [253].

No natural basis exists for optimization of ribozymes' primary structure, either, so as to yield pragmatic folds and three-dimensional shapes. Ribozymes are in a sense nothing more than combinatorial composomes. There is no reason a composome can't exist as a folded linear digital biopolymer. Prior to folding, ribozymes are combinatorial strings of "alphabetical characters" or token sequences. The fact that they start out as linearly arranged prior to folding does not disqualify them from being composomes. The alphabet consists of four possible "letter" or token options. Only one of four nucleotide options can be selected at each locus in the string. In the absence of selection in a materialistic world, the monomeric component would either be 1) random, 2) ordered to some degree by physicochemical law-like tendencies, or 3) ordered to some degree by unequal availability of each base in a given environment. Some bases like cytosine would have been extremely rare in a prebiotic environment. So any spontaneously forming composomal string of ribonucleotides would be considerably ordered by a greater abundance of some nucleotide options over others. This would greatly reduce the Shannon uncertainty and information potential of any string formation. If there were no physicodynamic preferences for any one base over another from the base-4 alphabet, and all nucleotides were equally available in the prebiotic environment, each potential selection at each locus would represent two bits of Shannon uncertainty.

No folding takes place until the ribonucleotide string is already polymerized with rigidly formed 3'5' phosphodiester bonds. The sequencing determines via thermodynamics the particular secondary and tertiary folding that eventually produces a functional ribozyme.

Natural selection is only eliminative, not constructive [239, 240]. Empirical evidence and prediction fulfillment are both sorely lacking in support of the contention that random variation (noise) can generate good new ideas, implementations, designs or engineering from which the environment could "select." This is true of RNA stochastic ensembles and any early RNA analogs forming in a pre-RNA world. It is all the more true of the derivation of current life's universal linear digital symbol system. These tokens represent and prescribe future primary structure (sequencing), folding, secondary and tertiary structure (three-dimensional catalysts, machines and nanocomputers), and the eventual molecular interactions between them (e.g., protein-protein interactions).

Natural selection does nothing more than *eliminate* second-rate phenotypic organisms arising from inferior programming [147]. Evolution contributes nothing to new programming [15, 250]. Duplication plus variation (random noise) has never been shown to generate a single new nontrivial program. A mechanism for programming potential new function using a linear digital material symbol system exists nowhere in the environmental selection paradigm. Natural selection has never been observed to prescribe a single new superior genome or metabolome from mere noise "variation." Instead of a natural process mechanism for spontaneous programming of genetic and genomic Prescriptive Information (PI), blind belief in the amazing, mystical powers of "variation" of duplications is propagated. Optimism still exists, however, that one day the prebiotic self-organization of an RNA World will be worked out [45, 63, 301-305]. But the RNA World remains to this day entirely imaginative [306]. Even if it ever existed, Blomberg, head of NASA's Astrobiology program at the time, admitted, "The RNA world may have been a great achievement, but it could hardly provide, in a direct way, the functions that were necessary for the final steps to the first organisms." [233]

5.2 Pre-RNA World and RNA World analogs

The Pre-RNA World model [133, 254, 307-313] arose out of necessity because spontaneous RNA chemistry in a prebiotic environment proved to be unrealistic. "Ribose is difficult to form selectively, and the addition of nucleobases to ribose is inefficient in the case of purines and does not occur at all in the case of the canonical pyrimidines."[256] Thus pre-RNA World models began to arise in which RNA analogs, alternative RNA-like molecules with backbones different from ribose-phosphate, and possibly even peptides, serve primarily as catalysts. The backbone of RNA analogs is made up phosphate-bonded non-ribose sugars [69, 71].

Early on, Orgel and Joyce realized the seriousness of problems with initial RNA chemistry in a prebiotic environment [314, 315]. They also investigated simpler nucleic acid polymers. Examples of specific RNA analogs are (L)-a-threofuranosyl oligonucleotides and TNAs [316]. These molecules have threose rather than ribose in their sugar-phosphate backbones and yet retain many of the properties of RNA including the ability to pair up in double helices. Anastasi, et al reviewed recent experimental work on the assembly of potential RNA precursors [303]. Powner, Gerland and Sutherland investigated the synthesis of activated pyrimidine ribonucleotides under plausible prebiotic conditions [256]. House discussed possible roles for dihydrouracils in the pre-RNA world [312]. Matray and Gryaznov investigated the synthesis and properties of RNA analogs-oligoribonucleotide N3'-->P5' phosphoramidates. [317] Kolb found many of the properties of urazole to make it a good potential precursor to uracil and guanazole a potential precursor to cytosine in a pre-RNA world [313].

Nielson investigated peptide nucleic acids (PNAs) using an amide rather than ribose backbone [318, 319]. Diederichsen investigated alternating D- and L- alanyl peptide nucleic acids (ANAs) [320].

Iris Fry [141] describes models such as Cairns-Smith's and de Duve's as "Preparatory metabolism" models. Both Cairns-Smith and de Duve find the prebiotic emergence of RNA implausible. But they both believe correctly that a genetic polymer is essential for open-ended evolution (OEE). Says Howard Pattee, "Separate description and construction components are necessary for complex systems that can adapt and evolve." [321, pg 261]. Fry argues that other models should also be considered "preparatory" rather than Metabolism First [141]. As usual, no explanation is provided as to how inanimate nature would go about "preparing" for eventual metabolic success. Failure to explain how so-called "natural process" physicodynamics could anticipate and pursue formal function impedes virtually all models of life origin, whether Protometabolic First or Information/Replication First. Appealing to a preRNA World does not solve this problem either. No motive or basis exists for selection of cooperative organization over disorganization. Only investigator minds imagine inanimate nature incorporating urazole and its ribosides, peptide RNA analogs, group II introns and other self-replicative ribozymes into auto-catalytic schemes. Nature does not scheme. All code-origin models are plagued by the same lack of environmental pursuit of pragmatism of any kind. Inanimate nature does not value or pursue "usefulness." Unaided physicodynamic constraints and laws do not generate controls that steer interactions toward formal function.

Even Pre-RNA World chemistry proved to be far more unrealistic than originally envisioned [152, 177, 178, 257, 258, 322-324]. Leslie Orgel was the premier life-origin investigator of the late twentieth and early twenty-first century. Orgel could find no basis for a Protometabolic World of self-organization:

> In my opinion, there is no basis in known chemistry for the belief that long sequences of reactions can organize spontaneously---and every reason to believe that they cannot. [155]

Thus Orgel had little choice but to hypothesize a simpler RNA analog scenario that could eventually evolve into an RNA World. He believed there had to have been alternate polymers, perhaps RNA analogs, in a preRNA World. But his optimism only seemed to wane near the end of his illustrious career. And opponents of the RNA World are no more impressed with the plausibility of useful RNA analog formation than that of RNA. It has been argued by opponents of the preRNA and RNA World that the spontaneous formation and self-organization of RNA analogs such as TNA, PNA, ANA and urazole in a prebiotic environment is not significantly more plausible than the RNA World model [150, 178, 303, 325]. In addition, it has not been established that any of these alternative backbones could provide the extensive Shannon uncertainty and Prescriptive Information (PI) potential provided by current life nucleic acids [15, 16, 18, 318, 326]. Nevertheless, adherence to the Pre-RNA World model remains strong in many circles [133, 175, 264, 311, 313, 327-331].

6. Early photosynthetic models

The earliest photosynthetic cells are thought to have been almost certainly anoxygenic. Oxygenic photosynthesis and the subsequent rise of atmospheric oxygen supposedly occurred around 2.4 billion years ago [332]. Complex biosynthetic pathways of carbon fixation would have been needed involving new photosynthetic cofactors, electron carriers, and pigments [141]. Endosymbiosis is thought to have played a role [333-337]. Schopf states,

> Fossil evidence of photosynthesis, documented in Precambrian sediments by microbially laminated stromatolites, cyanobacterial microscopic fossils, and carbon isotopic data consistent with the presence of Rubisco-mediated CO_2-fixation, extends from the present to ~3,500 million years ago. Such

data, however, do not resolve time of origin of O2-producing photoauto-trophy from its anoxygenic, bacterial, evolutionary precursor. Though it is well established that Earth's ecosystem has been based on autotrophy since its very early stages, the time of origin of oxygenic photosynthesis, more than 2,450 million years ago, has yet to be established.[338]

Grenfell et al [339] point out that the Sun was originally shining 20-25% less brightly than today. They maintain that earth would have been an ice ball without greenhouse-like conditions to warm the atmosphere. They thus conclude that greenhouse gases must have been present on early Earth to warm the planet. Argue Grenfell et al.:

> Evidence from the geological record indicates an abundance of the greenhouse gas CO(2). CH(4) was probably present as well; and, in this regard, methanogenic bacteria, which belong to a diverse group of anaerobic prokaryotes that ferment CO(2) plus H(2) to CH(4), may have contributed to modification of the early atmosphere. Molecular oxygen was not present, as is indicated by the study of rocks from that era, which contain iron carbonate rather than iron oxide. [339]

Mulkidjanian and Galperin believe that life started within photosynthe-sizing ZnS compartments [61, 340]. They contend that life could have evolved under the conditions of elevated levels of Zn2+ ions, byproducts of the ZnS-mediated photosynthesis. UV-rich solar radiation set the stage for a Zinc World. They envision precipitated zinc sulfide (ZnS) providing photosyntheti-cally active porous edifices similar to what is found near deep-sea hydrother-mal vents. They cite as evidence the roles of Zn2+ ions and possibly manga-nese sulfide in modern organisms, particularly in RNA and protein structures:

Under the high pressure of the primeval, carbon dioxide-dominated atmos-phere ZnS could precipitate at the surface of the first continents, within reach of solar light. It is suggested that the ZnS surfaces (1) used the solar radiation to drive carbon dioxide reduction, yielding the building blocks for the first biopolymers, (2) served as templates for the synthesis of longer biopolymers from simpler building blocks, and (3) prevented the first bi-opolymers from photo-dissociation, by absorbing from them the excess ra-diation. In addition, the UV light may have favoured the selective enrich-ment of photostable, RNA-like polymers. [61]

Baltscheffsky investigated stepwise molecular evolution of bacterial photosynthetic energy conversion [98]. Various other embryonic Photosynthetic models have been proposed [341-363]. All of them require considerable organization and pathway integration to be able to harness, transduce, store, and call up when needed energy in a usable form. Even if this organization were to spontaneously generate, innumerable mechanisms would have to simultaneously arise that could accomplish something creative using the transduced stored energy. Even the simplest protometabolism would had to have been masterfully organized to meet homeostatic metabolic needs necessary for life.

7. Code-origin Models

Much work has been done trying to elucidate how genetic code arose naturalistically [159, 184-186, 188, 298, 364-391]. Di Giulio, Wong, Yarus, Schimmel and Guimaraes have probably been the leading code-origin theorists. The conceptually ideal nature of genetic code [392-398], however, is difficult to explain working only with after-the-fact differential survival and reproduction of the best already-programmed organisms. Genetic programming must not only preceed, but prescribe the existence of any organism. This is not just true for the fittest organisms, but for any living organism. No basis exists within the metaphysical belief system of naturalism for selection of coded instructions prior to the realization of phenotypic superiority [147].

The Code Co-evolution Model of code origin was first proposed by Wong [387, 388] and updated 30 years later [184]. Wong's model suggested that genetic code co-evolved with biosynthetic pathways of AA. "Amino acid biosynthesis and hydrophobicity were important factors in shaping the genetic code, as the primitive code co-evolved with new varieties of amino acids generated by the expanding pathways of biosynthesis." [390] Ronneberg concluded that Wong's code co-evolution theory cannot adequately explain the structure of the genetic code [379]. Wong countered with a defense of his model [389, 390]. Others have entertained variations of Wong's approach that differ in important respects [368, 399] [159, 186, 400]. Lahav suggested a unique co-evolution of enzymes and ribozymes [128]. Guimaraes has published probably the most recently updated code origin model [159, 186, 400-402].

Yockey [403, pg 4-5] shows that the Central Dogma—the one-way-only flow of Prescriptive Information (PI) from the codon alphabet of 64 (or 61) block codes to the amino acid alphabet of 20—is a mathematical property arising from the redundancy (poorly termed "degeneracy") of the genetic code, not just a physicodynamic property of nucleic acids and amino acids per se. It is a mathematical impossibility for PI to flow in the opposite direction as is required by many code-origin models that propose slow growth of the code table over long periods of time. The PI in the redundant codon table had to have been there first. When told only that a pair of tossed die generated a total of 7, there is no way to recover the information of whether the two die showed a 1 and 6, 2 and 5, 3 and 4, 4 and 3, 5 and 2, or 6 and 1. Detailed information is permanently lost in the totaling of the two die. One cannot determine from knowing only the amino acid, without the codon table, which of the 1, 2, 3, 4 or 6 codons that prescribed that amino acid. Such information is crucial with many genetic diseases and with cancer mutations.

The genetic code is ideally arranged to minimize the effect of genetic noise in the Shannon channel [392, 395-397, 403, pg 104-107]. This optimal coding requires the additional information that would not have existed if the code were slowly constructed through evolution from amino acids to codons. Naturalistic code-origin models seem to consistently ignore this reality. The devil is in the details of which codon prescribed which amino acid in each unique situation. Some genetic diseases are caused by a point mutation of one of the three nucleotides in a codon even though the amino acid prescribed is unaffected!

8. Composome, Chemoton and RNA evolution would have been extremely limited.

Composomes, chemotons and ribozymes do NOT have separate description and construction components. The two components are one in the same. Sequencing directly determines structure and catalytic ability. This creates real problems for any abiogenic evolutionary model that hopes to maintain its optimized self-replicative sequencing, folds, structure and function while simultaneously mutating to a different sequence that is optimized for some other metabolic function other than self-replication. When the "genome" changes, the phenotypic "organism" itself changes in an immediate and direct way. The odds of the primary structure (sequencing) being the same for both optimized self-replicative function and a separate sophisticated metabolic function are miniscule at best. Any move towards an improved new metabolic function will simultaneously compromise the ribozyme's or protocell's hoped-for (never once observed) spontaneous self-replicative function.

Another unanswered question for RNA World advocates is, "What correlation exists between highly-optimized, error-free self-replication of sequences, and highly optimized metabolic function? How could one sequence simultaneously be optimized for both important functions? The reason ribozymes receive so much attention in life-origin science is their ability to simultaneously serve as catalysts and information-retaining linear digital strings. Their catalytic function and information retention depends upon their particular sequence of ribonucleotides. But in ribozymes, the exact same sequence that would be optimized for catalysis of self-replication is also the "heritable" sequence. For the heritable sequence to mutate necessarily involves the de-optimization of the already optimized catalytic sequence and function. In ribozymes the two sequences are one in the same with shared functions. Progress in one area compromises progress in another. No selection pressure for either function takes place upon polymerization of ribonucleotides with rigid bonds. The environment has no goal or desire to optimize either function, let alone organize both together into an integrated homeostatic metabolic scheme with accurate self-replication. The Functional Information (FI) [10-14] found in ribozyme's primary-structure sequencing would be largely limited to the sequencing that prescribes folding into that particular ribozyme. The ribonucleotide sequencing would first and foremost have to be optimized for self-replication. Simultaneous optimization of this same sequence for contribution to any holistic metabolic scheme would be almost impossible. Even in a protocell, hundreds if not thousands of metabolic functions other than the self-replication function would be needed to organize anything close to life.

Recently even Szathmáry's group has argued effectively that composomes cannot evolve [252]. Compositional "genomes" (assemblies of varying molecular species) are thought by some to be able to propagate evolvable chemical and structural information [162-165]. Szathmary's group calls these macromolecular aggregates "ensemble replicators." They are thought to be able to replicate three-dimensional structures that would support the idea of composome evolution. Vasas, Szathmáry and Santos found in a 2010 PNAS paper entitled "Lack of evolvability in self-sustaining autocatalytic networks constrains metabolism-first scenarios for the origin of life,"

> In sharp contrast with template-dependent replication dynamics, we demonstrate here that replication of compositional information is so inaccurate that fitter compositional genomes cannot be maintained by selection and, therefore, the system lacks evolvability (i.e., it cannot substantially depart from the asymptotic steady-state solution already built-in in the dynamical equations). We conclude that this fundamental limitation of ensemble replicators cautions against metabolism-first theories of the origin of life, although ancient metabolic systems could have provided a stable habitat within which polymer replicators later evolved. [252, pg. 1470]

Thus three-dimensional composomes cannot evolve. Composomes lack the dichotomization of genome from phenotype needed for evolution to occur

[404-408]. A buffer zone must exist between genome and phenotype to allow for the simultaneous existence of genetic drift and a relatively stable phenotypic life simultaneously.

Any mutation of an oligoribonucleotide sequence in the RNA World would tend to produce a much more immediate phenotypic change than seen with current organism DNA mutations. And that mutation would be almost certainly be far more deleterious to the already accidentally and barely achieved self-replicative function. Any movement towards a beneficial new metabolic function would be de-optimizing to the self-replicative function.

But even before that, no reason or mechanism was ever provided for how the sequencing was achieved that would produce self-replicative function. How did an inanimate nature pursue optimization of self-replicative sequencing in the first place?

The primary structure (the sequencing) is "written in stone" with rigid 3'5' phosphodiester bonds before any prescriptive function could be realized. And the environment has no desire or preference for replication over non-replication. No basis exists for selection for potential self-replicative function. Either a stochastic ensemble against all odds just happens to form that has self-replicative catalytic ability, or it doesn't. The finest RNA chemists in the world have had an extremely difficult time purposefully engineering ribozymes to auto-catalyze [62, 294]. The statistical prohibitiveness of a stochastic ensemble achieving this function cannot be circumvented in the absence of a selection mechanism or experimenter engineering.

But suppose by some miracle a primary structure stochastic ensemble formed spontaneously which catalyzed self-replication. Any mutation would not only not pursue a new function, it would compromise and de-optimize the already-existing self-replicative function. Self-replication would be sacrificed at the expense of selection for almost any other metabolic function.

Virtually all Protometabolism First and RNA World models lack organizational motive, ability and naturalistic explanation. No reason is provided from these purely physicodynamic interactions as to why inanimate nature would pursue the goal of formal integration and function.

The self-replicative optimization was itself already problematic because it would consume all resources in the mass production of the same self-replicator oligoribonucleotide sequence. Selfish auto-catalytic self-replications would completely exhaust the phase space resources so that none of the other scores (if not hundreds) of essential protometabolic contributors could have formed at the same place and time.

The vast combinatorial and structural phase space needed for other spontaneous stochastic evolutions of functions would be completely consumed before any "search" could begin. But of course *there is no search*. Environments don't search for anything. Evolution has no goal, especially not prebiotic evolution (for which there is zero observational evidence, or even plausible rational support). Function first has to exist for it to be secondarily preferred. The environment has no preference for function over non-function. Evolution is nothing more than elimination of lesser quality already-living organisms [239, 240].

The difficulty with which most of these needed organic contributors are chemically produced, their short-lived stability in most prebiotic environments, the innumerable cross reactions that occur, the exclusivity of left-handed amino acids and right-handed sugars with no straight forward means of homochiral production, are just a few of the many problems with Metabolism First models.

Rocha describes natural selection as a "statistical bias on the rates of reproduction of populations of individuals," but acknowledges that "this is as far as (statistical) dynamics can take us to describe this process" [243, pg. 11]. No reason is provided as to why a statistical bias might exist that would favor formal organization. No contrast is drawn between mere physicodynamic constraints and formal controls, between fixed laws and the arbitrary formal rules needed to generate formal pragmatic systems. Choices, not constraints—and rules, not laws—are needed to organize any formal utility. Until science fully acknowledges this objective reality, progress in life-origin studies will continue to encounter immovable road blocks.

Literal genetic algorithms, not figurative ones, prescribe and control life. Nucleotides function in an objective, not just a subjective human symbolic capacity. The particular symbol selection at each decision node of nucleotide polymerization is isolated from physicodynamic causation by a *dynamic discontinuity* [242, 243, 409]. The nonphysical instructions are physically instantiated into material symbol systems using physical symbol vehicles. The programming is fundamentally formal, not physical. "Semantic/semiotic/bioengineering function requires dynamically inert, resortable, physical symbol vehicles that *represent* time-independent, non-dynamic "meaning (e.g., codons)." [9] Physicodynamics cannot participate in representationalism. The latter is purely formal, not physical. No empirical or rational basis exists for granting to physics or chemistry such non-dynamic capabilities of functional sequencing. Neither chance nor necessity (fixed law) can program configurable switches to integrate circuits, write coded instructions, or organize formal utility.

The bottom line is that no naturalistic basis exists for optimization of ribozymes' primary structure. The environment is blind to isolated function.

The notion of molecular evolution has nothing to work with as a basis for selection other than self-replication. But self-replicative function is very different from metabolic function. No reason exists for a sequence optimized for self-replication to also be simultaneously optimized for hundreds of other metabolic catalytic functions needed for a protometabolism to organize. Even if all the individual ribozyme catalytic metabolic functions somehow got optimized at the same time as the self-replicative function, nothing would exist in a prebiotic environment to organize a living system. We would have nothing but a vesicle or micelle that was a "bag of ribozymes." If a "bag of enzymes" does not constitute a living cell, a "bag of ribozymes" certainly wouldn't. The capabilities of ribozymes are extremely limited with only minimally effective rate constants and sophistication compared to proteins.

A recent paper in The Quarterly Review of Biology shows that almost all of the very few supposedly "helpful" adaptive mutations involve either the loss or the modification of existing function [410]. It was already well-known that most mutations are outrightly deleterious. What is significant about this study is that it shows even the rare supposedly "beneficial" mutations degrade the genome. Genetic research seems to be consistently showing information decay from mutations, not improvement in the quality of sophisticated prescription [279].

Systems biology [411-417] has risen to the forefront as investigators have discovered ever more amazing degrees of metabolic organization and control. No basis exists for organization or control in a chance-and-necessity-only materialistic worldview. This problem is just as real, if not more acute, in primordial systems biology [418, 419].

Biosemiosis—communication within and between cells, is now firmly entrenched in our understanding of life. This includes the molecular biological level [420-424] [15, 326, 406-408, 425-449]. Communication of meaningful/functional messages is impossible without arbitrary selection of symbols or programming choices.

9. Panspermia

It has never been clear to this author why any astrobiologist would think that panspermia would help solve the life-origin problem on earth. The age of the cosmos is only three times that of the earth. How could the statistical prohibitiveness of spontaneous generation of life on earth possibly be helped significantly by multiplying that ridiculously low probability by a mere factor of 3?

Problems exist with the notion of panspermia from other solar systems. For a rock or ET spacecraft to overcome the gravity of an average source planet, it would need an escape velocity equivalent to around 16,000 mph. An asteroid hit would be required. In addition, any rock coming from around 40 light years away would take around a million years to get here. DNA would be destroyed by radiation even in spores. The temperature in space is nearly absolute zero. No nutrients or oxygen would exist during the trip. Entry into the earth's atmosphere would cook the rock. The impact of whatever superheated meteriorite was left wouldn't help much.

Recently Hoover [450] has claimed to have found microfossils in carbonaceous chondritic meteriorites. This is not the first time he has made this claim [451]. The astrobiological scientific community remains skeptical. Carbonaceous chondrites are meteorites that are thought to have formed in the early solar system. The metals in these meteorites are found as silicates, oxides and sulfides rather than in their free form. They can also contain considerable amounts of carbon and organic compounds, especially if they have never been heated to more than 50 degrees C. They are important because they seem to have formed in oxygen-rich environments, and many of them contain minerals that appear to have been exposed to water. Different kinds of carbonaceous chondrites have formed depending upon the planet and region of the early solar system from which they originated.

Hoover attempted to rule out contamination of his samples, but Brasier argues: "In terms of syngenicity, these samples have been sitting around in laboratories for between 205 and 73 years. It is well known that microbial contaminants can penetrate deep into such rocks, even during storage." [452]

Morphology is not very reliable in identifying such microfossils. In addition to filamentous bacteria varying a great deal in morphology, electron microscopy introduces many artifacts. Brasier [452] argues that such filamentous structures can form abiotically as ambient inclusion trails (AITs). These AITs greatly resemble cyanobacterial microfossils, and have compositions similar to those described by Hoover that are enriched in carbon, sulphur and silica-rich minerals in their filament margins. NASA itself is cautious about supporting Hoover's claims. Carbon enrichment is a major component of all the meteorites that Hoover studied. The evidence for life in Hoover's samples is probably no better than that for the Martian meteorite ALH840001.

After over 40 years of trying to detect an intelligent message from outer space, the Search for Extraterrestrial Intelligence (SETI) has come up empty handed.

For a discussion of the metaphysical notion of multiverse, see Chapter 11, Section 5.

10. Conclusions:

No evasion of the facts is any longer possible. The reality of Prescriptive Information's control of life is undeniable [281, 453-460]. The many recently discovered additional layers and dimensions of PI only make explaining the derivation of PI all the more daunting a task. The roles of micro RNAs, peptides and small polypeptides, in addition to regulatory proteins, have proven linear digital regulation of function within bona fide formal molecular biological systems. Even linear digital controls of development are now quite apparent.

Any kind of organization requires programming for potential function, arrangement of components, control and regulation of events prior to the existence of any pursued biofunction. But of course no prebiotic chemical system is capable of pursuing eventual function.

Eigen/Schuster hypercycle models and Ganti chemoton/stoichiometric models excite naturalistic explanatory hope. The reason is that they provide a vivid illusion of control and self-organization. In reality, no such formal control or organization is provided by mere constraints. Both positive and negative feedback circular constraints occur that affect reaction products deterministically. But no formally integrative and fine-tuned system of regulation and sophisticated function arises from circular physicodynamic constraints.

Any evolution of improved metabolic function would compromise its self-replicative function, as the two would almost always have two widely different optimized structures. Similarly, the molecular evolution of any new metabolically functional structure would compromise its self-replicative structure. Ribozymes, like any other composome, cannot evolve much without progressive loss of their already-optimized auto-catalytic tertiary structure. Not even the auto-catalytic function of ribozymes has ever been observed to occur without extensive human engineering. As mentioned above, so-called "directed evolution" is a nonsense term that violates the evolutionary requirement of being non-teleological.

In Leslie Orgel's last paper, entitled "The implausibility of metabolic cycles on the prebiotic earth" [153], Orgel emphasized that ". . . , solutions offered by supporters of geneticist or metabolist scenarios that are dependent on "if pigs could fly" hypothetical chemistry are unlikely to help."

Nobel laureate biologist George Wald stated without hesitancy that "one has only to contemplate the magnitude of [the] task to concede that the spontaneous generation of a living organism is impossible." [461]

References

1. Miller, S.L. 1953, A production of amino acids under possible primitive earth conditions, Science 117, (3046) 528-529.
2. Morowitz, H.J. 1999, The theory of biochemical organization, metabolic pathways, and evolution, Complexity, 4, 39-53.
3. Parker, E.T.; Cleaves, H.J.; Dworkin, J.P.; Glavin, D.P.; Callahan, M.; Aubrey, A.; Lazcano, A.; Bada, J.L. 2011, Primordial synthesis of amines and amino acids in a 1958 Miller H2S-rich spark discharge experiment, Proceedings of the National Academy of Sciences.
4. Cairns-Smith, A.G. 1990, *Seven Clues to the Origin of Life*. Canto ed., Cambridge University Press: Cambridge.
5. Cairns-Smith, A.G. 1966, The origin of life and the nature of the primitive gene, J Theor Biol, 10, (1) 53-88.
6. Cairns-Smith, A.G. 1977, Takeover mechanisms and early biochemical evolution, Biosystems, 9, (2-3) 105-109.
7. Cairns-Smith, A.G.; Ingram, P.; Walker, G.L. 1972, Formose production by minerals: possible relevance to the origin of life, J Theor Biol, 35, (3) 601-604.
8. Cairns-Smith, A.G.; Walker, G.L. 1974, Primitive metabolism, Curr Mod Biol, 5, (4) 173-186.
9. Abel, D.L.; Trevors, J.T. 2005, Three subsets of sequence complexity and their relevance to biopolymeric information., Theoretical Biology and Medical Modeling, 2, 29 Open access at http://www.tbiomed.com/content/2/1/29.
10. Teller, C.; Willner, I. 2010, Functional nucleic acid nanostructures and DNA machines, Curr Opin Biotechnol, 21, (4) 376-91.
11. McIntosh, A.C. 2010, Information And Entropy – Top-down Or Bottom-up Development In Living Systems?, International Journal of Design & Nature and Ecodynamics, 4, (4) 351-385.
12. Hazen, R.M.; Griffin, P.L.; Carothers, J.M.; Szostak, J.W. 2007, Functional information and the emergence of biocomplexity, Proc Natl Acad Sci U S A, 104 Suppl 1, 8574-81.
13. Carothers, J.M.; Oestreich, S.C.; Davis, J.H.; Szostak, J.W. 2004, Informational complexity and functional activity of RNA structures, J Am Chem Soc, 126, (16) 5130-7.
14. Szostak, J.W. 2003, Functional information: Molecular messages, Nature, 423, (6941) 689.
15. Abel, D.L. 2009, The biosemiosis of prescriptive information, Semiotica, 2009, (174) 1-19.
16. Abel, D.L. Prescriptive Information (PI) [Scirus Topic Page]. http://www.scitopics.com/Prescriptive_Information_PI.html (Last accessed September, 2011).
17. Abel, D.L. 2011, Moving 'far from equilibrium' in a prebitoic environment: The role of Maxwell's Demon in life origin. In *Genesis - In the Beginning: Precursors of Life, Chemical Models and Early Biological Evolution* Seckbach, J.Gordon, R., Eds. Springer: Dordrecht.
18. Abel, D.L.; Trevors, J.T. 2007, More than Metaphor: Genomes are Objective Sign Systems. In *BioSemiotic Research Trends*, Barbieri, M., Ed. Nova Science Publishers: New York, pp 1-15
19. Mann, S.; Perry, C.C. 1986, Structural aspects of biogenic silica. In *Silicon Biochemistry; CIBA Foundation Symposium 121*, Joyn Wiley & Sons: Sussex, UK, pp 40-58.
20. Trevors, J.T. 1997, Bacterial evolution and silicon, Antonie Van Leeuwenhoek, 71, (3) 271-276.
21. Williams, R.J.P. 1986, Introduction to silicon chemistry and biochemistry. In *Silicon Biochemistry: CIBA Foundation Symposium 121*, John Wiley & Sons: Sussex, UK, pp 24-39.
22. Vysotskii, Z.Z.; Danilov, V.I.; Strelko, V.V. 1967, [Properties of polysilicic acid gels and conditions of the origin and development of life], Usp Sovrem Biol, 63, (3) 362-79.
23. Trevors, J.T.; Abel, D.L. 2004, Chance and necessity do not explain the origin of life, Cell Biology International, 28, 729-739.
24. Ball, P. 2004, Water, water, everywhere?, Nature, 427, 19-20.
25. Good, W. 1973, The role of water in the origin of life and its function in the primitive gene, J Theor Biol, 39, (2) 249-276.
26. Papagiannis, M.D. 1992, What makes a planet habitable, and how to search for habitable planets in other solar systems, J Br Interplanet Soc, 45, (6) 227-230.
27. Burton, F.G.; Lohrmann, R.; Orgel, L.E. 1974, On the possible role of crystals in the origins of life. VII. The adsorption and polymerization of phosphoramidates by montmorillonite clay, J Mol Evol, 3, (2) 141-150.
28. Ertem, G.; Ferris, J.P. 1998, Formation of RNA oligomers on montmorillonite: site of catalysis, Orig Life Evol Biosph, 28, (4-6) 485-499.
29. Ertem, G.; Ferris, J.P. 2000, Sequence- and regio-selectivity in the montmorillonite-catalyzed synthesis of RNA, Orig Life Evol Biosph, 30, (5) 411-422.
30. Ferris, J.P.; Huang, C.H.; Hagan, W.J., Jr. 1988, Montmorillonite: a multifunctional mineral catalyst for the prebiological formation of phosphate esters, Orig Life Evol Biosph, 18, (1-2) 121-133.
31. Ferris, J.P.; Ertem, G.; Agarwal, V. 1989, Mineral catalysis of the formation of dimers of 5'-AMP in aqueous solution: the possible role of montmorillonite clays in the prebiotic synthesis of RNA, Orig Life Evol Biosph, 19, (2) 165-178.

32. Ferris, J.P.; Ertem, G. 1992, Oligomerization of ribonucleotides on montmorillonite: reaction of the 5'-phosphorimidazolide of adenosine, Science, 257, (5075) 1387-1389.

33. Friebele, E.; Shimoyama, A.; Hare, P.E.; Ponnamperuma, C. 1981, Adsorption of amino acid entantiomers by Na-montmorillonite, Orig Life, 11, (1-2) 173-184.

34. Miyakawa, S.; Ferris, J.P. 2003, Sequence- and regioselectivity in the montmorillonite-catalyzed synthesis of RNA, J Am Chem Soc, 125, (27) 8202-8208.

35. Paecht-Horowitz, M.; Eirich, F.R. 1988, The polymerization of amino acid adenylates on sodium-montmorillonite with preadsorbed polypeptides, Orig Life Evol Biosph, 18, (4) 359-387.

36. Ding, P.Z.; Kawamura, K.; Ferris, J.P. 1996, Oligomerization of uridine phosphorimidazolides on montmorillonite: a model for the prebiotic synthesis of RNA on minerals, Orig Life Evol Biosph, 26, (2) 151-171.

37. Ertem, G.; Ferris, J.P. 1996, Synthesis of RNA oligomers on heterogeneous templates, Nature, 379, (6562) 238-240.

38. Friebele, E.; Shimoyama, A.; Ponnamperuma, C. 1980, Adsorption of protein and non-protein amino acids on a clay mineral: a possible role of selection in chemical evolution, J Mol Evol, 16, (3-4) 269-278.

39. Huang, W.; Ferris, J.P. 2003, Synthesis of 35-40 mers of RNA oligomers from unblocked monomers. A simple approach to the RNA world, Chem Commun (Camb), 12, 1458-1459.

40. Kawamura, K.; Ferris, J.P. 1999, Clay catalysis of oligonucleotide formation: kinetics of the reaction of the 5'-phosphorimidazolides of nucleotides with the non-basic heterocycles uracil and hypoxanthine, Orig Life Evol Biosph, 29, (6) 563-591.

41. Ferris, J.P.; Hill, A.R., Jr.; Liu, R.; Orgel, L.E. 1996, Synthesis of long prebiotic oligomers on mineral surfaces, Nature, 381, (6577) 59-61.

42. Kerr, R.A. 2003, Geochemistry. Minerals cooked up in the laboratory call ancient microfossils into question, Science, 302, (5648) 1134.

43. Garcia-Ruiz, J.M.; Hyde, S.T.; Carnerup, A.M.; Christy, A.G.; Van Kranendonk, M.J.; Welham, N.J. 2003, Self-assembled silica-carbonate structures and detection of ancient microfossils, Science, 302, (5648) 1194-7.

44. Javaux, E.J.; Marshall, C.P.; Bekker, A. 2010, Organic-walled microfossils in 3.2-billion-year-old shallow-marine siliciclastic deposits, Nature, 463, (7283) 934-8.

45. Ricardo, A.; Carrigan, M.A.; Olcott, A.N.; Benner, S.A. 2004, Borate Minerals Stabilize Ribose, Science, 303, (5655) 196-.

46. Wächtershäuser, G. 1988, Before enzymes and templates: theory of surface metabolism, Microbiol Rev, 52, (4) 452-84.

47. Wächtershäuser, G. 1992, Groundworks for an evolutionary biochemistry: the iron-sulphur world, Prog Biophys Mol Biol, 58, (2) 85-201.

48. Wächtershäuser, G. 2007, On the Chemistry and Evolution of the Pioneer Organism, Chemistry & Biodiversity, 4, (4) 584-602.

49. Alpermann, T.; Rüdel, K.; Rüger, R.; Steiniger, F.; Nietzsche, S.; Filiz, V.; Förster, S.; Fahr, A.; Weigand, W. 2011, Polymersomes Containing Iron Sulfide (FeS) as Primordial Cell Model, Origins of Life and Evolution of Biospheres, 41, (2) 103-119.

50. Kundell, F.A. 2011, A suggested pioneer organism for the wachtershauser origin of life hypothesis, Orig Life Evol Biosph, 41, (2) 175-98.

51. Martin, W.; Baross, J.; Kelley, D.; Russell, M.J. 2008, Hydrothermal vents and the origin of life, Nat Rev Microbiol, 6, (11) 805-14.

52. Martin, W.; Russell, M.J. 2003, On the origins of cells: a hypothesis for the evolutionary transitions from abiotic geochemistry to chemoautotrophic prokaryotes, and from prokaryotes to nucleated cells, Philos Trans R Soc Lond B Biol Sci, 358, (1429) 59-83; discussion 83-85.

53. Russell, M.J. 2003, Geochemistry. The importance of being alkaline, Science, 302, (5645) 580-1.

54. Russell, M.J.; Hall, A.J. 1997, The emergence of life from iron monosulphide bubbles at a submarine hydrothermal redox and pH front., Journal Geological Society London,, 154, 377-402.

55. Russell, M.J.; Hall, A.J.; Martin, W. 2010, Serpentinization as a source of energy at the origin of life, Geobiology, 8, (5) 355-71.

56. Russell, M.J.; Martin, W. 2004, The rocky roots of the acetyl-CoA pathway, Trends in biochemical sciences, 29, (7) 358-63.

57. Russell, R.; Zhuang, X.; Babcock, H.P.; Millett, I.S.; Doniach, S.; Chu, S.; Herschlag, D. 2002, Exploring the folding landscape of a structured RNA, PNAS, 99, (1) 155-160.

58. Fox, S.W. 1995, Thermal synthesis of amino acids and the origin of life, Geochim Cosmochim Acta, 59, (6) 1213-4.

59. MacLeod, G.; McKeown, C.; Hall, A.J.; Russell, M.J. 1994, Hydrothermal and oceanic pH conditions of possible relevance to the origin of life, Orig Life Evol Biosph, 24, (1) 19-41.

60. Lane, N.; Allen, J.F.; Martin, W. 2010, How did LUCA make a living? Chemiosmosis in the origin of life, Bioessays, 32, (4) 271-80.

61. Mulkidjanian, A.Y. 2009, On the origin of life in the zinc world: 1. Photosynthesizing, porous edifices built of hydrothermally precipitated zinc sulfide as cradles of life on Earth, Biol Direct, 4, 26.

62. Attwater, J.; Wochner, A.; Pinheiro, V.B.; Coulson, A.; Holliger, P. 2010, Ice as a protocellular medium for RNA replication, Nature Communications, 1, (1 September).

63. Kanavarioti, A.; Monnard, P.A.; Deamer, D.W. 2001, Eutectic phases in ice facilitate nonenzymatic nucleic acid synthesis, Astrobiology, 1, (3) 271-81.

64. Miyakawa, S.; Cleaves, H.J.; Miller, S.L. 2002, The cold origin of life: B. Implications based on pyrimidines and purines produced from frozen ammonium cyanide solutions, Orig Life Evol Biosph, 32, (3) 209-18.

65. Miyakawa, S.; Cleaves, H.J.; Miller, S.L. 2002, The cold origin of life: A. Implications based on the hydrolytic stabilities of hydrogen cyanide and formamide, Orig Life Evol Biosph, 32, (3) 195-208.

66. Monnard, P.A.; Apel, C.L.; Kanavarioti, A.; Deamer, D.W. 2002, Influence of ionic inorganic solutes on self-assembly and polymerization processes related to early forms of life: implications for a prebiotic aqueous medium, Astrobiology, 2, (2) 139-52.

67. Monnard, P.A.; Kanavarioti, A.; Deamer, D.W. 2003, Eutectic phase polymerization of activated ribonucleotide mixtures yields quasi-equimolar incorporation of purine and pyrimidine nucleobases, J Am Chem Soc, 125, (45) 13734-40.

68. Orgel, L.E. 2004, Prebiotic adenine revisited: eutectics and photochemistry, Origins of life and evolution of the biosphere, 34, (4) 361-9.

69. Eschenmoser, A. 1999, Chemical etiology of nucleic acid structure, Science, 284, (5423) 2118-24.

70. Eschenmoser, A. 2007, On a Hypothetical Generational Relationship between HCN and Constituents of the Reductive Citric Acid Cycle, Chemistry & Biodiversity, 4, (4) 554-573.

71. Eschenmoser, A. 2007, The search for the chemistry of life's origin, Tetrahedron, 63, 12821-12844.

72. Rode, B.M. 1999, Peptides and the origin of life, Peptides, 20, (6) 773-86.

73. Carny, O.; Gazit, E. 2011, Creating prebiotic sanctuary: self-assembling supramolecular Peptide structures bind and stabilize RNA, Orig Life Evol Biosph, 41, (2) 121-32.

74. Hoenigsberg, H.F. 2007, From geochemistry and biochemistry to prebiotic evolution...we necessarily enter into Ganti's fluid automata, Genet Mol Res, 6, (2) 258-73.

75. Bender Christopher, J. 2009, Chemical Evolution and Biomimetic Chemistry. In *Chemical Evolution II: From the Origins of Life to Modern Society*, American Chemical Society: Washington DC, pp 313-331.

76. de Souza-Barros, F.; Vieyra, A. 2007, Mineral interface in extreme habitats: a niche for primitive molecular evolution for the appearance of different forms of life on earth, Comp Biochem Physiol C Toxicol Pharmacol, 146, (1-2) 10-21.

77. Huber, C.; Wachtershauser, G. 1998, Peptides by activation of amino acids with CO on (Ni,Fe)S surfaces: implications for the origin of life [see comments], Science, 281, (5377) 670-2.

78. Lazard, D.; Lahav, N.; Orenberg, J.B. 1987, The biogeochemical cycle of the adsorbed template. I: Formation of the template, Orig Life Evol Biosph, 17, (2) 135-48.

79. Lazard, D.; Lahav, N.; Orenberg, J.B. 1988, The biogeochemical cycle of the adsorbed template. II: Selective adsorption of mononucleotides on adsorbed polynucleotide templates, Orig Life Evol Biosph, 18, (4) 347-57.

80. Mojzsis, S.; Krishnamurthy, R.; Arrhenius, G. 1999, Before RNA and After: Geophysical and Geochemical Constraints on Molecular Evolution. In *The RNA World*, 2nd ed.; Mojzsis, S.; Krishnamurthy, R.Arrhenius, G., Eds. Cold Spring Harbor Laboratory Press.

81. Nutman, A.P.; Clark, R.L.F.; Bennett, V.C.; Wright, D.; Norman, M.D. 2010, ≥3700 Ma pre-metamorphic dolomite formed by microbial mediation in the Isua supracrustal belt (W. Greenland): Simple evidence for early life? , PreCambrian Research, 183, (4, 15 December) 725-737.

82. Marakushev, S.A.; Belonogovaa, O.g.V. 2009, The parageneses thermodynamic analysis of chemoautotrophic CO2 fixation archaic cycle components, their stability and self-organization in hydrothermal systems, Journal of Theoretical Biology, 257, (4, 21 April) 588-597

83. Bartsev, S.I.; Mezhevikin, V.V. 2005, Pre-biotic stage of life origin under non-photosynthetic conditions, Advances in space research, 35, (9) 1643-7.

84. Chernavskii, D.S.; Chernavskaya, N.M. 1975, Some theoretical aspects of the problem of life origin, J Theor Biol, 50, (1) 13-23.

85. Edwards, M.R. 1996, Metabolite channeling in the origin of life, J Theor Biol, 179, (4) 313-22.

86. Gánti, T. 1979, *A theory of biochemical supersystems and its application to problems of natural and artificial biogenesis*. University Park Press: Baltimore.

87. Gánti, T. 2003, *Chemoton theory*. Kluwer Academic/Plenum Publishers: New York.

88. Goldstein, J. 2003, The construction of emergent order, or, how to resist the temptation of hylozoism, Nonlinear dynamics, psychology, and life sciences, 7, (4) 295-314.

89. Kasting, J.F. 1990, Impacts and the origin of life, Earth Miner Sci, 59, (4) 37-42.
90. Lifson, S. 1997, On the crucial stages in the origin of animate matter, J Mol Evol, 44, (1) 1-8.
91. Miyakawa, S.; Yamanashi, H.; Kobayashi, K.; Cleaves, H.J.; Miller, S.L. 2002, Prebiotic synthesis from CO atmospheres: Implications for the origins of life, PNAS, 99, (23) 14628-14631.
92. Sowerby, S.J.; Petersen, G.B. 2002, Life before RNA, Astrobiology, 2, (3) 231-9.
93. Sylvester-Bradley, P.C. 1976, Evolutionary oscillation in prebiology: igneous activity and the origins of life, Orig Life, 7, (1) 9-18.
94. Trevors, J.T. 2011, Origin of microbial life: Nano- and molecular events, thermodynamics/entropy, quantum mechanisms and genetic instructions, J Microbiol Methods, 84, (3) 492-5.
95. Acevedo, O.L.; Orgel, L.E. 1987, Non-enzymatic transcription of an oligodeoxynucleotide 14 residues long, J Mol Biol, 197, (2) 187-93.
96. Baaske, P.; Weinert, F.M.; Duhr, S.; Lemke, K.H.; Russell, M.J.; Braun, D. 2007, Extreme accumulation of nucleotides in simulated hydrothermal pore systems, Proc Natl Acad Sci U S A, 104, (22) 9346-51.
97. Baltscheffsky, H. 1975, Protein structure and the molecular evolution of biological energy conversion, Biosystems, 6, (4) 217-23.
98. Baltscheffsky, H. 1981, Stepwise molecular evolution of bacterial photosynthetic energy conversion, Biosystems, 14, (1) 49-56.
99. Bartel, D.P. 1999, Creation and evolution of new ribozymes, Biol Bull, 196, (3) 322-3.
100. Joyce, G.F. 2002, Molecular evolution: booting up life, Nature, 420, (6913) 278-9.
101. Joyce, G.F.; Orgel, L.E. 1998, The origins of life--a status report, Am Biol Teach, 60, (1) 10-2.
102. Jukes, T.H. 1987, Transitions, transversions, and the molecular evolutionary clock, J Mol Evol, 26, (1-2) 87-98.
103. Jukes, T.H. 2000, The neutral theory of molecular evolution, Genetics, 154, (3) 956-8.
104. Jukes, T.H.; Kimura, M. 1984, Evolutionary constraints and the neutral theory, J Mol Evol, 21, (1) 90-2.
105. Kauffman, S.A. 1992, Applied molecular evolution, J Theor Biol, 157, (1) 1-7.
106. Kauffman, S.A.; Macready, W.G. 1995, Search strategies for applied molecular evolution, J Theor Biol, 173, (4) 427-40.
107. Kimura, M. 1969, The rate of molecular evolution considered from the standpoint of population genetics, Proc Natl Acad Sci U S A, 63, (4) 1181-8.
108. Kimura, M. 1983, *Neutral Theory of Molecular Evolution.* Cambridge University Press: N.Y., N.Y.
109. Lahav, N. 1985, The synthesis of primitive 'living' forms: definitions, goals, strategies and evolution synthesizers, Orig Life Evol Biosph, 16, (2) 129-49.
110. Miller, S.L.; Lazcano, A. 1995, The origin of life -- Did it occur at high temperatures?, Journal of Molecular Evolution, 41, 689-692.
111. Nei, M.; Niimura, Y.; Nozawa, M. 2008, The evolution of animal chemosensory receptor gene repertoires: roles of chance and necessity, Nat Rev Genet, 9, (12) 951-63.
112. Nei, M.; Suzuki, Y.; Nozawa, M. 2010, The Neutral Theory of Molecular Evolution in the Genomic Era, Annual Review of Genomics and Human Genetics, 11, (1) 265-289.
113. Oehlenschlager, F.; Eigen, M. 1997, 30 years later--a new approach to Sol Spiegelman's and Leslie Orgel's in vitro evolutionary studies. Dedicated to Leslie Orgel on the occasion of his 70th birthday, Orig Life Evol Biosph, 27, (5-6) 437-57.
114. Pennisi, E. 2001, MOLECULAR EVOLUTION: Genome Duplications: The Stuff of Evolution?, Science, 294, (5551) 2458-2460.
115. Perlovsky, L. 2002, Statistical limitations on molecular evolution., J Biomol Struct Dyn, Jun;19, (6) 1031-44.
116. Wahl, L.M.; Krakauer, D.C. 2000, Models of experimental evolution: the role of genetic chance and selective necessity, Genetics, 156, (3) 1437-48.
117. Weiner, A.M. 1999, Molecular evolution: aminoacyl-tRNA synthetases on the loose, Curr Biol, 9, (22) R842-4.
118. White, D.H. 1980, A theory for the origin of a self-replicating chemical system. I: Natural selection of the autogen from short, random oligomers, Journal of molecular evolution, 16, (2) 121-47.
119. White, D.H.; Raab, M.S. 1982, A theory for the origin of a self-replicating chemical system. II. Computer simulation of the autogen, Journal of molecular evolution, 18, (3) 207-16.
120. Wright, M.C.; Joyce, G.F. 1997, Continuous in vitro evolution of catalytic function, Science, 276, (5312) 614-7.
121. Yoshiya Ikawa, K.T., Shigeyoshi Matsumura, Shota Atsumi, and Tan Inoue 2003, Putative intermediary stages for the molecular evolution from a ribozyme to a catalytic RNP, Nucleic Acids Res., March 1; 31, 1488-1496.
122. Zeldovich, K.B.; Shakhnovich, E.I. 2008, Understanding Protein Evolution: From Protein Physics to Darwinian Selection, Annual Review of Physical Chemistry, 59, (1) 105-127.
123. Küppers, B. 1975, The general principles of selection and evolution at the molecular level, Prog Biophys Mol Biol, 30, (1) 1-22.
124. Küppers, B. 1979, Towards an experimental analysis of molecular self-organization and precellular Darwinian evolution, Naturwissenschaften, 66, (5) 228-43.

125. Küppers, B.-O. 1990, *Information and the Origin of Life*. MIT Press: Cambridge, MA.
126. Kurakin, A. 2006, Self-organization versus Watchmaker: molecular motors and protein translocation., Biosystems 84, (1) 15-23.
127. Kurakin, A. 2010, Order without design, Theoretical Biology and Medical Modelling, 7, (1) 12.
128. Lahav, N. 1991, Prebiotic co-evolution of self-replication and translation or RNA world?, J Theor Biol, 151, (4) 531-539.
129. Lahav, N. 1993, The RNA-world and co-evolution hypotheses and the origin of life: implications, research strategies and perspectives, Orig Life Evol Biosph, 23, (5-6) 329-44.
130. Lahav, N. 1999, *Biogenesis: Theories of Life's Origin*. Oxford University Press: Oxford.
131. Lahav, N.; Chang, S. 1982, The possible role of soluble salts in chemical evolution, J Mol Evol, 19, (1) 36-46.
132. Lazcano, A. 2010, Historical Development of Origins Research, Cold Spring Harbor Perspectives in Biology, -.
133. Lazcano, A.; Miller, S.L. 1996, The origin and early evolution of life: prebiotic chemistry, the pre-RNA world, and time, Cell, 85, (6) 793-8.
134. Lifson, S.; Lifson, H. 1999, A model of prebiotic replication: survival of the fittest versus extinction of the unfittest, J Theor Biol, 199, (4) 425-33.
135. Lifson, S.; Lifson, H. 2001, Coexistence and Darwinian selection among replicators: response to the preceding paper by Scheuring and Szathmary, J Theor Biol, 212, (1) 107-9.
136. Kiedrowski von, G. 1996, Origins of life. Primordial soup or crepes?, Nature, 381, (6577) 20-1.
137. Kiedrowski von, G. 2001, Chemistry: a way to the roots of biology, Chembiochem, 2, (7-8) 597-8.
138. Miller, S.L. 1997, Peptide nucleic acids and prebiotic chemistry [news] [see comments], Nat Struct Biol, 4, (3) 167-9.
139. Miller, S.L. 1998, The endogenous synthesis of organic compounds. In *The Molecular Origins of Life*, Brack, A., Ed. Oxford University Press: Oxford, U.K., pp 59-85.
140. Miller, S.L.; Schopf, J.W.; Lazcano, A. 1997, Oparin's "Origin of Life": sixty years later, J Mol Evol, 44, (4) 351-3.
141. Fry, I. 2011, The role of natural selection in the origin of life, OLEB, 41, 3-16.
142. Dyson, F., 1999, Life in the Universe: Is Life Digital or Analog? In *Life in the Universe: Is Life Digital or Analog?*, NASA Goddard Space Flight Center Colloquiem, Greenbelt, MD, 1999; Greenbelt, MD,
143. Dyson, F.J. 1998, *Origins of Life*. 2nd ed., Cambridge University Press: Cambridge.
144. Anet, F.A. 2004, The place of metabolism in the origin of life, Curr Opin Chem Biol, 8, (6) 654-9.
145. Kiedrowski von, G. 1986, A self-replicating hexadeoxy nucleotide, Angewandte Chemie Inernational Edition in English, 25, 932-935.
146. Gánti, T. 2003, *The Principles of Life*. Oxford University Press: Oxford, UK.
147. Abel, D.L. 2009, The GS (Genetic Selection) Principle, Frontiers in Bioscience, 14, (January 1) 2959-2969 Open access at http://www.bioscience.org/2009/v14/af/3426/fulltext.htm.
148. Gesteland, R.F.; Cech, T.R.; Atkins, J.F. 2006, *The RNA World*. 3 ed., Cold Spring Harbor Laboratory Press: Cold Spring Harbor.
149. Smith, E.; Morowitz, H.J. 2004, Universality in intermediary metabolism, Proc Natl Acad Sci USA, 101, 13168-13173.
150. de Duve, C. 2005, *Singularities landmarks on the pathways of life*. Cambridge Univ. Press: Cambridge, UK.
151. Kenyon, D.H.; Steinman, G. 1969, *Biochemical Predestination*. McGraw-Hill Book Co.: New York.
152. Shapiro, R. 2000, A replicator was not involved in the origin of life, IUBMB Life, 49, (3) 173-176.
153. Orgel, L.E. 2008, The implausibility of metabolic cycles on prebiotic earth, PLoS biology, 6, 5-13.
154. Pross, A. 2004, Causation and the origin of life. Metabolism or replication first?, Orig Life Evol Biosph, 34, (3) 307-21.
155. Orgel, L.E. 1998, The origin of life--a review of facts and speculations, Trends Biochem Sci, 23, (12) 491-5.
156. Dyson, F.J. 1982, A model for the origin of life, J Mol Evol, 18, (5) 344-50.
157. Orgel, L.E. 2000, Self-organizing biochemical cycles, Proc Natl Acad Sci U S A, 97, (23) 12503-7.
158. de Duve, C. 2007, Chance and necessity revisited, Cell Mol Life Sci, 64, (24) 3149-58.
159. Guimaraes, R.C. 2010, Metabolic Basis for the Self-Referential Genetic Code, Origins of life and evolution of the biosphere.
160. Lee, D.H.; Granja, J.R.; Martinez, J.A.; Severin, K.; Ghadri, M.R. 1996, A self-replicating peptide, Nature, 382, (6591) 525-8.
161. Isaac, R.; Chmielewski, J. 2002, Approaching exponential growth with a self-replicating peptide, J Am Chem Soc, 124, 6808-6809.
162. Shenhav, B.; Oz, A.; Lancet, D. 2007, Coevolution of compositional protocells and their environment, Philos Trans R Soc Lond B Biol Sci, 362, (1486) 1813-9.
163. Segre, D.; Lancet, D.; Kedem, O.; Pilpel, Y. 1998, Graded autocatalysis replication domain (GARD): kinetic analysis of self-replication in mutually catalytic sets, Orig Life Evol Biosph, 28, (4-6) 501-14.

164. Segre, D.; Ben-Eli, D.; Deamer, D.W.; Lancet, D. 2001, The lipid world, Orig Life Evol Biosph, 31, (1-2) 119-45.
165. Segre, D.; Ben-Eli, D.; Lancet, D. 2000, Compositional genomes: prebiotic information transfer in mutually catalytic noncovalent assemblies, Proc Natl Acad Sci U S A, 97, (8) 4112-7.
166. Segre, D.; Lancet, D. 2000, Composing Life, EMBO Rep, 1, (3) 217-222.
167. Lancet, D.; Shenhav, B. 2009, Compositional lipid protocells: reproduction without polynucleotides. In *Protocells: Bridging Nonliving and Living Matter*, Rasmussen, S.; Bedau , M. A.; Chen, L.; Deamer, D.; Krakauer, D. C.; Packard, N. H.Stadler, P. F., Eds. MIT Press: Cambridge, MA.
168. Maynard Smith, J. 1998, The units of selection, Novartis Found Symp, 213, 203-11.
169. Maynard Smith, J. 1999, The 1999 Crafoord Prize Lectures. The idea of information in biology, Q Rev Biol, 74, (4) 395-400.
170. Maynard Smith, J. 2000, The concept of information in biology, Philosophy of Science, 67, (June) 177-194 (entire issue is an excellent discussion).
171. Dawkins, R. 1995, *River out of eden : a Darwinian view of life*. Basic Books: New York, NY.
172. Sterelny, K.; Smith, K.; Dickison, M. 1996, The extended replicator, Biology and Philosophy, 11, 377-403.
173. Sterelny, K.; Griffiths, P.E. 1999, *Sex and Death*. Univ of Chicago Press: Chicago.
174. Pohorille, A. 2009, Early ancestors of existing cells. In *Protocells: Bridging Non-living and Living Matter*, Rasmussen, S.; Bedau , M. A.; Chen, L.; Deamer, D.; Krakauer, D. C.; Packard, N. H.Stadler, P. F., Eds. MIT Press: Cambridge, pp 563-581.
175. Yarus, M. 2011, Getting past the RNA world: the initial darwinian ancestor, Cold Spring Harb Perspect Biol, 3, (4).
176. Mielke, R.E.; Russell, M.J.; Wilson, P.R.; McGlynn, S.E.; Coleman, M.; Kidd, R.; Kanik, I. 2010, Design, fabrication, and test of a hydrothermal reactor for origin-of-life experiments, Astrobiology, 10, (8) 799-810.
177. Shapiro, R. 2006, Small molecule interactions were central to the origin of life, Quarterly Review of Biology, 81, 105-125.
178. Shapiro, R. 2007, A simpler origin of life, Scientific American, Feb 12.
179. Kauffman, S.A. 2010, On emergence, OLEB, 40, (4-5) 381-383.
180. Kauffman, S. 2007, Question 1: origin of life and the living state, Orig Life Evol Biosph, 37, (4-5) 315-22.
181. Morowitz, H.J.; Srinivasan, V.; Smith, E. 2010, Ligand field theory and the origin of life as an emergent feature of the periodic table of elements, Biol Bull, 219, (1) 1-6.
182. Deamer, D.; Szostak, J.W. 2010, The Origins of Life. Cold Spring Harbor Press: Cold Spring Harbor, NY.
183. Lindahl, P.A. 2004, Stepwise evolution of nonliving to living chemical systems, OLEB, 34, 371-389.
184. Wong, J.T. 2005, Coevolution theory of the genetic code at age thirty, Bioessays, 27, (4) 416-25.
185. Wong, J.T. 2007, Question 6: coevolution theory of the genetic code: a proven theory, Orig Life Evol Biosph, 37, (4-5) 403-8.
186. Guimaraes, R.C. 2004, The Genetic Code as a Self-Referential and Functional System, International Conference on Computation, Communications and Control Technologies, 7, 160-165.
187. Di Giulio, M. 2005, The origin of the genetic code: theories and their relationships, a review, Bio Systems, 80, (2) 175-84.
188. Di Giulio, M. 2008, An extension of the coevolution theory of the origin of the genetic code, Biol Direct, 3, 37.
189. Luisi, P.L.; Ferri, F.; Stano, P. 2006, Approaches to semi-synthetic minimal cells: a review, Naturwissenschaften, 93, 1-13.
190. Luisi, P.L. 2006, *The Emergence of Life: From chemical Origins to Synthetic Biology*. Cambridge Univ. Press: Cambridge.
191. Luisi, P.L.; Rasi, P.S.; Mavelli, F. 2004, A possible route to prebiotic vesicle reproduction, Artif Life, 10, (3) 297-308.
192. Luisi, P.L. 2002, Toward the engineering of minimal living cells, Anat Rec, 268, (3) 208-14.
193. Szostak, J.W.; Bartel, D.P.; Luisi, P.L. 2001, Synthesizing life, Nature, 409, (3, Jan 18) 387-90.
194. Fischer, A.; Oberholzer, T.; Luisi, P.L. 2000, Giant vesicles as models to study the interactions between membranes and proteins, Biochim Biophys Acta, 1467, (1) 177-88.
195. Deamer, D.W. 2010, Origin of life just got closer: Life After the synthetic cell, Nature Struc Biol, 465, (27 May).
196. Deamer, D. 2010, Introduction to Definition of Life Section, Astrobiology, 10, (10) 1001-1002.
197. Rasmussen, S.; Bedau, M.A.; Liaohai, C.; Deamer, D.; Krakauer, D.C.; Packard, N.H.; Stadler, P.F. 2008, *Protocells: Bridging Nonliving and Living Matter*. MIT Press: Cambridge.
198. Deamer, D.W. 2008, Origins of life: How leaky were primitive cells?, Nature, 454, (7200) 37-38.
199. Deamer, D.; Pohorille, A. 2008, Session 34. Synthetic Cells and Life's Origins, Astrobiology, 8, (2) 453-455.
200. Monnard, P.A.; Deamer, D.W. 2003, Preparation of vesicles from nonphospholipid amphiphiles, Methods Enzymol, 372, 133-51.

201. Pohorille, A.; Deamer, D. 2002, Artificial cells: prospects for biotechnology, Trends Biotechnol, 20, (3) 123-8.
202. Monnard, P.; Deamer, D. 2002, Membrane self-assembly processes: Steps toward the first cellular life., Anat Rec, 268, (3) 196-207.
203. Deamer, D.; Dworkin, J.P.; Sandford, S.A.; Bernstein, M.P.; Allamandola, L.J. 2002, The first cell membranes, Astrobiology, 2, (4) 371-81.
204. Apel, C.L.; Deamer, D.W.; Mautner, M.N. 2002, Self-assembled vesicles of monocarboxylic acids and alcohols: conditions for stability and for the encapsulation of biopolymers, Biochim Biophys Acta, 1559, (1) 1-9.
205. Mansy, S.S. 2010, Membrane Transport in Primitive Cells, Cold Spring Harbor Perspectives in Biology, -.
206. Mansy, S.S.; Schrum, J.P.; Krishnamurthy, M.; Tobe, S.; Treco, D.A.; Szostak, J.W. 2008, Template-directed synthesis of a genetic polymer in a model protocell, Nature, 454, (7200) 122-5.
207. Mansy, S.S.; Szostak, J.W. 2008, Thermostability of model protocell membranes, Proc Natl Acad Sci U S A, 105, (36) 13351-5.
208. Mansy, S.S.; Szostak, J.W. 2009, Reconstructing the Emergence of Cellular Life through the Synthesis of Model Protocells, Cold Spring Harb Symp Quant Biol.
209. Chen, I.A.; Roberts, R.W.; Szostak, J.W. 2004, The emergence of competition between model protocells., Science, 305, 1474-1476.
210. Luisi, P.L. 2007, The problem of macromolecular sequences: the forgotten stumbling block, Orig Life Evol Biosph, 37, (4-5) 363-5.
211. Abel, D.L. 2007, Complexity, self-organization, and emergence at the edge of chaos in life-origin models, Journal of the Washington Academy of Sciences, 93, (4) 1-20.
212. Cronhjort, M.B. 1995, Hypercycles versus parasites in the origin of life: model dependence in spatial hypercycle systems, Orig Life Evol Biosph, 25, (1-3) 227-33.
213. Eigen, M. 1971, Self-organization of matter and the evolution of biological macromolecules, Naturwissenschaften, 58, (In German) 465-523.
214. Eigen, M.; Biebricher, C.K.; Gebinoga, M.; Gardiner, W.C. 1991, The hypercycle. Coupling of RNA and protein biosynthesis in the infection cycle of an RNA bacteriophage, Biochemistry, 30, (46) 11005-18.
215. Eigen, M.; Gardiner, W.C., Jr.; Schuster, P. 1980, Hypercycles and compartments. Compartments assists--but do not replace--hypercyclic organization of early genetic information, J Theor Biol, 85, (3) 407-411.
216. Eigen, M.; Schuster, P. 1977, The hypercycle. A principle of natural self-organization. Part A: Emergence of the hypercycle, Naturwissenschaften, 64, (11) 541-65.
218. Eigen, M.; Schuster, P. 1981, Comments on "growth of a hypercycle" by King (1981), Biosystems, 13, (4) 235.
219. Eigen, M.; Schuster, P.; Sigmund, K.; Wolff, R. 1980, Elementary step dynamics of catalytic hypercycles, Biosystems, 13, (1-2) 1-22.
220. Hecht, R.; Happel, R.; Schuster, P.; Stadler, P.F. 1997, Autocatalytic networks with intermediates. I: Irreversible reactions, Math Biosci, 140, (1) 33-74.
221. Igamberdiev, A.U. 1999, Foundations of metabolic organization: coherence as a basis of computational properties in metabolic networks, Bio Systems, 50, (1) 1-16.
222. Kaneko, K. 2003, Recursiveness, switching, and fluctuations in a replicating catalytic network, Phys Rev E Stat Nonlin Soft Matter Phys, 68, (3 Pt 1) 031909.
223. Lee, D.H.; Severin, K.; Yokobayashi, Y.; Ghadiri, M.R. 1997, Emergence of symbiosis in peptide self-replication through a hypercyclic network [published erratum appears in Nature 1998 Jul 2;394(6688):101], Nature, 390, (6660) 591-4.
224. Scheuring, I.; Czaran, T.; Szabo, P.; Karolyi, G.; Toroczkai, Z. 2003, Spatial models of prebiotic evolution: soup before pizza?, Orig Life Evol Biosph, 33, (4-5) 319-55.
225. Smith, J.M. 1979, Hypercycles and the origin of life, Nature, 280, (5722) 445-446.
226. Tsokolov, S. 2010, A Theory of Circular Organization and Negative Feedback: Defining Life in a Cybernetic Context, Astrobiology, 10, (10) 1031-1042.
227. Morowitz, H.J. 1992, *Beginnings of cellular life*. Yale University Press: New Haven.
228. New, M.; Pohorille, A. 2000, An inherited efficiences model of non-genomic evolution, Simulat Pract Theor, 8, 99-108.
229. Shenhav, B.; Segre, D.; Lancet, D. 2003, Mesobiotic emergence: molecular and ensemble complexity in early evolution Adv. Complex Syst, 6, (15-35).
230. Smith, E.; Morowitz, H.J.; Copley, S. 2009, Core metabolisma as a self-organized system. In *Protocells: Bridging Living and Non-Living Matter*, Rasmussen, S.; Bedau , M. A.; Chen, L.; Deamer, D.; Krakauer, D. C.; Packard, N. H.Stadler, P. F., Eds. MIT Press: Cambridge, pp 433-460.
231. Eigen, M. 1992, *(with Winkler-Oswatitsch, R.), Steps Toward Life: A Perspective on Evolution.* Oxford University Press: Oxford, UK.
232. Swetina, J.; Schuster, P. 1982, Self-replication with errors. A model for polynucleotide replication, Biophysical chemistry, 16, (4) 329-45.

233. Blomberg, C. 1997, On the appearance of function and organisation in the origin of life, J Theor Biol, 187, (4) 541-54.
234. Gánti, T. 1975, Organization of chemical reactions into dividing and metabolizing units: the chemotons, Biosystems, 7, (1) 15-21.
235. Gánti, T. 1980, On the organizational basis of the evolution, Acta Biol, 31, (4) 449-59.
236. Gánti, T. 1997, Biogenesis itself, J Theor Biol, 187, (4) 583-93.
237. Gánti, T. 2002, On the early evolutionary origin of biological periodicity, Cell Biol Int, 26, (8) 729-35.
238. Abel, D.L.; Trevors, J.T. 2006, Self-Organization vs. Self-Ordering events in life-origin models, Physics of Life Reviews, 3, 211-228.
239. Johnson, D.E. 2010, *Probability's Nature and Nature's Probability (A call to scientific integrity)*. Booksurge Publishing: Charleston, S.C.
240. Johnson, D.E. 2010, *Programming of Life*. Big Mac Publishers: Sylacauga, Alabama.
241. Abel, D.L. 2000 To what degree can we reduce "life" without "loss of life"?, Workshop on Life: a satellite meeting before the Millenial World Meeting of University Professors, Modena, Italy,
242. Rocha, L.M. 2000, Syntactic autonomy: or why there is no autonomy without symbols and how self-organizing systems might evolve them, Annals of the New York Academy of Sciences, 207-223.
243. Rocha, L.M. 2001, Evolution with material symbol systems, Biosystems, 60, 95-121.
244. Weiss, O.; Jimenez-Montano, M.A.; Herzel, H. 2000, Information content of protein sequences, J Theor Biol, 206, (3) 379-86.
245. Costanzo, G.; Pino, S.; Ciciriello, F.; Di Mauro, E. 2009, RNA: Processing and Catalysis: Generation of Long RNA Chains in Water, J. Biol. Chem., 284, 33206-33216.
246. Abel, D.L. 2010, Constraints vs. Controls, Open Cybernetics and Systemics Journal, 4, 14-27 Open Access at http://www.bentham.org/open/tocsj/articles/V004/14TOCSJ.pdf.
247. Behe, M.J. 1996, *Darwin's Black Box*. Simon & Shuster: The Free Press: New York.
248. Stegmann, U.E. 2005, Genetic Information as Instructional Content, Phil of Sci, 72, 425-443.
249. Glass, J.I.; Assad-Garcia, N.; Alperovich, N.; Yooseph, S.; Lewis, M.R.; Maruf, M.; Hutchison, C.A., III; Smith, H.O.; Venter, J.C. 2006, Essential genes of a minimal bacterium, PNAS, 0510013103.
250. Abel, D.L. 2008, 'The Cybernetic Cut': Progressing from description to prescription in systems theory, The Open Cybernetics and Systemics Journal, 2, 234-244 Open access at www.bentham.org/open/tocsj/articles/V002/252TOCSJ.pdf
251. Abel, D.L. 2009, The capabilities of chaos and complexity, Int. J. Mol. Sci., 10, (Special Issue on Life Origin) 247-291 Open access at http://mdpi.com/1422-0067/10/1/247
252. Vasas, V.; Szathmáry, E.; Santos, M. 2010, Lack of evolvability in self-sustaining autocatalytic networks constrains metabolism-first scenarios for the origin of life, Proceedings of the National Academy of Sciences, 107, (4) 1470-1475.
253. Kovac, L.; Nosek, J.; Tomaska, L. 2003, An overlooked riddle of life's origins: energy-dependent nucleic acid unzipping, J Mol Evol, 57 Suppl 1, S182-9.
254. Larralde, R.; Robertson, M.P.; Miller, S.L. 1995, Rates of decomposition of ribose and other sugars: implications for chemical evolution, Proc Natl Acad Sci U S A, 92, (18) 8158-60.
255. Szostak, J.W. 2009, Origins of life: Systems chemistry on early Earth, Nature, 459, (7244) 171-2.
256. Powner, M.W.; Gerland, B.; Sutherland, J.D. 2009, Synthesis of activated pyrimidine ribonucleotides in prebiotically plausible conditions, Nature, 459, (7244) 239-42.
257. Shapiro, R. 1999, Prebiotic cytosine synthesis: a critical analysis and implications for the origin of life, Proc Natl Acad Sci U S A, 96, (8) 4396-4401.
258. Shapiro, R. 2002, Comments on `Concentration by Evaporation and the Prebiotic Synthesis of Cytosine', Origins Life Evol Biosph, 32, (3) 275-278.
259. Joyce, G.F.; Orgel, L.E. 1999, Prospects for understanding the origin of the RNA World. In *The RNA World*, Second ed.; Gesteland, R. F.; Cech, T. R.Atkins, J. F., Eds. Cold Spring Harbor Laboratory Press: Cold Spring Harbor, NY, pp 49-78.
260. Orgel, L. 2003, Some consequences of the RNA world hypothesis., Orig Life Evol Biosph, 33, (2) 211-8.
261. Lehman, N.; Jukes, T.H. 1988, Genetic code development by stop codon takeover, J Theor Biol, 135, (2) 203-14.
262. Lehmann, J. 2000, Physico-chemical constraints connected with the coding properties of the genetic system, J Theor Biol, 202, (2) 129-44.
263. Lehmann, J.; Riedo, B.; Dietler, G. 2004, Folding of small RNAs displaying the GNC base-pattern: implications for the self-organization of the genetic system, J Theor Biol, 227, (3) 381-95.
264. Lehmann, K.; Schmidt, U. 2003, Group II introns: structure and catalytic versatility of large natural ribozymes, Crit Rev Biochem Mol Biol, 38, (3) 249-303.
265. Szathmary, E. 1999, The origin of the genetic code: amino acids as cofactors in an RNA world, Trends Genet, 15, (6) 223-9.
266. Szathmary, E. 2000, The evolution of replicators, Philos Trans R Soc Lond B Biol Sci, 355, (1403) 1669-76.

267. Szathmary, E. 2001, Biological information, kin selection, and evolutionary transitions, Theoretical Population Biology, 59, 11-14.
268. Szathmary, E. 2003, Why are there four letters in the genetic alphabet?, Nat Rev Genet, 4, (12) 995-1001.
269. Szathmary, E. 2006, The origin of replicators and reproducers, Philos Trans R Soc Lond B Biol Sci, 361, (1474) 1761-76.
270. Szathmary, E. 2007, Coevolution of metabolic networks and membranes: the scenario of progressive sequestration, Philos Trans R Soc Lond B Biol Sci, 362, (1486) 1781-7.
271. Szathmary, E.; Demeter, L. 1987, Group selection of early replicators and the origin of life, J Theor Biol, 128, (4) 463-86.
272. Szathmary, E.; Gladkih, I. 1989, Sub-exponential growth and coexistence of non-enzymatically replicating templates, J Theor Biol, 138, (1) 55-8.
273. Szathmáry, E.; Jordan, F.; Pal, C. 2001, Can genes explain biological complexity?, Science, 292, 1315-1316.
274. Szathmary, E.; Maynard Smith, J. 1997, From replicators to reproducers: the first major transitions leading to life, J Theor Biol, 187, (4) 555-71.
275. Szathmary, E.; Santos, M.; Fernando, C. 2005, Evolutionary potential and requirements for minimal protocells, Top Curr Chem, 259, 167-211.
276. Szathmary, E.; Smith, J.M. 1995, The major evolutionary transitions, Nature, 374, (6519) 227-32.
277. Szathmary, E.; Zintzaras, E. 1992, A statistical test of hypotheses on the organization and origin of the genetic code, J Mol Evol, 35, (3) 185-9.
278. Ycas, M. 1999, Codons and hypercycles, Orig Life Evol Biosph, 29, (1) 95-108.
279. Zintzaras, E.; Santos, M.; Szathmary, E. 2002, "Living" Under the Challenge of Information Decay: The Stochastic Corrector Model vs. Hypercycles, J Theor Biol, 217, (2) 167-181.
280. Ruiz-Mirazo, K.; Pereto, J.; Moreno, A. 2004, A universal definition of life: autonomy and open-ended evolution, Orig Life Evol Biosph, 34, (3) 323-46.
281. Ruiz-Mirazo, K.; Umerez, J.; Moreno, A. 2008, Enabling conditions for 'open-ended evolution', Biol Phil, 23, (1) 67-85.
282. Abel, D.L. The GS (Genetic Selection) Principle [Scirus Topic Page]. http://www.scitopics.com/The_GS_Principle_The_Genetic_Selection_Principle.html (Last accessed September, 2011).
283. Johnston, W.K.; Unrau, P.J.; Lawrence, M.S.; Glasner, M.E.; Bartel, D.P. 2001, RNA-catalyzed RNA polymerization: accurate and general RNA-templated primer extension, Science, 292, (5520) 1319-25.
284. Bartel, D.P.; Szostak, J.W. 1993, Isolation of new ribozymes from a large pool of random sequences, Science, 261, (5127) 1411-8.
285. Paul, N.; Joyce, G.F. 2002, Inaugural Article: A self-replicating ligase ribozyme, PNAS, 99, (20) 12733-12740.
286. Paul, N.; Joyce, G.F. 2003, Self-replication, Curr Biol, 13, (2) R46.
287. Zaher, H.S.; Unrau, P.J. 2007, Selection of an improved RNA polymerase ribozyme with superior extension and fidelity, RNA, 13, (7) 1017-26.
288. Ma, W.; Yu, C. 2006, Intramolecular RNA replicase: possibly the first self-replicating molecule in the RNA world, Orig Life Evol Biosph, 36, (4) 413-20.
289. Hud, N.V.; Anet, F.A. 2000, Intercalation-mediated synthesis and replication: a new approach to the origin of life, J Theor Biol, 205, (4) 543-62.
290. Hud, N.V.; Jain, S.S.; Li, X.; Lynn, D.G. 2007, Addressing the problems of base pairing and strand cyclization in template-directed synthesis: a case for the utility and necessity of 'molecular midwives' and reversible backbone linkages for the origin of proto-RNA, Chem Biodivers, 4, (4) 768-83.
291. Hud, N.V.; Lynn, D.G. 2004, From life's origins to a synthetic biology, Curr Opin Chem Biol, 8, (6) 627-8.
292. Hud, N.V.; Plavec, J. 2003, A unified model for the origin of DNA sequence-directed curvature, Biopolymers, 69, (1) 144-58.
293. Hud, N.V.; Polak, M. 2001, DNA-cation interactions: The major and minor grooves are flexible ionophores, Curr Opin Struct Biol, 11, (3) 293-301.
294. Cheng, L.K.L.; Unrau, P.J. 2010, Closing the Circle: Replicating RNA with RNA, Cold Spring Harb Perspect Biol.
295. Engelhart, A.E.; Hud, N.V. 2010, Primitive Genetic Polymers, Cold Spring Harbor Perspectives in Biology, -.
296. Schrum, J.P.; Zhu, T.F.; Szostak, J.W. 2010, The Origins of Cellular Life, Cold Spring Harbor Perspectives in Biology, 2, (9) -.
297. JANAS, T.; YARUS, M. 2003, Visualization of membrane RNAs, RNA, 9, (11) 1353-1361.
298. Yarus, M. 2002, PRIMORDIAL GENETICS: Phenotype of the Ribocyte, Annu Rev Genet, 36, 125-51.
299. Macia, J.; Sole, R.V. 2007, Protocell self-reproduction in a spatially extended metabolism-vesicle system, J Theor Biol, 245, (3) 400-10.
300. Chen, I.A. 2006, GE Prize-winning essay. The emergence of cells during the origin of life, Science, 314, (5805) 1558-9.

301. Klussmann, M.; Blackmond Donna, G. 2009, Origin of Homochirality. In *Chemical Evolution II: From the Origins of Life to Modern Society*, American Chemical Society: Washington DC, pp 133-145.

302. Klussmann, M.; Iwamura, H.; Mathew, S.P.; Wells, D.H., Jr.; Pandya, U.; Armstrong, A.; Blackmond, D.G. 2006, Thermodynamic control of asymmetric amplification in amino acid catalysis, Nature, 441, (7093) 621-3.

303. Anastasi, C.; Buchet, F.F.; Crowe, M.A.; Parkes, A.L.; Powner, M.W.; Smith, J.M.; Sutherland, J.D. 2007, RNA: prebiotic product, or biotic invention?, Chem Biodivers, 4, (4) 721-39.

304. Pino, S.; Ciciriello, F.; Costanzo, G.; Di Mauro, E. 2008, Nonenzymatic RNA ligation in water, JBC, 283, 36494-36503.

305. Breslow, R.; Cheng, Z.L. 2009, On the origin of terrestrial homochirality for nucleosides and amino acids, Proc Natl Acad Sci U S A, 106, (23) 9144-6.

306. Wu, M.; Higgs, P.G. 2009, Origin of self-replicating biopolymers: autocatalytic feedback can jump-start the RNA world, J Mol Evol, 69, (5) 541-54.

307. Kamioka, S.; Ajami, D.; Rebek, J., Jr. 2010, Autocatalysis and organocatalysis with synthetic structures, Proc Natl Acad Sci U S A, 107, (2) 541-4.

308. Spitzer, J.; Poolman, B. 2009, The role of biomacromolecular crowding, ionic strength, and physicochemical gradients in the complexities of life's emergence, Microbiol Mol Biol Rev, 73, (2) 371-88.

309. Cairns-Smith, A.G. 2008, Chemistry and the missing era of evolution, Chemistry, 14, (13) 3830-9.

310. Strasdeit, H. 2005, New studies on the Murchison meteorite shed light on the pre-RNA world, Chembiochem, 6, (5) 801-3.

311. Dworkin, J.P.; Lazcano, A.; Miller, S.L. 2003, The roads to and from the RNA world, J Theor Biol, 222, (1) 127-34.

312. House, C.H.; Miller, S.L. 1996, Hydrolysis of dihydrouridine and related compounds, Biochemistry, 35, (1) 315-20.

313. Kolb, V.M.; Dworkin, J.P.; Miller, S.L. 1994, Alternative bases in the RNA world: the prebiotic synthesis of urazole and its ribosides, J Mol Evol, 38, 549-57.

314. Joyce, G.F.; Inoue, T.; Orgel, L.E. 1984, Non-enzymatic template-directed synthesis on RNA random copolymers. Poly(C, U) templates, J Mol Biol, 176, (2) 279-306.

315. Joyce, G.F.; Orgel, L. 2006, Progress toward understanding the origin of the RNA world. In *RNA World*, 3rd ed.; Gesteland, R.; Cech, T.Atkins, J. F., Eds. Cold Spring Harbor Laboratory: Cold Spring Harbor, pp 23-56.

316. Schoning, K.; Scholz, P.; Guntha, S.; Wu, X.; Krishnamurthy, R.; Eschenmoser, A. 2000, Chemical etiology of nucleic acid structure: the alpha-threofuranosyl-(3'-->2') oligonucleotide system, Science, 290, (5495) 1347-51.

317. Matray, T.J.; Gryaznov, S.M. 1999, Synthesis and properties of RNA analogs-oligoribonucleotide N3'-->P5' phosphoramidates. In Vol. 27, pp 3976-3985.

318. Nielsen, P. 2009, Peptide nucleic acids as prebiotic and abiotic genetic material. In *Protocells: Bridging Nonliving and Living Matter*, Rasmussen, S.; Bedau , M. A.; Chen, L.; Deamer, D.; Krakauer, D. C.; Packard, N. H.Stadler, P. F., Eds. MIT Press: Cambridge.

319. Nielsen, P.E. 1993, Peptide nucleic acid (PNA): a model structure for the primordial genetic material?, Orig Life Evol Biosph, 23, (5-6) 323-7.

320. Diederichsen, U. 1996, Pairing properties of alanyl peptide nucleic acids containing an amino acid backbone with alternating configuration, Angew Chem Int Ed Engl, 35, 445-448.

321. Pattee, H.H. 1977, Dynamic and linguistic modes of complex systems, Int. J. General Systems, 3, 259-266.

322. Shapiro, R. 1984, The improbability of prebiotic nucleic acid synthesis, Orig Life, 14, (1-4) 565-570.

323. Shapiro, R. 1987, *Origins : A Skeptic's Guide to the Creation of Life on Earth*. Bantam: New York.

324. Shapiro, R. 1988, Prebiotic ribose synthesis: a critical analysis, Orig Life Evol Biosph, 18, (1-2) 71-85.

325. Orgel, L.E. 2004, Prebiotic chemistry and the origin of the RNA world, Critical reviews in biochemistry and molecular biology, 39, (2) 99-123.

326. Abel, D.L.; Trevors, J.T. 2006, More than metaphor: Genomes are objective sign systems, Journal of BioSemiotics, 1, (2) 253-267.

327. Kamioka, S.; Ajami, D.; Rebek, J. 2009, Autocatalysis and organocatalysis with synthetic structures, Proceedings of the National Academy of Sciences,

328. Perry, R.S.; Kolb, V.M. 2004, On the applicability of Darwinian principles to chemical evolution that led to life, International Journal of Astrobiology, 3, 45-53.

329. Rodin, S.; Ohno, S.; Rodin, A. 1993, Transfer RNAs with complementary anticodons: could they reflect early evolution of discriminative genetic code adaptors?, Proc Natl Acad Sci U S A, 90, (10) 4723-7.

330. Schuster, P. 1984, Polynucleotide evolution, hypercycles and the origin of the genetic code, Adv Space Res, 4, (12) 143-151.

331. Stadler, B.M.; Stadler, P.F.; Schuster, P. 2000, Dynamics of autocatalytic replicator networks based on higher-order ligation reactions, Bull Math Biol, 62, (6) 1061-86.

332. Hohmann-Marriott, M.F.; Blankenship, R.E. 2010, Evolution of Photosynthesis, Annu Rev Plant Biol.
333. Margulis, L. 1981, *Symbiosis in Cell Evolution: Life and Its Environment on the Early Earth*. W.H. Freeman: San Francisco, CA.
334. Palmer, J.D. 1997, Organelle Genomes--Going, Going, Gone!, Science, 275, (5301) 790-.
335. Fukui, S.; Fukatsu, T.; Ikegami, T.; Shimada, M. 2007, Endosymbiosis as a compact ecosystem with material cycling: parasitism or mutualism?, Journal of theoretical biology, 246, (4) 746-54.
336. Kutschera, U.; Niklas, K.J. 2008, Macroevolution via secondary endosymbiosis: a Neo-Goldschmidtian view of unicellular hopeful monsters and Darwin's primordial intermediate form, Theory Biosci, 127, (3) 277-89.
337. Kleine, T.; Maier, U.G.; Leister, D. 2009, DNA Transfer from Organelles to the Nucleus: The Idiosyncratic Genetics of Endosymbiosis, Annual Review of Plant Biology, 60, (1) 115-138.
338. Schopf, W.J. 2011, The paleobiological record of photosynthesis, Photosynth Res, 107, (1) 87-101.
339. Grenfell, J.L.; Rauer, H.; Selsis, F.; Kaltenegger, L.; Beichman, C.; Danchi, W.; Eiroa, C.; Fridlund, M.; Henning, T.; Herbst, T.; Lammer, H.; Leger, A.; Liseau, R.; Lunine, J.; Paresce, F.; Penny, A.; Quirrenbach, A.; Rottgering, H.; Schneider, J.; Stam, D.; Tinetti, G.; White, G.J. 2010, Co-evolution of atmospheres, life, and climate, Astrobiology, 10, (1) 77-88.
340. Mulkidjanian, A.Y.; Galperin, M.Y. 2009, On the origin of life in the zinc world. 2. Validation of the hypothesis on the photosynthesizing zinc sulfide edifices as cradles of life on Earth, Biol Direct, 4, 27.
341. Blankenship, R.E.; Hartman, H. 1998, The origin and evolution of oxygenic photosynthesis, Trends Biochem Sci, 23, (3) 94-97.
342. Broda, E. 1975, The begginning of photosynthesis, Orig Life, 6, (1-2) 247-51.
343. Buick, R. 1992, The antiquity of oxygenic photosynthesis: evidence from stromatolites in sulphate-deficient Archaean lakes, Science, 255, (5040) 74-7.
344. Castresana, J.; Saraste, M. 1995, Evolution of energetic metabolism: the respiration-early hypothesis, Trends Biochem Sci, 20, (11) 443-448.
345. Hartman, H. 1998, Photosynthesis and the origin of life, Orig Life Evol Biosph, 28, (4-6) 515-21.
346. Jennings, R.C.; Belgio, E.; Casazza, A.P.; Garlaschi, F.M.; Zucchelli, G. 2007, Entropy consumption in primary photosynthesis, Biochim Biophys Acta, 1767, (10) 1194-7; discussion 1198-9.
347. Gupta, R.S.; Mukhtar, T.; Singh, B. 1999, Evolutionary relationships among photosynthetic prokaryotes (Heliobacterium chlorum, Chloroflexus aurantiacus, cyanobacteria, Chlorobium tepidum and proteobacteria): implications regarding the origin of photosynthesis, Mol Microbiol, 32, (5) 893-906.
348. Margulis, L.; Obar, R. 1985, Heliobacterium and the origin of chrysoplasts, Biosystems, 17, (4) 317-25.
349. Morowitz, H.J.; Heinz, B.; Deamer, D.W. 1988, The chemical logic of a minimum protocell, Orig Life Evol Biosph, 18, (3) 281-7.
350. Sundstrom, V. 2008, Femtobiology, Annual Review of Physical Chemistry, 59, (1) 53-77.
351. Suzuki, J.Y.; Bauer, C.E. 1995, A prokaryotic origin for light-dependent chlorophyll biosynthesis of plants, Proc Natl Acad Sci U S A, 92, (9) 3749-53.
352. Tamulis, A.; Tamulis, V.; Graja, A. 2006, Quantum mechanical modeling of self-assembly and photoinduced electron transfer in PNA-based artificial living organisms, J Nanosci Nanotechnol, 6, (4) 965-73.
353. Volkov, A.G.; Gugeshashvili, M.I.; Deamer, D.W. 1995, Energy conversion at liquid/liquid interfaces: artificial photosynthetic systems, Electrochim Acta, 40, (18) 2849-68.
354. Xiong, J.; Fischer, W.M.; Inoue, K.; Nakahara, M.; Bauer, C.E. 2000, Molecular evidence for the early evolution of photosynthesis, Science, 289, (5485) 1724-30.
355. Ham, M.-H.; Choi, J.H.; Boghossian, A.A.; Jeng, E.S.; Graff, R.A.; Heller, D.A.; Chang, A.C.; Mattis, A.; Bayburt, T.H.; Grinkova, Y.V.; Zeiger, A.S.; Van Vliet, K.J.; Hobbie, E.K.; Sligar, S.G.; Wraight, C.A.; Strano, M.S. 2010, Photoelectrochemical complexes for solar energy conversion that chemically and autonomously regenerate, Nat Chem, advance online publication.
356. Foyer, C.H.; Bloom, A.J.; Queval, G.; Noctor, G. 2009, Photorespiratory Metabolism: Genes, Mutants, Energetics, and Redox Signaling, Annual Review of Plant Biology, 60, (1) 455-484.
357. Trevors, J.T. 2002, The subsurface origin of microbial life on the Earth, Res Microbiol, 153, (8) 487-91.
358. Leebens-Mack, J.; DePamphilis, C. 2002, Power analysis of tests for loss of selective constraint in cave crayfish and nonphotosynthetic plant lineages, Molecular biology and evolution, 19, (8) 1292-302.
359. Dismukes, G.C.; Klimov, V.V.; Baranov, S.V.; Kozlov, Y.N.; DasGupta, J.; Tyryshkin, A. 2001, The origin of atmospheric oxygen on Earth: The innovation of oxygenic photosynthesis, Proc. Natl. Acad. Sci. USA, February 27; 98, (2170-2175).
360. Cavalier-Smith, T. 2001, Obcells as proto-organisms: membrane heredity, lithophosphorylation, and the origins of the genetic code, the first cells, and photosynthesis, J Mol Evol, 53, (4-5) 555-95.
361. Pereto, J.G.; Velasco, A.M.; Becerra, A.; Lazcano, A. 1999, Comparative biochemistry of CO2 fixation and the evolution of autotrophy, Int Microbiol, 2, (1) 3-10.
362. Fridlyand, L.E.; Scheibe, R. 1999, Regulation of the Calvin cycle for CO2 fixation as an example for general control mechanisms in metabolic cycles, Bio Systems, 51, (2) 79-93.

363. Volkov, A.G.; Deamer, D.W. 1997, Redox chemistry at liquid/liquid interfaces, Prog Colloid Polym Sci, 103, 21-8.

364. Crick, F.H.C. 1968, The Origin of the Genetic Code, J. Mol. Biol, 38, 367-379.

365. Di Giulio, M. 1997, On the origin of the genetic code, J Theor Biol, 187, (4) 573-581.

366. Di Giulio, M. 2000, Genetic code origin and the strength of natural selection, J Theor Biol, 205, (4) 659-661.

367. Di Giulio, M. 2001, The Non-universality of the Genetic Code: the Universal Ancestor was a Progenote, Journal of Theoretical Biology, 209, (3, April) 345-349.

368. Di Giulio, M. 2001, A blind empiricism against the coevolution theory of the origin of the genetic code, J Mol Evol, 53, (6) 724-732.

369. Di Giulio, M. 2002, Genetic code origin: are the pathways of type Glu-tRNA(Gln) --> Gln-tRNA(Gln) molecular fossils or not?, J Mol Evol, 55, (5) 616-622.

370. Di Giulio, M. 2003, The early phases of genetic code origin: conjectures on the evolution of coded catalysis., Orig Life Evol Biosph, 33, (4-5) 479-489.

371. Di Giulio, M.; Medugno, M. 1998, The historical factor: the biosynthetic relationships between amino acids and their physicochemical properties in the origin of the genetic code, J Mol Evol, 46, (6) 615-621.

372. Di Giulio, M.; Medugno, M. 1999, Physicochemical optimization in the genetic code origin as the number of codified amino acids increases, J Mol Evol, 49, (1) 1-10.

373. Di Giulio, M.; Medugno, M. 2000, The robust statistical bases of the coevolution theory of genetic code origin, J Mol Evol, 50, (3) 258-263.

374. Di Giulio, M.; Medugno, M. 2001, The level and landscape of optimization in the origin of the genetic code, J Mol Evol, 52, (4) 372-382.

375. Higgs, P.G.; Pudritz, R.E. 2009, A Thermodynamic Basis for Prebiotic Amino Acid Synthesis and the Nature of the First Genetic Code, Astrobiology, 9, (5) 483-490.

376. Ikehara, K. 2002, Origins of gene, genetic code, protein and life: comprehensive view of life systems from a GNC-SNS primitive genetic code hypothesis., J Biosci, 27, (2) 165-86.

377. Legiewicz, M.; Yarus, M. 2005, A More Complex Isoleucine Aptamer with a Cognate Triplet, J. Biol. Chem., 280, (20) 19815-19822.

378. Ribas de Pouplana, L.; Turner, R.J.; Steer, B.A.; Schimmel, P. 1998, Genetic code origins: tRNAs older than their synthetases?, Proc Natl Acad Sci U S A, 95, (19) 11295-11300.

379. Ronneberg, T.A.; Landweber, L.F.; Freeland, S.J. 2000, Testing a biosynthetic theory of the genetic code: Fact or artifact?, PNAS, 97, (25) 13690-13695.

380. Sanchez, R.; Grau, R. 2006, A novel algebraic structure of the genetic code over the galois field of four DNA bases, Acta biotheoretica, 54, (1) 27-42.

381. Schimmel, P.; Ribas de Pouplana, L. 1999, Genetic code origins: experiments confirm phylogenetic predictions and may explain a puzzle [published erratum appears in Proc Natl Acad Sci U S A 1999 May 11;96(10):5890], Proc Natl Acad Sci U S A, 96, (2) 327-328.

382. Seligmann, H.; Amzallag, G.N. 2002, Chemical interactions between amino acid and RNA: multiplicity of the levels of specificity explains origin of the genetic code, Naturwissenschaften, 89, (12) 542-551.

383. Yarus, M.; Christian, E.L. 1989, Genetic code origins, Nature, 342, (6248) 349-350.

384. Schultz, D.W.; Yarus, M. 1996, On malleability in the genetic code, J Mol Evol, 42, (5) 597-601.

385. Yarus, M. 1991, An RNA-amino acid complex and the origin of the genetic code, New Biol, 3, (2) 183-9.

386. Yarus, M.; Schultz, D.W. 1997, Further comments on codon reassignment. Response, J Mol Evol, 45, (1) 3-6.

387. Wong, J.T. 1975, A co-evolution theory of the genetic code, Proc Natl Acad Sci U S A, 72, (5) 1909-1912.

388. Wong, J.T. 1976, The evolution of a universal genetic code, Proc Natl Acad Sci U S A, 73, (7) 2336-40.

389. Wong, J.T. 1980, Role of minimization of chemical distances between amino acids in the evolution of the genetic code, Proc Natl Acad Sci U S A, 77, (2) 1083-6.

390. Wong, J.T. 1988, Evolution of the genetic code, Microbiol Sci, 5, (6) 174-181.

391. Wong, J.T. 1991, Origin of genetically encoded protein synthesis: a model based on selection for RNA peptidation, Orig Life Evol Biosph, 21, (3) 165-76.

392. Labouygues, J.M.; Figureau, A. 1984, The logic of the genetic code: synonyms and optimality against effects of mutations, Orig Life, 14, (1-4) 685-692.

393. Szathmary, E. 1991, Four letters in the genetic alphabet: a frozen evolutionary optimum?, Proc R Soc Lond B Biol Sci, 245, (1313) 91-9.

394. Szathmary, E. 1992, What is the optimum size for the genetic alphabet?, Proc Natl Acad Sci U S A, 89, (7) 2614-8.

395. Bradley, D. 2002, Informatics. The genome chose its alphabet with care, Science, 297, (5588) 1789-1791.

396. Freeland, S.J.; Hurst, L.D. 1998, The genetic code is one in a million, Journal of Molecular Evolution, 47, 238-248.

397. Itzkovitz, S.; Alon, U. 2007, The genetic code is nearly optimal for allowing additional information within protein-coding sequences, Genome Res, 17, (4) 405-12.

398. Segal, E.; Fondufe-Mittendorf, Y.; Chen, L.; Thastrom, A.; Field, Y.; Moore, I.K.; Wang, J.P.; Widom, J. 2006, A genomic code for nucleosome positioning, Nature, 442, (7104) 772-8.

399. Di Giulio, M. 2004, The coevolution theory of the origin of the genetic code, Physics of Life Reviews.

400. Guimaraes, R.C. 2001, Two punctuation systems in the genetic code. In *First steps in the origin of life in the universe*, Chela-Flores, J.; Owen, T.Raulin, F., Eds. Kluwer Acad. Publ: Dordrecht, NL.

401. Farias, S.T.; Moreira, C.H.; Guimaraes, R.C. 2007, Structure of the genetic code suggested by the hydropathy correlation between anticodons and amino acid residues, Orig Life Evol Biosph, 37, (1) 83-103.

402. Guimaraes, R.C.; Moreira, C.H.; de Farias, S.T. 2008, A self-referential model for the formation of the genetic code, Theory Biosci, 127, (3) 249-70.

403. Yockey, H.P. 1992, *Information Theory and Molecular Biology*. Cambridge University Press: Cambridge.

404. Pattee, H.H. 2001, The physics of symbols: bridging the epistemic cut, Biosystems, 60, (1-3) 5-21.

405. Pattee, H.H. 2001, Irreducible and complementary semiotic forms, Semiotica, 134, 341-358.

406. Pattee, H.H. 2005, The physics and metaphysics of Biosemiotics, Journal of Biosemiotics, 1, 303-324.

407. Pattee, H.H. 2007, The necessity of biosemiotics: Matter-symbol complementarity. In *Introduction to Biosemiotics: The New Biological Synthesis*, Springer: Dordrecht, The Netherlands, pp 115-132.

408. Pattee, H.H.; Kull, K. 2009, A biosemiotic conversation: Between physics and semiotics, Sign Systems Studies 37, (1/2).

409. Rocha, L.M. 1998, Selected self-organization and the semiotics of evolutionary systems. In *Evolutionary Systems: Biological and Epistemological Perspectives on Selection and Self-Organization*, Salthe, S.; van de Vijver, G.Delpos, M., Eds. Kluwer: The Netherlands, pp 341-358.

410. Behe, M. 2010, Experimental Evolution, Loss-of-Function Mutations, and "The First Rule of Adaptive Evolution", The Quarterly Review of Biology, 85, (4 December).

411. Sun, W.; Julie Li, Y.-S.; Huang, H.-D.; Shyy, J.Y.-J.; Chien, S. 2010, microRNA: A Master Regulator of Cellular Processes for Bioengineering Systems, Annual Review of Biomedical Engineering, 12, (1) 1-27.

412. Sun, L.; Becskei, A. 2010, Systems biology: The cost of feedback control, Nature, 467, (7312) 163-164.

413. Cotterell, J.; Sharpe, J. 2010, An atlas of gene regulatory networks reveals multiple three-gene mechanisms for interpreting morphogen gradients, Molecular Systems Biology, 6, (425).

414. Chuang, H.-Y.; Hofree, M.; Ideker, T. 2010, A Decade of Systems Biology, Annual Review of Cell and Developmental Biology, 26, (1) 721-744.

415. Buchanan, M.; Caldarelli, G.; De Los Rios, P.; Rao, F.; Vendruscolo, M. 2010, Networks in Cell Biology. Cambridge University Press: Cambridge.

416. Ferrell, J. 2009, Q&A: Systems biology, Journal of Biology, 8, (1) 2.

417. Danesi, M. 2005, Modeling systems theory, Journal of Biosemiotics, 1, 213-225.

418. Krishnamurthy, L.; Nadeau, J.; Ozsoyoglu, G.; Ozsoyoglu, M.; Schaeffer, G.; Tasan, M.; Xu, W. 2003, Pathways database system: an integrated system for biological pathways, Bioinformatics, 19, (8) 930-937.

419. Kuhn, H. 1972, Self-organization of molecular systems and evolution of the genetic apparatus, Angew Chem Int Ed Engl, 11, (9) 798-820.

420. Kull, K. 1993, Semiotic paradigm in theoretical biology. In *Lectures in Theoretical Biology: The second Stage*, Kull, K.Tiivel, T., Eds. Estonian Academy of Sciences: Tallinn, pp 52-62.

421. Kull, K. 1998, Organism as a self-reading text: Anticipation and semiosis, International Journal of Computing Anticipatory Systems, 1, 93-104.

422. Kull, K. 1999, Biosemiotics in the twentieth century: A view from biology., Semiotica, 127, 385-414.

423. Kull, K. 2005, A brief history of biosemiotics, Journal of Biosemiotics, 1, 1-36.

424. Kull, K. 2006, Biosemiotics and biophysics--The fundamental approaches to the study of life. In *Introduction to Biosemiotics: The New Biological Synthesis*, Barbieri, M., Ed. Springer-Verlag New York, Inc. : Dordrecht, The Netherlands; Secaucus, NJ, USA

425. Artmann, S. 2005, Biosemiotics as a structural science, Journal of Biosemiotics, 1, 247-285.

426. Barbieri, M. 2005, Life is 'artifact-making', Journal of Biosemiotics, 1, 113-142.

427. Barbieri, M. 2006, Is the Cell a Semiotic System? In *Introduction to Biosemiotics: The New Biological Synthesis*, Barbieri, M., Ed. Springer-Verlag New York, Inc. : Secaucus, NJ, USA

428. Barbieri, M. 2006, Introduction to Biosemiotics: The New Biological Synthesis. Springer-Verlag   Dordrecht, The Netherlands.

429. Barbieri, M. 2007, Has biosemiotics come of age? In *Introduction to Biosemiotics: The New Biological Synthesis*, Barbieri, M., Ed. Springer: Dorcrecht, The Netherlands, pp 101-114.

430. Barbieri, M. 2007, Is the cell a semiotics system? In *Introduction to Biosemiotics: The New Biological Synthesis*, Barbieri, M., Ed. Springer: Dordrecht, The Netherlands, pp 179-208.

431. Barbieri, M. 2007, BioSemiotic Research Trends Nova Science Publishers, Inc.: New York.

432. Barbieri, M. 2007, The Codes of Life: The Rules of Macroevolution (Biosemiotics). Springer: Dordrecht, The Netherlands.

433. Barbieri, M. 2008, Biosemiotics: a new understanding of life, Naturwissenschaften, 95, 577-599.

434. Barbieri, M. 2008, Cosmos and History: Life is semiosis; The biosemiotic view of nature, The Journal of Natural and Social Philosophy, 4, (1-2) 29-51.
435. Battail, G. 2005, Genetics as a communication process involving error-correcting codes, Journal of Biosemiotics, 1, 143-193.
436. Chang, H.-l. 2005, Natural history or natural systems? Encoding the textual sign, Journal of Biosemiotics, 1, 227-245.
437. Deely, J. 1992, Semiotics and biosemiotics: are sign-science and life-science coextensive? In *Biosemiotics: The Semiotic Web 1991*, Sebeok, T. A.Umiker-Sebeok, J., Eds. Mouton de Gruyter: Berlin/N.Y., pp 46-75.
438. Emmeche, C. 1999, The Sarkar challenge to biosemiotics: Is there any information in a cell?, Semiotica, 127, 273-293.
439. Favareau, D. 2006, The Evolutionary History of Biosemiotics. In *Introduction to Biosemiotics: The New Biological Synthesis*, Barbieri, M., Ed. Springer-Verlag New York, Inc. : Dordrecht, The Netherlands; Secaucus, NJ, USA pp 1-68.
440. Florkin, M. 1974, Concepts of molecular biosemiotics and of molecular evolution. In *Comprehensive Biochemistry*, Florkin, M.Stoltz, E. H., Eds. Elsevier: Amsterdam, pp 1-124.
441. Hoffmeyer, J. 1997, Biosemiotics: Towards a new synthesis in biology, European Journal for Semiotic Studies, 9, 355-376.
442. Hoffmeyer, J. 2006, Semiotic scaffolding of living systems. In *Introduction to Biosemiotics: The New Biological Synthesis*, Barbieri, M., Ed. Springer-Verlag New York, Inc. : Dordrecht, The Netherlands; Secaucus, NJ, USA pp 149-166.
443. Hoffmeyer, J.; Emmeche, C. 2005, Code-Duality and the Semiotics of Nature, Journal of Biosemiotics, 1, 37-91.
444. Jämsä, T. 2006, Semiosis in evolution. In *Introduction to Biosemiotics: The New Biological Synthesis*, Barbieri, M., Ed. Springer-Verlag New York, Inc. : Dordrecht, The Netherlands; Secaucus, NJ, USA
445. Kawade, Y. 1996, Molecular biosemiotics: molecules carry out semiosis in living systmes, Semiotica, 111, 195-215.
446. Nöth, W. 2005, Semiotics for biologists, Journal of Biosemiotics, 1, 195-211.
447. Salthe, S.N. 2006, What is the scope of biosemiotics? Information in living systems. . In *Introduction to Biosemiotics: The New Biological Synthesis* Barbieri, M., Ed. Springer-Verlag New York, Inc.: Dordrecht, The Netherlands; Secaucus, NJ, USA
448. Sharov, A. 1992, Biosemiotics. A functional-evolutionary approach to the analysis of the sense of evolution. In *Biosemiotics: The Semiotic Web 1991*, Sebeok, T. A.Umiker-Sebeok, J., Eds. Mouton de Gruyter: Berlin, pp 345-373.
449. Vehkavaara, T. 1998, Extended concept of knowledge for evolutionary epistemology and for biosemiotics. In *Emergence, Complexity, Hierarchy, Organization. Selected and edited papers from Echo III*, Farre, G. L.Oksala, T., Eds. Espoo: Helsinki, pp 207-216.
450. Hoover, R.B. 2011, Fossils of Cyanobacteria in CI1 Carbonaceous Meteorites NASA/Marshall Space Flight Center Journal of Cosmology, 13, (March).
451. Hoover, R.B. 1997, Meteorites, microfossils and exobiology. . In *Instruments, Methods, and Missions for the Investigation of Extraterrestrial Microorganisms SPIE, 3111*, Hoover, R. B., Ed. pp 115–136.
452. Brasier, M.D. 2011, Life in CI1 Carbonaceous Chondrites?, Journal of Cosmology, 13, (March).
453. Ruiz-Mirazo, K.; Moreno, A. 2006, On the origins of information and its relevance for biological complexity, Biological Theory, 1, (3) 227-229.
454. Ruiz-Mirazo, K.; Moreno, A.; Moran, F. 1998, Merging the Energetic and the Relational-Constructive Logic of Life. . In *Artificial Life VI*, Adami, C.; R., B.; Kitano, H.Taylor, C., Eds. MIT Bradford Books: Cambridge MA, pp 448-451.
455. Konopka, A.K. 1994, Sequences and Codes: Fundamentals of Biomolecular Cryptology. In *Biocomputing: Informatics and Genome Projects*, Smith, D., Ed. Academic Press: San Diego, pp 119-174.
456. Konopka, A.K. 1997, Theoretical Molecular Biology. In *Encyclopedia of Molecular Biology and Molecular Medicine*, Meyers, R. A., Ed. VCH Publishers: Weinheim, Vol. 6, pp 37-53.
457. Konopka, A.K. 2001, Towards understanding life itself, Comput Chem, 25, (4) 313-5.
458. Konopka, A.K. 2003, Systems biology: aspects related to genomics. In *Nature Encyclopidia of the Human Genome*, Cooper, D. N., Ed. Nature Publishing Group Reference: London, Vol. 5, pp 459-465.
459. Konopka, A.K. 2003, Information theories in molecular biology and genomics. In *Nature Encyclopedia of teh Human Genome*, Cooper, D. N., Ed. Nature Publishing Group Reference: London, Vol. 3, pp 464-469.
460. Konopka, A.K. 2003, Sequence complexity and composition. In *Nature Encyclopedia of the Human Genome. Vol. 5.*, Cooper, D. N., Ed. Nature Publishing Group Reference: London, pp 217-224.
461. Wald, G. 1954, The Origin of Life, Scientific American, 191 44-53.

The First Gene, David L. Abel, Editor, 2011,   pp 287-304   ISBN: 978-0-9657988-9-1

10.   What Might Be a Protocell's Minimal "Genome?"

Donald E. Johnson

(Ph.D: Computer & Information Science;   Ph.D: Chemistry)

Science Integrity, 5002 Holly Tree Road, Wilmington, NC 28409

Abstract.   The origin of life's biggest mystery is the origin of the genome which contains the information to cybernetically control all aspects of cellular life today.   Without formal control, no life would exist.   The genetics-first and metabolism-first models will be examined, each having characteristics that strain scientific credibility.   Major physical science limitations and the formidable information science problems are examined. These problems usually result in over-simplifications in speculative scenarios.   More   serious   are   the   11   peer-reviewed   scientific   null hypotheses that require falsification before any of the naturalistic scenarios can be considered as serious science.   Assuming the problems can be resolved, the requirements for a minimal "genome" can be discussed in the areas of initial generation of programmed controls, replication of the genome and needed components that make it useful, regulation of "life's" processes, and evolvability.   Life is an intersection of the physical sciences of chemistry and physics and the nonphysical formalism of information science.   Each domain must be investigated using that domain's principles. Yet most scientists have been attempting to use physical science to explain life's nonphysical information domain, a practice that has no scientific justification.

Correspondence/Reprint request: Donald E. Johnson, Science Integrity, 5002 Holly Tree Road, Wilmington, NC 28409   Email:. donjohnson@ieee.org

Introduction: Pseudo-Scientific Speculations or Science?

A hundred years ago, the title's question wouldn't have been needed since a cell was thought to be bag of plasm [1] originating in a "warm little pond" [2].   Fifty years ago, protein and DNA structures had been determined so science "knew" the secrets of the genome.   With the Miller/Urey [3] synthesis, many thought that the origin of life explanation was near.   Fifteen years ago, it started to be realized that "junk DNA" was a misnomer.   Five years ago, epigenetic control systems largely determined by non-coding DNA began to be discovered.   As new knowledge of functional complexity is revealed, we realize that our knowledge of that complexity has been increasing exponentially, with no end in sight.   As one layer is pealed back, a new level of functional complexity is exposed. Rather than getting simpler, the more we know, the more we know we don't know! "As sequencing and other new technologies spew forth data, the complexity of biology has seemed to grow by orders of Magnitude" [4]. There seems to be an exponential increase in knowledge, with the target of understanding the origin receding ever faster.

The origin of life (OOL) is unknown and is obscured by the lack of knowledge of the prebiotic conditions that existed as life "developed." "Most of the (bio)chemical processes found within all the living organisms are well understood at the molecular level, whereas the origin of life remains one of the most vexing issues in chemistry, biology, and philosophy" [5].   "The origin of life remains one of the humankind's last great unanswered questions, as well as one of the most experimentally challenging research areas" [6]. Any speculation inevitably involves science as we don't know it.   It is metaphysically presumed that since life obviously exists, there must have been a time when non-life developed into life through natural mechanisms.   It is also presumed (with no substantiating reasons) that Pasteur's Law of Biogenesis, all life is from life ("Omne vivum ex vivo" [7]), must not have been applicable during life's formation from inanimate material.   Pasteur's warning that "Spontaneous generation is a dream" ("La génération spontanée est une chimère" [8]) is perhaps appropriate to consider with the various speculations.   It is important to realize that "we don't yet know, but the answers will be coming" isn't a scientific statement, but rather expresses faith in naturalism-of-the-gaps, which is no more scientific than the god-of-the-gaps explanation that most scientists would dismiss out-of-hand.

Speculation on a particular aspect of life may not prove fruitful since all known life is a carefully-orchestrated cybernetic system [9-12]. Without consideration of the origin of cybernetic processes, they are "systems and processes that interact with themselves and produce themselves from themselves" [13]. Michael Polanyi argued that life is not reducible to physical and chemical principles, but rather that, "the information content of a biological whole exceeds that of the sum of its parts" [14]. "A whole exists when it acts like a whole, when it produces combined effects that the parts cannot produce alone" [15]. "Understanding the origin of life requires knowledge not only of the origin of biological molecules such as amino acids, nucleotides and their polymers, but also the manner in which those molecules are integrated into the organized systems that characterize cellular life" [16].

It should be noted that speculation is important within science, since that is the way that new lines of thought are proposed in order to test scenarios for possibility and feasibility [17]. While the dream of becoming a Nobel laureate may encourage wide dissemination of a speculation, it seems appropriate to warn about spreading such speculations outside the scientific community. The public too often views a scientist's speculation as validated science, so that the speculative nature is overlooked. The public may value a scientist's view in much the same way that a movie star's endorsement of a product is seen as important. There seems to be a wide-spread belief in chemical predestination, even though its chief promoter [18] has repudiated its possibility. For example, when signs of water on Mars were discovered, the media proclaimed that there must be life then. Our collective preoccupation with the Search for Extraterrestrial Intelligence [19] illustrates the belief in the inevitability of life.

1. Overview

The approach of this essay will be to consider scenarios for developing the minimal replication and control information ("proto-genome") for a protocell, since even "protolife" would require self-replication and control ability. Note that the ability to use the "genomic" information for functionality is also critical. Metabolic cycles [20], homochirality [21-24], cell membranes [25-26], and other required components will not be the primary thrust, even though all are indirectly controlled by today's genome.   An excellent review of the organic chemistry for biomolecular origin is available [27].   Each proponent's scenario will be briefly highlighted, with the primary arguments against the scenario coming from proponents of an alternative scenario, typically as quotes.   Finally, we'll examine principles that are almost universally ignored in OOL scenarios, but are in critical need of scientific explanation.

1.1 RNA (Genetics) First Scenarios

A ribosome, "a molecular fossil" [28], can join amino acids without additional enzymes except for those that are imbedded in the ribosome itself to make it a ribozyme (enzymes needed to manufacture tRNAs presumably developed later).   "An appeal of the RNA world hypothesis is that it solves the 'chicken and egg' problem; it shows that in an earlier, simplified biota the genotype/replicator and pheno-type/catalyst could have been one and the same molecule" [29] (but the RNA/enzyme of a ribozyme is another chicken/egg problem).   "RNA appears well suited to have served as the first replicative polymer on this planet" [30].   The origin of the RNA World by stringing together optimistic extrapolations of prebiotic chemistry achievements and experimenter-directed RNA "evolution" (a misnomer) has been described as "the 'Molecular Biologists' Dream ... [and]   the prebiotic chemist's nightmare" [31].   An excellent review [32] describes the potential and problems of the RNA world.   The "difficulties in nucleobase ribosylation can be overcome with directing, blocking, and activating groups on the nucleobase and ribose ... These molecular interventions are synthetically ingenious, but serve to emphasize the enormous difficulties that must be overcome if ribonucleosides are to be efficiently produced by nucleobase ribosylation under prebiotically plausible conditions. This impasse has led many scientists to abandon the idea that a RNA "genome" might have assembled abiotically, and has prompted a search for potential pre-RNA informational molecules" [33].   It has been pointed out that "what is essential, therefore, is a reasonably detailed description, hopefully supported by experimental evidence, of how an evolvable family of cycles might operate. The scheme should not make unreasonable demands on the efficiency and specificity of the various external and internally generated catalysts that are supposed to be involved.   Without such a description, acceptance of the possibility of complex non-enzymatic cyclic organizations that are capable of evolution can only be based on faith, a notoriously dangerous route to scientific progress" [20].   The experimenter-directed "side products would have amounted to a fatal and committed step in the synthesis of a nascent proto-RNA. This problem illustrates a difficulty in non-enzymatic polymerization that must be taken into account when considering how the nature of the synthetic routes to and structural identities of early genetic polymers: irreversible linkages are adaptive for an informational polymer only when mechanisms exist to make them conditionally reversible [34].

"No physical law need be broken for spontaneous RNA formation to happen, but the chances against it are so immense, that the suggestion implies that the non-living world had an innate desire to generate RNA

There is no reason to presume that an indifferent nature would not combine units at random, producing an immense variety of hybrid short, terminated chains, rather than the much longer one of uniform backbone geometry needed to support replicator and catalytic functions" [35]. "The RNA molecule is too complex, requiring assembly first of the monomeric constituents of RNA, then assembly of strings of monomers into polymers. As a random event without a highly structured chemical context, this sequence has a forbiddingly low probability and the process lacks a plausible chemical explanation, despite considerable effort to supply one" [36].   "It has been challenging to identify possible prebiotic chemistry that might have created RNA.   Organic molecules, given energy, have a well-known propensity to form multiple products, sometimes referred to collectively as 'tar' or 'tholin.'   These mixtures appear to be unsuited to support Darwinian processes, and certainly have never been observed to spontaneously yield a homochiral genetic polymer.   To date, proposed solutions to this challenge either involve too much direct human intervention to satisfy many in the community, or generate molecules that are unreactive 'dead ends' under standard conditions of temperature and pressure" [27].

Some [27, 33] believe that inorganic crystals or clay served as a template for the original RNA.   The "replication of clay 'information' has remained hypothetical, and transfer of replicated clay properties to nucleic acids even more so" [29].   Crystals contain a very small quantity of information in their regular structures, so that any significant information would have to be in irregularities.   How would inanimate nature produce those irregularities to serve as templates for functional information in replicative polymers?

"The reaction system... is a purified reconstituted system in which all of the components and their concentrations are defined. The number of components is amazingly large, yet this is one of the simplest encapsulated systems for carrying out protein translation and RNA replication. With regard to the origin of life, the first living systems would have had functionally identical translation and replication systems, but they must have been simpler and contained machinery for nutrient transport. The complexity of our system implies that extant translation machinery has become highly sophisticated during the evolutionary process" [16].

1.2   Metabolism-First Scenarios

Metabolism-first scenarios involve development of a self-replicating, self-sustaining chemical system that is able to capture energy and that is contained within a protocell [24] or geothermal vent [38-39]. Perhaps energy transfer used an "osmosis first" paradigm [40, 26]. Unlike RNA first, there is no nucleotide genome to control replication or component construction so that selection would have favored "not the best replicator, but the reaction that sucked in fuel the quickest, denying energy to other chemical processes" [41]. The "bag of chemicals" (composome) presumably would grow until it reaches a size that enables it to divide, with each "daughter" inheriting about half the chemical contents. "The origin of life was marked when a rare few protocells happened to have the ability to capture energy from the environment to initiate catalyzed heterotrophic growth directed by heritable genetic information in the polymers ... The origin of life occurred when a subset of these molecules was captured in a compartment and could interact with one another to produce the properties we associate with the living state" [39]. There have been simulations [42-43] in which the composomes "undergo mutation-like compositional changes" that are claimed to illustrate evolution, but these have never been experimentally observed.

Although metabolism-first avoids the infeasibility of forming functional RNA by chance, "replication of compositional information is so inaccurate that fitter compositional genomes cannot be maintained by selection and, therefore, the system lacks evolvability (i.e., it cannot substantially depart from the asymptotic steady-state solution already built-in in the dynamical equations). We conclude that this fundamental limitation of ensemble replicators cautions against metabolism-first theories of the origin of life" [44]. Concerning the chemical cycles required, "These are chemically very difficult reactions ... One needs, therefore, to postulate highly specific catalysts for these reactions. It is likely that such catalysts could be constructed by a skilled synthetic chemist, but questionable that they could be found among naturally occurring minerals or prebiotic organic molecules...The lack of a supporting background in chemistry is even more evident in proposals that metabolic cycles can evolve to 'life-like' complexity. The most serious challenge to proponents of metabolic cycle theories—the problems presented by the lack of specificity of most non-enzymatic catalysts—has, in general, not been appreciated. If it has, it has been ignored. Theories of the origin of life based on metabolic cycles cannot be justified by the inadequacy of competing theories: they must stand on their own" [20].

2. Major Unresolved Difficulties

Nearly all scenarios presented as science during this author's education using the American Chemical Society's "From Molecules to Man" have been shown to be incorrect by today's science. Scientists need to use much caution during speculative dreaming about mechanisms that can be considered as explanations for the observations that are currently available. Some of the major difficulties requiring scientific explanation will be highlighted in this section

2.1 Physical Science Limitations

What natural interactions produced homochilarity, α-linkage only amino acids, and non-enzymatic peptide bonds and other dehydration reactions in aqueous solutions to produce proteins and RNAs? What physical laws could integrate biochemical pathways and cycles into a formal protometabolic scheme? How did the enzymes required to level life's 10^{19} range of uncatalyzed reactions [45] spontaneously polymerize and self-assemble?

2.2 Formidable Information Science Problems

"Biological information is not a substance ... biological information is not identical to genes or to DNA (any more than the words on this page are identical to the printers ink visible to the eye of the reader). Information, whether biological or cultural, is not a part of the world of substance" [46].   "All the equations of physics taken together cannot describe, much less explain, living systems.   Indeed, the laws of physics do not even contain any hints regarding cybernetic processes or feedback control" [10].   The argument for abiogenesis "simply says it happened.   As such, it is nothing more than blind belief.   Science must provide rational theoretical mechanism, empirical support, prediction fulfillment, or some combination of these three.   If none of these three are available, science should reconsider that molecular evolution of genetic cybernetics is a proven fact and press forward with new research approaches which are not obvious at this time" [47].   "The challenge for an undirected origin of such a cybernetic complex interacting computer system is the need to demonstrate that the rules, laws, and theories that govern electronic computing systems and information don't apply to the even more complex digital information systems that are in living organisms.   Laws of chemistry and physics, which follow exact statistical, thermodynamic, and spatial laws, are totally inadequate for generating complex functional information or those systems that process that information using prescriptive algorithmic information" [48].

It is important to realize that data generated by regular fluctuations (such as seasons or light/dark cycles) have extremely low information content, offering no explanation for life's functional information. Communication of information requires that both sender and receiver know the arbitrary protocol determined by rules, not law.   A functioning protocell would have needed formal organization, not redundant order.   Organization requires control, which requires formalism as a reality [Chapt 1].   Each protein is currently the result of the execution of a real computer program running on the genetic operating system.   How did inanimate nature write those programs and operating systems?   The genome would be useless without the processing systems needed to carry out its prescriptive instructions.

2.3 Over-Simplification of Information Requirements

"Whatever the source of life (which is scientifically unknowable), the alphabet involved with the origin of life, by the necessary conditions of information theory, had to be at least as symbolically complex as the current codon alphabet.   If intermediate alphabets existed (as some have speculated), each predecessor also would be required to be at least as complex as its successor, or Shannon's Channel Capacity [49] would be exceeded for information transfer between the probability space of alphabets with differing Shannon capacity.   Therefore, life's original alphabet must have used a coding system at least as symbolically complex as the current codon alphabet.   There has been no feasible natural explanation proposed to produce such an alphabet since chance or physicality cannot produce functional information or a coding system, let alone a system as complex as that in life" [50].   Coded information has never been observed to originate from physicality. "Due to the abstract character of function and sign systems [semiotics -- symbols and their meaning], life is not a subsystem of natural laws. This suggests that our reason is limited in respect to solving the problem of the origin of life and that we are left accepting life as an axiom... Life express[es] both function and sign systems, which indicates that it is not a subsystem of the [physical] universe, since chance and necessity cannot explain sign systems, meaning, purpose, and goals" [51].   "The reductionist approach has been to regard information as arising out of matter and energy.   Coded information systems such as DNA are regarded as accidental in terms of the origin of life and that these then led to the evolution of all life forms as a process of increasing complexity by natural selection operating on mutations on these first forms of life" [52].   "From the information perspective, the genetic system is a pre-existing operating system of unknown origin that supports the storage and execution of a wide variety of specific genetic programs (the genome applications), each program being stored in DNA.   DNA is a storage medium, not a computer, that specifies all information needed to support the growth, metabolism, parts manu-facturing, etc. for a specific organism via gene subprograms" [50].

There are many features in current life that are extremely difficult to envision as arising from a protocell.   The smallest genome (though not autonomous) found so far is in "the psyllid symbiont Carsonella ruddii, which consists of a circular chromosome of 159,662 base pairs... The genome has a high coding density (97%) with many overlapping genes and reduced gene length" [53]. "The origin and evolution of overlapping genes are still unknown" [54]. Since they are prevalent in the simplest known genome, a big question is how and why did overlapping genes arise? Recently, sub-coded information [55] and a second genetic code [56] characterizing alternative splicing have been discovered. Various transcribed RNAs are mixed and matched and spliced into mRNAs for specifying protein construction and other controls. MicroRNAs regulate large networks of genes by acting as master control switches [57]. Tiny polypeptides (with 11-32 amino acids) can function as "micro-protein" gene expression regulators [58]. Were these features required initially, or by what interactions of nature did they arise later?

Scientists are investigating "the organization of information in genomes and the functional roles that non-protein coding RNAs play in the life of the cell. The most significant challenges can be summarized by two points: a) each cell makes hundreds of thousands of different RNAs and a large percent of these are cleaved into shorter functional RNAs demonstrating that each region of the genome is likely to be multifunctional and b) the identification of the functional regions of a genome is difficult because not only are there many of them but because the functional RNAs can be created by taking sequences that are not near each other in the genome and joining them together in an RNA molecule. The order of these sequences that are joined together need not be sequential. The central mystery is what controls the temporal and coordinated expression of these RNAs" [59]. "It is very difficult to wrap your head around how big the genome is and how complicated ... It's very confusing and intimidating ... The coding parts of genes come in pieces, like beads on a string, and by splicing out different beads, or exons, after RNA copies are made, a single gene can potentially code for tens of thousands of different proteins, although the average is about five ... It's the way in which genes are switched on and off, though, that has turned out to be really mind-boggling, with layer after layer of complexity emerging" [60]. When and how did these features arise? Were any present in the first life?

2.4 Scientific Hypotheses Requiring Falsification

In addition to falsifying Shannon Capacity Theorem [49] if a proposed original information system isn't as complex as today's codon-based system, the following testable null hypotheses (proposed in peer-reviewed papers) may require falsification. No scenario should be accepted as science if it violates one or more of these unfalsified null hypotheses [60-61, 11-12].

#1   Stochastic ensembles of physical units cannot program algorithmic/cybernetic function.

#2   Dynamically-ordered sequences of individual physical units (physicality patterned by natural law causation) cannot program algorithmic/cybernetic function.

#3   Statistically weighted means (e.g., increased availability of certain units in the polymerization environment) giving rise to patterned (compressible) sequences of units cannot program algorithmic/cybernetic function.

#4   Computationally successful configurable switches cannot be set by chance, necessity, or any combination of the two, even over large periods of time.

#5   Self-ordering phenomena cannot generate cybernetic organization.

#6   Randomness cannot generate cybernetic organization.

#7   PI (prescriptive information [12]) cannot be generated from/by the chance and necessity of inanimate physicodynamics.

#8   PI cannot be generated independent of formal choice contingency.

#9   Formal algorithmic optimization, and the conceptual organization that results, cannot be generated independent of PI.

#10 Physicodynamics cannot spontaneously traverse The Cybernetic Cut [11]: physicodynamics alone cannot organize itself into formally functional systems requiring algorithmic optimization computational halting, and circuit integration.

3. Could a Protocell Live and Reproduce Without a "Genome?"

Assuming that the problems highlighted in the previous sections can be overcome (including falsifications of 2.4), this section will discuss the key topic of this essay. The protocell will be assumed to have an appropriate boundary (membrane, microtubule, etc.) that separates the "living" protocell from its environment. This section will highlight what would be required of a "proto-genome," without regard as to whether such a "genome" is feasible (not operationally falsified). "There seems to be little general agreement as to how the molecular apparatus needed to implement genetics within a cell could have come about. In fact, there seems to be nothing but puzzlement on such questions with virtually no chemically founded suggestions being made at all" [63]. We will be examining the functional requirements of the proto-genome, as opposed to hypothetical implementations. A proto-genome may have little resemblance to today's DNA-based genome since it will be assumed that life's origin didn't involve DNA. Consequently, we will be attempting to examine life as we don't know it, an exercise that should always be accompanied by healthy scientific skepticism.

It is important to realize that John von Neumann proposed and proved the requirements for a self-replicating automaton long before the discovery of DNA's information [64]. A self-reproducing system must contain the necessary components of any computer system, as well as the program for its own construction with the hardware needed to accomplish that construction.

3.1 Replication Requirements

A mechanism is needed to divide the protocell into two approximately equal daughters with each daughter being capable of growth and eventual division for exponential population potential. The "proto-genome" with its processing system must replicate itself, along with all cellular controls (functional information and senders/receivers/processors) into each daughter. Unless the "proto-genome" has replisome capabilities included in the "proto-genome" rather than a separate enzyme, the self-contained capability is required to duplicate all other needed components for "life" with high fidelity. Each daughter also needs a replicated (or split) cell boundary.

Science knows that the current replication hardware and software requires all the components to be fully functional for replication to occur at all. All known errors during replication result in a decrease of both Shannon and functional information [65], usually producing a cell that is no longer able to reproduce. Reliable replication is fundamental to life, a characteristic lacking in composomes [44].

3.2 Control Requirements

Controlled chemical metabolic networks are needed that can selectively admit "fuel" (redox, heat, photons, etc.) into the cell and process the "fuel" to harness the energy for growth, reproduction, manufacturing of needed components that can't migrate in, and other useful work. Both sender and receiver of the each control signal are needed, along with knowledge of the protocol rules for correct communication.   The manufacturing control for needed cellular components would probably require enzymatic functionality for polymerization, along with producing homochiral components.   In addition, control is required for cell division.   Without control, organization (as opposed to self-ordering) is impossible, and func-tionality would disintegrate, with "tar" a likely result.

Cellular control is a cybernetic process, so all of the requirements of the first eight chapters need instantiation into the protocell.   While the proto-genome may contain the control instructions, those instruc-tions must be read by other components (unless the proto-genome has expanded capabilities so that it can read itself), and communicated reliably (using "agreed upon" arbitrary protocols between sender and receiver, source and destination) to the components effecting the control operations.   This is not an easily-dismissed prerequisite since control in known life is critical to make the chemical components "alive."   In addition, mere physicodynamic constraints cannot generate formal biological controls [66].

3.3   Evolvability Requirements

The system would have to be capable of accurate duplication, but capable of gradual changes that would permit evolution to life-as-we-know-it.   A robust information structure that can be self-maintained (including error-correction), such as in a long genetic polymer, would be required.   The feasibility of formation of such a polymer has yet to be shown with any prebiotic mixture proposed to date.   The enzyme- and template-independent 120-mer polymers recently generated in water at high temperatures [67] are non-informational homopolymers similar to those adsorbed onto montmorillonite clay surfaces [68].   The aqueous polymers are also cyclic and require some experimenter engineering to achieve 120 mer length.

The proto-genome would also need to be able to effect highly accurate duplication of the entire proto-cell, with only an occasional "error" that could produce a very similar proto-cell, still possessing all three of the requirements in section 3. The proto-genome, along with all the proto-cell components, would need to have a feasible path to eventually produce cells with the functional complexity of today's life. It does little good to speculate a "simple" initial system unless there are feasible scenarios that can traverse from the proposed initial system to life as we know it, including coded information and other features highlighted previously. For example, one could envision dipping a finger into a bottle of ink and flicking the ink toward a white sheet would eventually produce a pattern that looks like an English letter. That would not explain the formal rules and meaningful syntax of letters that you are currently observing in this book, however.

4. Conclusions

While scenarios for the first cell can be envisioned purely from physicality, a "proto-genome" introduces cybernetic aspects that can have no origination from inanimate material. In particular, organization, prescriptive information, and control require traversing The Cybernetic Cut on a one-way CS (Configurable Switch) Bridge [11] that allows traffic only from formalism to physicality. Just as formalism needs recognition as reality, it is also critical to recognize the limits of physical science, such as physics and chemistry, whose spontaneous inanimate mass/energy interaction behavior is constrained by laws, not formal controls. Initial starting constraints chosen by an experimenter become controls for an experiment, but those chosen constraints are instantiations into physicality of nonphysical formalisms.

Life is an intersection of physical science and information science. Both domains are critical for any life to exist, and each must be investigated using that domain's principles. Yet most scientists have been attempting to use physical science to explain life's information domain, a practice that has no scientific justification. Since the chemistry and physics of life are controlled by prescriptive information (not just constrained by laws), biology is really an information science, not a physical science.

References

1. Haeckel E 2004, The Evolution of Man, Translation (original 1911), p49.

2. Darwin C 1871,
   http://darwin-online.org.uk/content/frameset?viewtype=side&itemID=F1452.3&pageseq=30

3. Miller SL 1953, "Production of Amino Acids Under Possible Primitive Earth Conditions," Science: 117, p528.

4. Hayden E 2010, "Human Genome at Ten: Life is Complicated," Nature 464, 3/31/10, p664-667.

5. Stochel G, Brindell M, Macyk W, Stasicka W, & Szacilowski K 2009, Bioinorganic Photochemistry (West Sussex, England: John Wiley & Sons,), p109.

6. Miller SL & Cleaves HJ 2007, "Prebiotic Chemistry on the Primitive Earth" in Isidore Rigoutsos and Gregroy Stephanopoulos, eds., Systems Biology Volume 1: Genomics (New York: Oxford University Press), 3.

7. Encyclopaedia Britannica 1902, 10th Edition.

8. Pasteur L 1862, http://www.pasteur.fr/ip/easysite/go/03b-00000j-0e7/institut-pasteur/histoire/

9. Corning P & Kline S 2000, "Thermodynamics, Information and Life Revisited, Part I: to Be or Entropy," Systems Research, 4/7/00, p273-295.

10. Corning P 2005, Holistic Darwinism, p330.

11. Abel DL 2008, "The 'Cybernetic Cut': Progressing from Description to Prescription in Systems Theory," The Open Cybernetics and Systemics Journal (2), p252-262.

12. Abel DL 2009, "The Biosemiosis of Prescriptive Information," Semiotica, 1/4/09, p1-19.

13. Kauffman L 2007, CYBCON discusstion group 18, 9/20/07, p15.

14. Polanyi M, Quote originally at Michael Polanyi Center site, now at nostalgia.wikipedia.org/wiki/Michael_Polanyi

15. Corning P 2002, "The Re-emergence of 'Emergence': A Venerable Concept in Search of a Theory," Complexity 7(6), p18-30.

16. Ichihashi N, Matsuura T, Kita H, SunamiT, Suzuki H, & Yomo T 2010, "Constructing Partial Models of Cells," Cold Spring Harb Perspect Biol, 4/28/10, 2:a004945.

17. Abel DL 2009, "The Universal Plausibility Metric (UPM) & Principle (UPP)," Theoretical Biology and Medical Modelling, 12/3/09, 6:27.

18. Kenyon D 1969, Biochemical Predestination.

19. SETI Website, setiathome.ssl.berkeley.edu/

20. Orgel L 2008, "The Implausibility of Metabolic Cycles on the Prebiotic Earth," PLoS Biol 6(1), e18.

21. Toxvaerd S 2009, "Origin of Homochirality in Biosystems," Int. J. Mol. Sci., 3/18/09, p1290-1299.

22. Blackmond D 2010, "The Origin of Biological Homochirality," Cold Spring Harb Perspect Biol, 3/10/10, 2:a002147.

23. Lee T & Lin Y 2010, "The Origin of Life and the Crystallization of Aspartic Acid in Water," Crystal Growth & Design 10 (4), p1652-1660.

24. Chown M 2010, "Did Exploding Stars Shatter Life's Mirror?," New Scientist, 5/19/10.

25. Mansy S 2010, "Membrane Transport in Primitive Cells," Cold Spring Harb Perspect Biol, 4/21/10 doi: 10.1101/cshperspect.a002188.

26. Lane N, Allen J, & Martin W 2010, "How Did LUCA Make a Living? Chemiosmosis in the Origin of Life," BioEssays 32, p271–280.

27. Benner S, Kim H, Kim M, & Ricardo A 2010, "Planetary Organic Chemistry and the Origins of Biomolecules," Cold Spring Harb Perspect Biol, 6/1/10.

28. Johnson D & Wang L 2010, "Imprints of the Genetic Code in the Ribosome," The Scripps Research Institute, 3/18/10.

29. Yarus M 2010, "Getting Past the RNA World: The Initial Darwinian Ancestor," 4/21/10, Cold Spring Harb Perspect Biol doi: 10.1101.

30. Cheng L & Unrau P 2010, "Closing the Circle: Replicating RNA with RNA," Cold Spring Harb Perspect Biol, 6/16/10, doi: 10.1101/cshperspect.a002204.

31. Joyce G & Orgel L 1999, "Prospects for Understanding the Origin of the RNA World," in The RNAWorld, 2nd ed., p49–77.

32. Orgel L 2004, "Prebiotic Chemistry and the Origin of the RNA World," Crit.l Rev. in Biochemistry and Molecular Biology: 39, p99–123.

33. Sutherland J 2010, "Ribonucleotides," Cold Spring Harb Perspect Biol, 3/10/10, 2:a005439.

34. Engelhart A & Hud N 2010, "Primitive Genetic Polymers," Cold Spring Harb Perspect Biol, 5/12/10, doi: 10.1101/cshperspect.a002196.

35. Shapiro R 2007, "A Simpler Origin for Life," Scientific American, 2/12/07.

36. Trefil J, Morowitz H, Smith E 2009, "The Origin of Life," American Scientist 97(2), 3-4/09, p206-210.

37. Hazen R & Sverjensky D 2010, "Mineral Surfaces, Geochemical Complexities, and the Origins of Life," Cold Spring Harb Perspect Biol, 4/14/10, 2:a002162.

38. Schrum J, Zhu T, & Szostak J 2010, 5/19/10, "The Origins of Cellular Life," Cold Spring Harb Perspect Biol doi: 10.1101/cshperspect.a002212.

39. Deamer D & Weber A 2010, "Bioenergetics and Life's Origins," Cold Spring Harb Perspect Biol, 1/13/10, 2:a004929.

40. Luisi, P, Ferri F, et al. 2006, "Approaches to semi-synthetic minimal cells: a review," Naturwissenschaften 93, p1-13.

41. Russell M 2009, as reported by John Whitfield in "Nascence man," Nature 459, 5/21/09.

42. Segre D, Ben-Eli D, & Lancet D 2000, "Compositional Genomes: Prebiotic Information Transfer in Mutually Catalytic Noncovalent Assemblies," PNAS 97 (8), 4/11/00, p4112–4117.

43. Shenhav B, Oz A, & Lancet D 2007, "Coevolution of Compositional Protocells and Their Environment," Phil. Trans. R. Soc. B 362, p1813-1819.

44. Vasas V, Szathmáry E, & Santos M 2010, "Lack of Evolvability in Self-sustaining Autocatalytic Networks Constrains Metabolism-first Scenarios for the Origin of Life," PNAS, 2/11/10, www.pnas.org/content/107/4/1470.

45. Lad C, Williams N, & Wolfenden R 2003, "The Rate of Yydrolysis of Phosphomonoester Dianions and the Exceptional Catalytic Proficiencies of Protein and Inositol Phosphatases," PNAS: 100 (10), 5/13/03, p5607-5610.

46. Hoffmeyer J & Emmeche C 2005, "Code-Duality and the Semiotics of Nature," J. Biosemiotics (1), p37-64.

47. Trevors JT & Abel DL 2004, "Chance and Necessity Do Not Explain the Origin of Life," Cell Biology International: 28, p729-739.

48. Johnson DE 2009/2010, Probability's Nature and Nature's Probability: A Call to Scientific Integrity, p48.

49. Shannon C 1948, "A Mathematical Theory of Communication," Bell System Technical Journal: 27, July & October, p379-423 & 623-656.

50. Johnson DE 2010, Programming of Life, p37&49.

51. Voie A 2006, "Biological Function and the Genetic Code are Interdependent," Chaos, Solitons and Fractals: 28(4), p1000-1004.

52. McIntosh A 2009, "Entropy, Free Energy and Information in Living Systems," International Journal of Design & Nature and Ecodynamics 4 (4), p351-385.

53. Nakabachi A, Yamashita A, Toh H, Ishikawa H, Dunbar H, Moran N, & Hattori M 2006, "The 160-Kilobase Genome of the Bacterial Endosymbiont Carsonella," Science: 314 (5797), 10/13/06, p267.

54. Veeramachaneni V, Makalowski1 W, Galdzicki M, Sood R, & Makalowska I 2004, "Mammalian Overlapping Genes: The Comparative Perspective," Genome Res.: 14, p280-286.

55. Cannarozzi G, Schraudolph N, Faty M, von Rohr P, Friberg M, Roth A, Gonnet P, Gonnet G, & Barral Y 2010, "Role for Codon Order in Translation Dynamics," Cell 141, 4/16/10, p355-367.

56. Barash Y, Calarco J, Gao W, Pan Q, Wang X, Shai O, Blencowe B, & Frey B 2010, "Deciphering the Splicing Code," Nature, 5/6/10, p53-9.

57. Lieberman J 2010, "Master of the Cell," The Scientist, 4/1/10, p42.

58. Kondo T, Plaza S, Zanet S, Benrabah E, Valenti P, Hashimoto Y, Kobayashi S, Payre F, Kageyama Y 2010, "Small Peptides Switch the Transcriptional Activity of Shavenbaby During Drosophila Embryogenesis," Science 329, 7/16/10, p336-339.

59. Gingeras T, www.desdeelexilio.com/2010/06/28/epigenetica- entrevista-a-thomas-gingeras/

60. Le Page M 2010, "Genome at 10: A Dizzying Journey into Complexity," New Scientist, 6/16/10.

61. Abel D & Trevors J 2005, "Three subsets of sequence complexity and their relevance to biopolymeric information," Theoretical Biology and Medical Modelling, 8/11/05, 2:29.

62. Abel D & Trevors J 2006, "Self-Organization vs Self-Ordering Events in Life-Origin Models," Physics of Life Reviews (3), p211-228.

63. Yana K, Fanga G, Bhardwaja N, Alexandera R, & Gersteinb M 2010, "Comparing Genomes to Computer Operating Systems in Terms of the Topology and Evolution of Their Regulatory Control Networks," www.pnas.org/cgi/doi/10.1073/pnas.0914771107.

64. von Neumann J, Theory of Self-Reproducing Automata, 1966.

65. Abel DL 2009, "The GS (genetic selection) Principle," Frontiers in Bioscience (14), 1/1/09, p2959-2969.

66. Abel, DL 2010, "Constraints vs. Controls," Open Cybernetics and Systemics Journal, 4, p14-27.

67. Costanzo G, Pino S, Ciciriello F, Di Mauro E 2009, "RNA: Processing and Catalysis: Generation of Long RNA Chains in Water," J. Biol. Chem., 284, p33206-33216.

68. Ferris JP 2002, "Montmorillonite Catalysis of 30-50 Mer Oligonucleotides: Laboratory Demonstration of Potential Steps in the Origin of the RNA World," Origins of Life and Evolution

The First Gene, David L. Abel, Editor, 2011, pp 305-324   ISBN: 978-0-9657988-9-1

Chapter 11: The Universal Plausibility Metric (UPM)

& Principle (UPP)*

David L. Abel

Department of ProtoBioCybernetics/ProtoBioSemiotics
Director, The Gene Emergence Project
The Origin-of-Life Science Foundation, Inc.
113 Hedgewood Dr. Greenbelt, MD 20770-1610 USA

Abstract: Mere possibility is not an adequate basis for asserting scientific plausibility. A precisely defined universal bound is needed beyond which the assertion of *plausibility,* particularly in life-origin models, can be considered operationally falsified. But can something so seemingly relative and subjective as plausibility ever be quantified? Amazingly, the answer is, "Yes."   A method of objectively measuring the plausibility of any chance hypothesis (The Universal Plausibility Metric [UPM]) is presented. A numerical inequality is also provided whereby any chance hypothesis can be definitively falsified when its UPM metric of ξ is < 1 (The Universal Plausibility Principle [UPP]). Both UPM and UPP pre-exist and are independent of any experimental design and data set.   No low-probability hypothetical plausibility assertion should survive peer-review without subjection to the UPP inequality standard of formal falsification ($\xi < 1$).

Correspondence/Reprint request: Dr. David L. Abel, Department of ProtoBioCybernetics/ProtoBioSemiotics
The Origin-of-Life Science Foundation, Inc., 113 Hedgewood Dr. Greenbelt, MD 20770-1610 USA   E-mail: life@us.net

*This chapter is reprinted with permission from Abel, D.L. 2009, The Universal Plausibility Metric (UPM) & Principle (UPP), Theor Biol Med Model, 6, (1) 27.

Introduction: The seemingly subjective liquidity of "plausibility"

Are there any *objective* standards that could be applied to evaluate the seemingly subjective notion of *plausibility*? Can something so psychologically relative as plausibility ever be quantified?

Our skepticism about defining a precise, objective Universal Plausibility Metric (UPM) stems from a healthy realization of our finiteness [1], subjectivity [2], presuppositional biases [3, 4], and epistemological problem [5]. We are rightly wary of absolutism.   The very nature of probability theory emphasizes gray-scales more than the black and white extremes of p = 0 or 1.0. Our problem is that extremely low probabilities can only asymptotically approach impossibility. An extremely unlikely event's probability always remains at least slightly > 0.   No matter how many orders of magnitude is the negative exponent of an event's probability, that event or scenario technically cannot be considered impossible. Not even a Universal Probability Bound [6-8] seems to establish absolute theoretical impossibility. The fanatical pursuit of absoluteness by finite subjective knowers is considered counterproductive in post modern science. Open-mindedness to all possibilities is encouraged [9].

But at some point our reluctance to exclude *any* possibility becomes stultifying to operational science [10]. Falsification is critical to narrowing down the list of serious possibilities [11]. Almost all hypotheses are possible. Only a few of them wind up being helpful and scientifically productive. Just because a hypothesis is possible should not grant that hypothesis scientific respectability. More attention to the concept of "infeasibility" has been suggested [12]. Millions of dollars in astrobiology grant money have been wasted on scenarios that are possible, but plausibly bankrupt. The question for scientific methodology should *not* be, "Is this scenario possible?" The question should be, "Is this possibility a *plausible* scientific hypothesis?" One chance in 10^{200} is theoretically possible, but given maximum cosmic probabilistic resources, such a possibility is hardly plausible. With funding resources rapidly drying up, science needs a foundational principle by which to falsify a myriad of theoretical possibilities that are not worthy of serious scientific consideration and modeling.

Proving a theory is considered technically unachievable [11].   Few bench scientists realize that falsification has also been shown by philosophers of science to be at best technically suspect [13].   Nevertheless, operational science has no choice but to proceed primarily by a process of elimination through practical falsification of competing models and theories.

Which model or theory best corresponds to the data? [14 (pg. 32-98)] [8]. Which model or theory best predicts future interactions? Answering these questions is made easier by eliminating implausible possibilities from the list of theoretical possibilities. Great care must be taken at this point, especially given the many non-intuitive aspects of scientifically addressable reality. But operational science must proceed on the basis of best-thus-far tentative knowledge. The human epistemological problem is quite real. But we cannot allow it to paralyze scientific inquiry.

If it is true that we cannot *know* anything for certain, then we have all the more reason to proceed on the basis of the greatest "*plausibility* of belief" [15-19]. If human mental constructions cannot be equated with objective reality, we are all the more justified in pursuing the greatest likelihood of correspondence of our knowledge to the object of that knowledge—presumed ontological being itself. Can we prove that objectivity exists outside of our minds? No. Does that establish that objectivity does not exist outside of our minds? No again. Science makes its best progress based on the axioms that 1) an objective reality independent of our minds does exist, and 2) scientists' collective knowledge can progressively correspond to that objective reality. The human epistemological problem is kept in its proper place through a) double-blind studies, b) groups of independent investigators all repeating the same experiment, c) prediction fulfillments, and d) the application of pristine logic (taking linguistic fuzziness into account), and e) the competition of various human ideas for best correspondence to repeated independent observations.

The physical law equations and the deductive system of mathematical rules that govern the manipulations of those equations are all formally absolute. But the axioms from which formal logic theory flows, and the decision of when to consider mathematical equations universal "laws" are not absolute. Acceptance of mathematical axioms is hypothetico-deductively relative. Acceptance of physical laws is inductively relative. The *pursuit* of correspondence between presumed objective reality and our knowledge of objective reality is laudable in science. But not even the axioms of mathematics or the laws of physics can be viewed as absolute. Science of necessity proceeds tentatively on the basis of best-thus-far subjective knowledge. At some admittedly relative point, the scientific community agrees by consensus to declare certain formal equations to be reliable descriptors and predictors of future physicodynamic interactions. Eventually the correspondence level between our knowledge and our repeated observations of presumed objective reality is considered *adequate* to make a tentative *commitment* to the veracity of an axiom or universal law until they are proven otherwise.

The same standard should apply in falsifying ridiculously implausible life-origin assertions. Combinatorial imaginings and hypothetical scenarios can be endlessly argued simply on the grounds that they are theoretically possible. But there is a point beyond which arguing the plausibility of an absurdly low probability becomes *operationally* counterproductive. That point can actually be quantified for universal application to all fields of science, not just astrobiology. Quantification of a Universal *Plausibility* Metric (UPM) and application of the Universal *Plausibility* Principle (UPP) inequality test to that specific UPM provides for definitive, unequivocal falsification of scientifically unhelpful and functionally useless hypotheses. When the UPP is violated, declaring falsification of that highly implausible notion is just as justified as the firm commitment we make to any mathematical axiom or physical "law" of motion.

1. Universal Probability Bounds

"Statistical prohibitiveness" in probability theory and the physical sciences has remained a nebulous concept for far too long. The importance of probabilistic resources as a context for consideration of extremely low probabilities has been previously emphasized [20 (pg. 13-17)] [6-8, 21]. Statistical prohibitiveness cannot be established by an exceedingly low probability alone [6]. Rejection regions and probability bounds need to be established independent of (preferably prior to) experimentation in any experimental design. But the setting of these zones and bounds is all too relative and variable from one experimental design to the next. In the end, however, probability is not the critical issue. The plausibility of hypotheses is the real issue. Even more important is the question of whether we can ever operationally falsify a preposterous but theoretically possible hypothesis.

The Universal Probability Bound (UPB) [6, 7] quantifies the maximum cosmic probabilistic resources (Ω, upper case omega) as the context of evaluation of any extremely low probability event. Ω corresponds to the maximum number of possible probabilistic trials (quantum transitions or physicochemical interactions) that could have occurred in cosmic history. The value of Ω is calculated by taking the product of three factors:

1) The number of seconds that have elapsed since the Big Bang (10^{17}) assumes a cosmic age of around 14 billion years. 60 sec/min X 60 min/hr X 24 hrs/day X 365 days per year X 14 billion years = 4.4 x 10^{17} seconds since the Big Bang.

2) The number of possible quantum events/transitions per second is derived from the amount of time it takes for light to traverse the minimum unit of distance. The minimum unit of distance (a quantum of space) is Planck length (10^{-33} centimeters). The minimum amount of

time required for light to traverse the Plank length is Plank time (10^{-43} seconds) [6, 7, 8, pg 215-217]. Thus a maximum of 10^{43} quantum transitions can take place per second. Since 10^{17} seconds have elapsed since the Big Bang, the number of possible quantum transitions since the Big Bang would be $10^{43} \times 10^{17} = 10^{60}$.

3) Sir Arthur Eddington's estimate of the number of protons, neutrons and electrons in the observable cosmos (10^{80}) [22] has been widely respected throughout the scientific literature for decades.

Some estimates of the total number of elementary particles have been slightly higher. The Universe is 95 billion light years (30 gigaparsecs) across. We can convert this to cubic centimeters using the equation for the volume of a sphere (5×10^{86} cc). If we multiply this times 500 particles (100 neutrinos and 400 photons) per cc, we would get 2.5×10^{89} elementary particles in the visible universe.

A Universal [6, 7]Probability Bound could therefore be calculated by the product of these three factors: $10^{17} \times 10^{43} \times 10^{80} = 10^{140}$

If the highest estimate of the number of elementary particles in the Universe is used (e.g., 10^{89}), the UPB would be 10^{149}.

The UPB's discussed above are *the highest calculated universal probability bounds ever published by many orders of magnitude* [7, 8]. They are the most permissive of (favorable to) extremely low-probability plausibility assertions in print [6] [8 (pg. 216-217)]. All other proposed metrics of probabilistic resources are far less permissive of low-probability chance-hypothesis plausibility assertions. Emile Borel's limit of cosmic probabilistic resources was only 10^{50} [23 (pg. 28-30)]. Borel based this probability bound in part on the product of the number of observable stars (10^{9}) times the number of possible human observations that could be made on those stars (10^{20}). Physicist Bret Van de Sande at the University of Pittsburgh calculates a UPB of 2.6×10^{92} [8, 24]. Cryptographers tend to use the figure of 10^{94} computational steps as the resource limit to any cryptosystem's decryption [25]. MIT's Seth Lloyd has calculated that the universe could not have performed more than 10^{120} bit operations in its history [26].
Here we must point out that a discussion of the number of cybernetic or cryptographic "operations" is totally inappropriate in determining a prebiotic UPB. Probabilistic combinatorics has nothing to do with "operations." Opera-

tions involve choice contingency [27-29]. Bits are "Yes/No" question *opportunities* [30 (pg. 66)], each of which could potentially reduce the total number of combinatorial possibilities (2^{NH} possible biopolymers: see Appendix 1) by half. But of course asking the right question and getting an answer is not a spontaneous physicochemical phenomenon describable by mere probabilistic uncertainty measures [31-33]. Any binary "operation" involves a bona fide decision node [34-36]. An operation is a *formal choice-based function.* Shannon uncertainty measures do not apply to specific choices [37-39]. Bits measure only the number of non-distinct, generic, *potential* binary choices, not actual specific choices [37]. Inanimate nature cannot ask questions, get answers, and exercise choice contingency at decision nodes in response to those answers. Inanimate nature cannot optimize algorithms, compute, pursue formal function, or program configurable switches to achieve integration and shortcuts to formal utility [28]. Cybernetic operations therefore have no bearing whatever in determining universal probability bounds for chance hypotheses.

Agreement on a sensible UPB in advance of (or at least totally independent of) any specific hypothesis, suggested scenario, or theory of mechanism is critical to experimental design. No known empirical or rational considerations exist to preclude acceptance of the above UPB. The only exceptions in print seem to come from investigators who argue that the above UPB is *too permissive* of the chance hypothesis [8, 12]. Faddish acceptance prevails of hypothetical scenarios of extremely low probability simply because they are in vogue and are theoretically possible. Not only a UPB is needed, but a fixed universal mathematical standard of *plausibility* is needed. This is especially true for complex hypothetical scenarios involving joint and/or conditional probabilities. Many imaginative hypothetical scenarios propose constellations of highly cooperative events that are theorized to self-organize into holistic formal schemes. Whether joint, conditional or independent, multiple probabilities must be factored into an overall *plausibility* metric. In addition, a universal plausibility bound is needed to eliminate overly imaginative fantasies from consideration for the best inference to causation.

2. The Universal Plausibility Metric (UPM)

To be able to definitively falsify ridiculously implausible hypotheses, we need first a Universal *Plausibility* Metric (UPM) to assign a numerical plausibility value to each proposed hypothetical scenario. Second, a Universal *Plausibility* Principle (UPP) inequality is needed as plausibility bound of this measurement for falsification evaluation. We need a cut-off point beyond which no extremely low probability scenario can be considered a "*scientifical-*

ly respectable" possibility. What is needed more than a probability bound is a *plausibility* bound. Any "possibility" that exceeds the ability of its probabilistic resources to generate should immediately be considered a *"functional non-possibility,"* and therefore an implausible scenario. While it may not be a theoretically absolute impossibility, if it exceeds its probabilistic resources, it is a gross understatement to declare that such a proposed scenario is simply not worth the expenditure of serious scientific consideration, pursuit, and resources. Every field of scientific investigation, not just biophysics and life-origin science, needs the application of the same independent test of credibility to judge the plausibility of its hypothetical events and scenarios. The application of this standard should be an integral component of the scientific method itself for all fields of scientific inquiry.

To arrive at the UPM, we begin with the maximum available probabilistic resources discussed above (Ω, upper case Omega) [6, 7]. But Ω could be considered from a quantum or a classical molecular/chemical perspective. Thus this paper proposes that the Ω quantification be broken down first according to the Level (L) or perspective of physicodynamic analysis ($^L\Omega$), where the perspective at the quantum level is represented by the superscript "q" ($^q\Omega$) and the perspective at the classical level is represented by "c" ($^c\Omega$). Each represents the maximum probabilistic resources available at each level of physical activity being evaluated, with the total number of quantum transitions being much larger than the total number of "ordinary" chemical reactions since the Big Bang.

Second, the maximum probabilistic resources $^L\Omega$ ($^q\Omega$ for the quantum level and $^c\Omega$ for classical molecular/chemical level) can be broken down even further according to the astronomical subset being addressed using the general subscript "A" for Astronomical: $^L\Omega_A$ (representing both $^q\Omega_A$ and $^c\Omega_A$). The maximum probabilistic resources can then be measured for each of the four different specific environments of each $^L\Omega$, where the general subscript A is specifically enumerated with "u" for universe, "g" for our galaxy, "s" for our solar system, and "e" for earth:

Universe $\quad ^L\Omega_u$
Galaxy $\quad\quad ^L\Omega_g$
Solar System $\; ^L\Omega_s$
Earth $\quad\quad\; ^L\Omega_e$ ($^L\Omega_e$ excludes meteorite and panspermia inoculations)

To include meteorite and panspermia inoculations in the earth metrics, we use the Solar System metrics $^L\Omega_s$ ($^q\Omega_s$ and $^c\Omega_s$).

As examples, for quantification of the maximum probabilistic resources at the quantum level for the astronomical subset of our galactic phase space, we would use the $^q\Omega_g$ metric. For quantification of the maximum probabilistic resources at the ordinary classical molecular/chemical reaction level in our solar system, we would use the $^c\Omega_s$ metric.

The most permissive UPM possible would employ the probabilistic resources symbolized by $^q\Omega_u$ where both the quantum level perspective and the entire universe are considered.

The sub division between the $^L\Omega_A$ for the quantum perspective (quantified by $^q\Omega_A$) and that for the classical molecular/chemical perspective (quantified by $^c\Omega_A$), however, is often not as clear and precise as we might wish. Crossovers frequently occur. This is particularly true where quantum events have direct bearing on "ordinary" chemical reactions in the "everyday" classical world. If we are going to err in evaluating the plausibility of any hypothetical scenario, let us err in favor of maximizing the probabilistic resources of $^L\Omega_A$. In cases where quantum factors seem to directly affect chemical reactions, we would want to use the four quantum level metrics of $^q\Omega_A$ ($^q\Omega_u$ $^q\Omega_g$, $^q\Omega_s$ and $^q\Omega_e$) to preserve the plausibility of the lowest-probability explanations.

3. Quantification of the Universal *Plausibility* Metric (UPM)

We keep italicizing plausibility because of prior experience with readers confusing the UPM with a probability measure. The UPM is *not* a probability measure. It is a plausibility measure. The computed Universal Plausibility Metric (UPM) objectively quantifies the level of plausibility of any chance hypothesis or theory. The UPM employs the symbol ξ (Xi, pronounced *zai* in American English, s*ai* in UK English, *ksi* in modern Greek) to represent the computed UPM according to the following equation:

$$\xi = \frac{f\,^L\Omega_A}{\omega}$$

Equation 1

where f represents the number of *functional* objects/events/scenarios that are known to occur out of all possible combinations (lower case omega, ω) (e.g., the number [f] of functional protein family members of varying sequence known to occur out of sequence space [ω]), and $^L\Omega_A$ (upper case Omega, Ω) represents the total probabilistic resources for any particular probabilistic context. The "L" superscript context of Ω describes which perspective of analysis, whether quantum (q) or a classical (c), and the "A" subscript context of Ω enumerates which subset of astronomical phase space is being evaluated: "u" for universe, "g" for our galaxy, "s" for our solar system, and "e" for earth. Note that the basic generic UPM (ξ) equation's form remains constant despite changes in the variables of levels of perspective (L: whether q or c) and astronomic subsets (A: whether u, g, s, or e).

The calculations of probabilistic resources in $^L\Omega_A$ can be found in Appendix 2. Note that the upper and lower case omega symbols used in this equation are *case sensitive* and each represents a completely different phase space.

The UPM from both the quantum ($^q\Omega_A$) and classical molecular/chemical ($^c\Omega_A$) perspectives/levels can be quantified by Equation 1. This equation incorporates *the number of possible transitions or physical interactions that could have occurred since the Big Bang*. Maximum quantum-perspective probabilistic resources $^q\Omega_u$ were enumerated above in the discussion of a UPB [6, 7] [8 (pg. 215-217)]. Here we use basically the same approach with slight modifications to the factored probabilistic resources that comprise Ω.

Let us address the quantum level perspective (q) first for the entire universe (u) followed by three astronomical subsets: our galaxy (g), our solar system (s) and earth (e).

Since approximately 10^{17} seconds have elapsed since the Big Bang, we factor that total time into the following calculations of quantum perspective probabilistic resource measures. Note that the difference between the age of the earth and the age of the cosmos is only a factor of 3. A factor of 3 is rather negligible at the high order of magnitude of 10^{17} seconds since the Big Bang (versus age of the earth). Thus, 10^{17} seconds is used for all three astronomical subsets:

$^q\Omega_u$ = Universe = 10^{43} trans/sec X 10^{17} secs X 10^{80} protons, neutrons & electrons = 10^{140}

$^q\Omega_g$ = Galaxy = 10^{43} X 10^{17} X 10^{67} = 10^{127}

$^q\Omega_s$ = Solar System = 10^{43} X 10^{17} X 10^{57} = 10^{117}

$^q\Omega_e$ = Earth = 10^{43} X 10^{17} X 10^{42} = 10^{102}

These above limits of probabilistic resources exist within the only known universe that we can repeatedly observe—the only universe that is scientifically addressable. Wild metaphysical claims of an infinite number of cosmoses may be fine for cosmological imagination, religious belief, or superstition. But such conjecturing has no place in hard science. Such claims cannot be empirically investigated, and they certainly cannot be falsified. They violate Ockham's (Occam's) Razor [40]. No prediction fulfillments are realizable.

They are therefore nothing more than blind beliefs that are totally inappropriate in peer-reviewed scientific literature. Such cosmological conjectures are far closer to metaphysical or philosophic enterprises than they are to bench science.

From a more classical perspective at the level of ordinary molecular/chemical reactions, we will again provide metrics first for the entire universe (u) followed by three astronomical subsets, our galaxy (g), our solar system (s) and earth (e).

The classical molecular/chemical perspective makes two primary changes from the quantum perspective. With the classical perspective, *the number of atoms* rather than the number of protons, neutrons and electrons is used. In addition, *the total number of classical chemical reactions that could have taken place since the Big Bang* is used rather than transitions related to cubic light-Planck's. The shortest time any transition requires before a chemical reaction can take place is 10 femtoseconds [41-46]. A femtosecond is 10^{-15} seconds. Complete chemical reactions, however, rarely take place faster than the picosecond range (10^{-12} secs). Most biochemical reactions, even with highly sophisticated enzymatic catalysis, take place no faster than the nano (10^{-9}) and usually the micro (10^{-6}) range. To be exceedingly generous (perhaps overly permissive of the capabilities of the chance hypothesis), we shall use 100 femtoseconds as the shortest chemical reaction time. 100 femtoseconds is 10^{-13} seconds. Thus 10^{13} simple and fastest chemical reactions could conceivably take place per second in the best of theoretical pipe-dream scenarios. The four $^c\Omega_A$ measures are as follows:

$$^c\Omega_u = \text{Universe} = 10^{13} \text{ reactions/sec X } 10^{17} \text{ secs X } 10^{78} \text{ atoms} = 10^{108}$$

$$^c\Omega_g = \text{Galaxy} = 10^{13} \text{ X } 10^{17} \text{ X } 10^{66} = 10^{96}$$

$$^c\Omega_s = \text{Solar System} = 10^{13} \text{ X } 10^{17} \text{ X } 10^{55} = 10^{85}$$

$$^c\Omega_e = \text{Earth} = 10^{13} \text{ X } 10^{17} \text{ X } 10^{40} = 10^{70}$$

Remember that $^L\Omega_e$ excludes meteorite and panspermia inoculations. To include meteorite and panspermia inoculations, we use the metric for our solar system $^c\Omega_s$.

These maximum metrics of the limit of probabilistic resources are based on the best-thus-far estimates of a large body of collective scientific investigations. We can expect slight variations up or down of our best guesses of the number of elementary particles in the universe, for example. But the basic formula presented as the Universal Plausibility Metric (PM) will never change. The Universal *Plausibility* Principle (UPP) inequality presented below is also immutable and worthy of law-like status. It affords the ability to objectively once and for all falsify not just highly improbable, but ridiculously implausible scenarios. Slight adjustments to the factors that contribute to the value of each $^L\Omega_A$ are straightforward and easy for the scientific community to update through time.

Most chemical reactions take longer by many orders of magnitude than what these exceedingly liberal maximum probabilistic resources allow. Biochemical reactions can take years to occur in the absence of highly sophisticated protein enzymes not present in a prebiotic environment. Even humanly engineered ribozymes rarely catalyze reactions by an enhancement rate of more than 10^5[47-51]. Thus the use of the fastest rate known for any complete chemical reaction (100 femtoseconds) seems to be the most liberal/forgiving probability bound that could possibly be incorporated into the classical chemical probabilistic resource perspective $^c\Omega_A$. For this reason, we should be all the more ruthless in applying the UPP test of falsification presented below to seemingly "far-out" metaphysical hypotheses that have no place in responsible science.

4. Falsification using The Universal Plausibility Principle (UPP)

The Universal Plausibility Principle (UPP) states that *definitive operational falsification* of any chance hypothesis is provided by the inequality of:

$$\xi < 1 \hspace{4cm} \text{Inequality 1}$$

This definitive operational falsification holds for hypotheses, theories, models, or scenarios at any level of perspective (q or c) and for any astronomical subset (u, g, s, and e). The UPP inequality's falsification is valid whether the hypothesized event is singular or compound, independent or conditional. Great care must be taken, however, to eliminate errors in the calculation of complex probabilities. Every aspect of the hypothesized scenario must have its probabilistic components factored into the one probability (p) that is used in the UPM (See equation 2 below). Many such combinatorial possibilities are joint or conditional. It is not sufficient to factor only the probabilities of each reactant's formation, for example, while omitting the probabilistic aspects of each reactant being presented at the same place and time, becoming available in the required reaction order, or being able to react at all (activated vs. not activated). Other factors must be included in the calculation of probabilities: optical isomers, non-peptide bond formation, many non-biological amino acids that also react [8]. The exact calculation of such probabilities is often not straightforward. But in many cases it becomes readily apparent that whatever the exact multi-factored calculation, the probability "p" of the entire scenario easily crosses the plausibility bound provided by the UPP inequality. This provides a definitive objective standard of falsification. When $\xi < 1$, immediately the notion should be considered "not a *scientifically plausible* possibility."   A ξ value < 1 should serve as an unequivocal operational falsification of that hypothesis. The hypothetical scenario or theory generating that ξ metric should be excluded from the differential list of possible causes. The hypothetical notion should be declared to be outside the bounds of scientific respectability. It should be flatly rejected as the equivalent of superstition.

f / ω in Equation 1 is in effect the probability of a particular *functional* event or object occurring out of all possible combinations. Take for example an RNA-World model. 23 different functional ribozymes in the same family might arise out of 10^{15} stochastic ensembles of 50-mer RNAs. This would reduce to a probability p of roughly 10^{-14} of getting a stochastic ensemble that manifested some degree of that ribozyme family's function.

Thus f / ω in Equation 1 reduces to the equivalent of a probability p:

$$UPM = \xi = p\, ^{c}\Omega_e \qquad\qquad \text{Equation 2}$$

where "p" represents an extremely low probability of any chance hypothesis that is *asserted to be plausible* given $^{L}\Omega_A$ probabilistic resources, in this particular case $^{c}\Omega_e$ probabilistic resources.

As examples of attempts to falsify, suppose we have three different chance hypotheses, each with its own low probability (p), all being evaluated from the quantum perspective at the astronomical level of the entire universe ($^{q}\Omega_u$). Given the three different probabilities (p) provided below, the applied UPP inequality for each $\xi = p\, ^{q}\Omega_u$ of each hypothetical scenario would establish definitive operational falsification for one of these three hypothetical scenarios, and fail to falsify two others:

$p = 10^{-140} \times 10^{140} = 10^0 = 1$ giving a ξ which is NOT < 1, so NOT falsified

$p = 10^{-130} \times 10^{140} = 10^{10}$ giving a $\xi > 1$, so NOT falsified

$p = 3.7 \times 10^{-151} \times 10^{140} = 3.7 \times 10^{-11}$ giving a $\xi < 1$, so Falsified

Let us quantify an example of the use of the UPM and UPP to attempt falsification of a chance hypothetical scenario:

Suppose 10^3 *biofunctional* polymeric sequences of monomers (f) exist out of 10^{17} possible sequences in sequence space (ω) all of the same number (N) of monomers. That would correspond to one chance in 10^{14} of getting a functional sequence by chance ($p = 10^3/10^{17} = 1/10^{14} = 10^{-14}$ of getting a functional sequence). If we were measuring the UPM from the perspective of a classical chemical view on earth over the last 5 billion years ($^c\Omega_e = 10^{70}$), we would use the following UPM equation (#1 above) with substituted values:

$$\xi = \frac{f \, ^c\Omega_e}{\omega} = \frac{10^3 \, x \, 10^{70}}{10^{17}}$$

$$\xi = \frac{10^{73}}{10^{17}} = 10^{56}$$

Since $\xi > 1$, this particular chance hypothesis is shown unequivocally to be plausible and worthy of further scientific investigation.

As one of the reviewers of this manuscript has pointed out, however, we might find the sequence space ω, and therefore the probability space f/ω, to be radically different for abiogenesis than for general physico-chemical reactions. The sequence space ω must include factors such as heterochirality, unwanted non-peptide-bond formation, and the large number of non-biological amino acids present in any prebiotic environment [8, 12]. This greatly increases ω, and would tend to substantially reduce the probability p of naturalistic abiogenesis. Spontaneously biofunctional stochastic ensemble formation was found to be only 1 in 10^{64} when TEM-1 β-lactamase's working domain of around 150 amino acids was used as a model [52]. Function was related to the hydropathic signature necessary for proper folding (tertiary structure). The ability to confer any relative degree of beta-lactam penicillin-like antibiotic resistance to bacteria was considered to define "biofunctional" in this study. Axe further measured the probability of a random 150-residue primary structure producing *any* short protein, despite many allowable monomeric substitutions, to be 10^{-74}. This probability is an example of a scientifically determined p that should be incorporated into any determination of the UPM in abiogenesis models.

5. Don't multiverse models undermine The UPP?

Multiverse models imagine that our universe is only one of perhaps countless parallel universes [53-55]. Appeals to the Multiverse worldview are becoming more popular in life-origin research as the statistical prohibitiveness of spontaneous generation becomes more incontrovertible in a finite universe [56-58]. The term "notion," however, is more appropriate to refer to multiverse speculation than "theory." The idea of multiple parallel universes cannot legitimately qualify as a testable scientific hypothesis, let alone a mature theory. Entertaining multiverse "thought experiments" almost immediately takes us beyond the domain of responsible science into the realm of pure metaphysical belief and conjecture. The dogma is literally "beyond physics and astronomy," the very meaning of the word "metaphysical."

The notion of multiverse has no observational support, let alone repeated observations. Empirical justification is completely lacking. It has no testability: no falsification potential exists. If provides no prediction fulfillments. The non-parsimonious construct of multiverse grossly violates the principle of Ockham's (Occam's) Razor [40]. No logical inference seems apparent to support the strained belief other than a perceived need to rationalize what we know is statistically prohibitive in the only universe that we *do* experience. Multiverse fantasies tend to constitute a back-door fire escape for when our models hit insurmountable roadblocks in the observable cosmos. When none of the facts fit our favorite model, we conveniently create imaginary extra universes that are more accommodating. This is not science. Science is interested in falsification within the only universe that science can address. Science cannot operate within mysticism, blind belief, or superstition. A multiverse may be fine for theoretical metaphysical models. But no justification exists for inclusion of this "dream world" in the sciences of physics and astronomy.

It could be argued that multiverse notions arose only in response to the severe time and space constraints arising out of Hawking, Ellis and Penrose's singularity theorems [59-61]. Solutions in general relativity involve singularities wherein matter is compressed to a point in space and light rays originate from a curvature. These theorems place severe limits on time and space since the Big Bang. Many of the prior assumptions of limitless time and sample space in naturalistic models were eliminated by the demonstration that time and space in the cosmos are quite finite, not infinite. For instance, we only have 10^{17}-10^{18} seconds at most to work with in any responsible cosmological universe model since the Big Bang. Glansdorff makes the point, "Conjectures about emergence of life in an infinite multiverse should not confuse probability with possibility." [62]

Even if multiple physical cosmoses existed, it is a logically sound deduction that linear digital genetic instructions using a representational material symbol system (MSS) [63] cannot be programmed by the chance and/or fixed laws of physicodynamics [27-29, 32, 33, 36-39, 64, 65]. This fact is not only true of the physical universe, but would be just as true in any imagined physical multiverse. Physicality cannot generate nonphysical Prescriptive Information (PI) [29]. Physicodynamics cannot practice formalisms (The Cybernetic Cut) [27, 34]. Constraints cannot exercise formal control unless those constraints are themselves chosen to achieve formal function [28, 66]. Environmental selection cannot select at the genetic level of arbitrary symbol sequencing (e.g., the polymerization of nucleotides and codons). (The GS Principle [Genetic Selection Principle]) [36, 64]. Polymeric syntax (sequencing; primary structure) prescribes future (potential; not-yet-existent) folding and formal function of small RNAs and even DNA. Symbol systems and configurable switch-settings can only be programmed with choice contingency, not chance contingency or fixed law, if nontrivial coordination and formal organization are expected [29, 38]. The all-important determinative sequencing of monomers is completed with rigid covalent bonds before any transcription, translation, or three-dimensional folding begins. Thus, imagining multiple physical universes or infinite time does not solve the problem of the origin of *formal* (nonphysical) biocybernetics and biosemiosis using a linear digital representational symbol system. The source of Prescriptive Information (PI) [29, 35] in a metaphysically presupposed material-only world is closely related to the problem of gene emergence from physicodynamics alone. The latter hurdles remain the number-one enigmas of life-origin research [67].

The main subconscious motivation behind multiverse conjecture seems to be, "Multiverse models can do anything we want them to do to make our models work for us." We can argue Multiverse models ad infinitum because their potential is limitless. The notion of Multiverse has great appeal because it can explain everything (and therefore nothing). Multiverse models are beyond scientific critique, falsification, and prediction fulfillment verification. They are purely metaphysical.

Multiverse imaginings, therefore, offer no scientific threat whatever to the universality of the UPM and UPP in the only cosmic reality that science knows and investigates.

6. Conclusions

Mere possibility is not an adequate basis for asserting scientific plausibility. Indeed, the practical need exists in science to narrow down lists of possibilities on the basis of objectively quantifiable plausibility.

A numerically defined Universal *Plausibility* Metric (UPM = ξ) has been provided in this paper. A numerical inequality of $\xi < 1$ establishes definitive operational falsification of any chance hypothesis (The Universal Plausibility Principle [UPP]). Both UPM and UPP pre-exist and are independent of any experimental design and data set. No low-probability plausibility assertion should survive peer-review without subjection to the UPP inequality standard of formal falsification ($\xi < 1$).

The use of the UPM and application of the UPP inequality to each specific UPM will promote clarity, efficiency and decisiveness in all fields of scientific methodology by allowing operational falsification of ridiculously implausible plausibility assertions. The UPP is especially important in astrobiology and all areas of life-origin research where mere theoretical possibility is often equated erroneously with plausibility. The application of The Universal Plausibility Principle (UPP) precludes the inclusion in scientific literature of wild metaphysical conjectures that conveniently ignore or illegitimately inflate probabilistic resources to beyond the limits of observational science. The UPM and UPP together prevent rapidly shrinking funding and labor resources from being wasted on preposterous notions that have no legitimate place in science. At best, notions with $\xi < 1$ should be considered not only operationally falsified hypotheses, but bad metaphysics on a plane equivalent to blind faith and superstition.

7. Appendix 1

2^{NH} is the "practical" number (high probability group), measured in bits, rather than the erroneous theoretical n^N as is usually published, of all possible biopolymeric sequences that could form, where

N = the number of loci in the string (or monomers in polymer)

n = the number of possible alphabetical symbols that could be used at each locus (4 nucleotides, 64 codons, or 20 amino acids)

H = the Shannon uncertainty at each locus

For a 100 mer biopolymeric primary structure, the number of sequence combinations is actually only 2.69×10^{-6} of the theoretically possible and more intuitive measure of n^N sequences. The reason derives from the Shannon-McMillan-Breiman Theorem [68-71] which is explained in detail by Yockey [72, pg 73-76].

8. Appendix 2

For best estimates of the number of atoms, protons, neutrons and electrons in the universe and its astronomical subsets, see [73].

Simple arithmetic is needed for many of these calculations. For example, the mass of our galaxy is estimated to be around 10^{12} solar masses. The mass of "normal matter" in our galaxy is around 10^{11} solar masses. The mass of the sun is about 2×10^{30} kg. The mass of our solar system is surprisingly not much more than the mass of the sun, still about 2×10^{30} kg. (The Sun contains 99.85% of all the matter in the Solar System, and the planets contain only 0.136% of the mass of the solar system.) The mass of a proton or neutron is 1.7×10^{-27} kg. Thus the number of protons & neutrons in our solar system is around $2 \times 10^{30} / 1.7 \times 10^{-27} = 1.2 \times 10^{57}$. The number of electrons is about half of that, or 0.6×10^{57}. The number of protons, neutrons and electrons in our solar system is therefore around 1.8×10^{57}. The number of protons, neutrons and electrons in our galaxy is around 1.8×10^{68}. We have crudely estimated a total of 100 protons, neutrons and electrons on average per atom. All of these estimates will of course vary some through time as consensus evolves. But adjustments to $^L\Omega_A$ are easily updated with absolutely no change in the Universal Plausibility Metric (UPM) equation or the Universal Plausibility Principle (UPP) inequality. Definitive operational falsification still holds when $\xi < 1$.

Acknowledgements

This author claims no originality or credit for some of the referenced technical probabilistic concepts incorporated into this paper, especially the application of a universal *probability* bound in evaluating low probabilities [6, 7].

References:

1. Emmeche, C. 2000, Closure, function, emergence, semiosis, and life: the same idea? Reflections on the concrete and the abstract in theoretical biology, Ann N Y Acad Sci, 901, 187-97.
2. Baghramian, M. 2004, *Relativism*. Routledge: London
3. Balasubramanian, P. 1984, *The concept of presupposition: a study*. Radhakrishnan Institute for Advanced Study in Philosophy, University of Madras: [Madras].
4. Beaver, D.I. 2001, *Presupposition and assertion in dynamic semantics*. CSLI Publications; FoLLI: Stanford, Calif.
5. Bohr, N. 1949, Discussion with Einstein on epistemological problems in atomic physics. In *Albert Einstein: Philosopher-Scientist*, Schilpp, P. A., Ed. Library of Living Philosophers: Evanston, IL.
6. Dembski, W. 1998, *The Design Inference: Eliminating Chance Through Small Probabilities*. Cambridge University Press: Cambridge.
7. Dembski, W.A. 2002, *No Free Lunch*. Rowman and Littlefield: New York.
8. Meyer, S.C. 2009, *Signature in the Cell*. Harper Collins: New York.
9. Kuhn, T.S. 1970, *The Structure of Scientific Revolutions*. 2nd 1970 ed., The University of Chicago Press: Chicago.
10. Sokal, A.; Bricmont, J. 1998, *Fashionable Nonsense*. Picador: New York, NY.
11. Popper, K.R. 1972, *The logic of scientific discovery*. 6th impression revised. ed., Hutchinson: London.
12. Johnson, D.E. 2010, *Probability's Nature and Nature's Probability (A call to scientific integrity)*. Booksurge Publishing: Charleston, S.C.
13. Slife, B.; Williams, R. 1995, Science and Human Behavior. In *What's Behind the Research? Discovering Hidden Assumptions in the Behavioral Sciences* Slife, B.Williams, R., Eds. SAGE Publications: Thousand Oaks, CA, Vol. 167–204.
14. Lipton, P. 1991, *Inference to the Best Explanation*. Routledge: New York.
15. Press, S.J.; Tanur, J.M. 2001, *The Subjectivity of Scientists and the Bayesian Approach*. John Wiley & Sons: New York.
16. Congdon, P. 2001, *Bayesian Statistical Modeling*. John Wiley and Sons: New York.
17. Bandemer, H. 1992, *Modeling uncertain data*. 1st ed., Akademie Verlag: Berlin.
18. Corfield, D.; Williamson, J. 2001, Foundations of Bayesianism. Kluwer Academic Publishers: Dorcrecht, p 428.
19. Slonim, N.; Friedman, N.; Tishby, N. 2006, Multivariate Information Bottleneck, Neural Comput., 18, (8) 1739-1789.
20. Fisher, R.A. 1935, *The Design of Experiments*. Hafner: New York.
21. Fisher, R.A. 1956, *Statistical Methods and Statistical Inference*. Oliver and Boyd: Edinburgh.
22. Eddington, A. 1928, *The Nature of the Physical World*. Macmillan: New York.
23. Borel, E. 1962, *Probabilities and Life*. Dover: New York.
24. van de Sande, B. 2006 Measuring complexity in dynamical systems, RAPID II, Biola University, May.
25. Dam, K.W.; Lin, H.S. 1996, Cryptography's Role in Securing the Information Society. National Academy Press: Washington, D.C.
26. Lloyd, S. 2002, Computational capacity of the universe, Phys Rev Lett, 88, 237901-8.
27. Abel, D.L. 2008, 'The Cybernetic Cut': Progressing from description to prescription in systems theory, The Open Cybernetics and Systemics Journal, 2, 234-244 Open access at www.bentham.org/open/tocsj/articles/V002/252TOCSJ.pdf
28. Abel, D.L. 2009, The capabilities of chaos and complexity, Int. J. Mol. Sci., 10, (Special Issue on Life Origin) 247-291 Open access at http://mdpi.com/1422-0067/10/1/247
29. Abel, D.L. 2009, The biosemiosis of prescriptive information, Semiotica, 2009, (174) 1-19.
30. Adami, C. 1998, *Introduction to Artificial Life*. Springer/Telos: New York.
31. Abel, D.L. 2002, Is Life Reducible to Complexity? In *Fundamentals of Life*, Palyi, G.; Zucchi, C.Caglioti, L., Eds. Elsevier: Paris, pp 57-72.
32. Abel, D.L. 2006 Life origin: The role of complexity at the edge of chaos, Washington Science 2006, Headquarters of the National Science Foundation, Arlington, VA
33. Abel, D.L. 2007, Complexity, self-organization, and emergence at the edge of chaos in life-origin models, Journal of the Washington Academy of Sciences, 93, (4) 1-20.
34. Abel, D.L. The Cybernetic Cut [Scirus Topic Page]. http://www.scitopics.com/The_Cybernetic_Cut.html (Last accessed May, 2011).
35. Abel, D.L. Prescriptive Information (PI) [Scirus Topic Page]. http://www.scitopics.com/Prescriptive_Information_PI.html (Last accessed May, 2011).
36. Abel, D.L. 2009, The GS (Genetic Selection) Principle, Frontiers in Bioscience, 14, (January 1) 2959-2969 Open access at http://www.bioscience.org/2009/v14/af/3426/fulltext.htm.

37. Abel, D.L.; Trevors, J.T. 2005, Three subsets of sequence complexity and their relevance to biopolymeric information., Theoretical Biology and Medical Modeling, 2, 29 Open access at http://www.tbiomed.com/content/2/1/29.

38. Abel, D.L.; Trevors, J.T. 2006, Self-Organization vs. Self-Ordering events in life-origin models, Physics of Life Reviews, 3, 211-228.

39. Abel, D.L.; Trevors, J.T. 2007, More than Metaphor: Genomes are Objective Sign Systems. In *BioSemiotic Research Trends*, Barbieri, M., Ed. Nova Science Publishers: New York, pp 1-15

40. Vitányi, P.M.B.; Li, M. 2000, Minimum Description Length Induction, Bayesianism and Kolmogorov Complexity, IEEE Transactions on Information Theory, 46, (2) 446 - 464.

41. Zewail, A.H. 1990, The Birth of Molecules, Scientific American December, 40-46.

42. Zewail, A.H. 1999, The Nobel Prize in Chemistry. For his studies of the transition states of chemical reactions using femtosecond spectroscopy: Press Release. In

43. Xia, T.; Becker, H.-C.; Wan, C.; Frankel, A.; Roberts, R.W.; Zewail, A.H. 2003, The RNA-protein complex: Direct probing of the interfacial recognition dynamics and its correlation with biological functions, PNAS, 1433099100.

44. Sundstrom, V. 2008, Femtobiology, Annual Review of Physical Chemistry, 59, (1) 53-77.

45. Schwartz, S.D.; Schramm, V.L. 2009, Enzymatic transition states and dynamic motion in barrier crossing, Nat Chem Biol, 5, (8) 551-558.

46. Pedersen, S.; Herek, J.L.; Zewail, A.H. 1994, The Validity of the "Diradical" Hypothesis: Direct Femtoscond Studies of the Transition-State Structures, Science, 266, (5189) 1359-1364.

47. Wiegand, T.W.; Janssen, R.C.; Eaton, B.E. 1997, Selection of RNA amide synthases, Chem Biol, 4, (9) 675-83.

48. Emilsson, G.M.; Nakamura, S.; Roth, A.; Breaker, R.R. 2003, Ribozyme speed limits, RNA, 9, (8) 907-18.

49. Robertson, M.P.; Ellington, A.D. 2000, Design and optimization of effector-activated ribozyme ligases, Nucleic Acids Res, 28, (8) 1751-9.

50. Hammann, C.; Lilley, D.M. 2002, Folding and activity of the hammerhead ribozyme, Chembiochem, 3, (8) 690-700.

51. Breaker, R.R.; Emilsson, G.M.; Lazarev, D.; Nakamura, S.; Puskarz, I.J.; Roth, A.; Sudarsan, N. 2003, A common speed limit for RNA-cleaving ribozymes and deoxyribozymes, Rna, 9, (8) 949-57.

52. Axe, D.D. 2004, Estimating the prevalence of protein sequences adopting functional enzyme folds, J Mol Biol, 341, (5) 1295-315.

53. Barrau, A. 2007, Physics in the multiverse. In CERN Courier, Vol. December.

54. Carr, B. 2007, Universe or Multiverse? Cambridge University Press: Cambridge.

55. Garriga, J.; Vilenkin, A. 2008, Prediction and explanation in the multiverse. In Phys.Rev.D 77:043526,2008.

56. Axelsson, S. 2003, Perspectives on handedness, life and physics, Med Hypotheses, 61, (2) 267-74.

57. Koonin, E.V. 2007, The Biological Big Bang model for the major transitions in evolution, Biol Direct, 2, 21.

58. Koonin, E.V. 2007, The cosmological model of eternal inflation and the transition from chance to biological evolution in the history of life, Biol Direct, 2, 15.

59. Hawking, S.; Ellis, G.F.R. 1973, *The Large Scale Structure of Space-Time*. Cambridge University Press. : Cambridge.

60. Hawking, S. 1988, *A Brief History of Time*. Bantam Books: New York.

61. Hawking, S.; Penrose, R. 1996, *The Nature of Space and Time*. Princeton U. Press: Princeton, N.J.

62. Glansdorff, N.; Xu, Y.; Labedan, B. 2009, The origin of life and the last universal common ancestor: do we need a change of perspective?, Res Microbiol, 160, (7) 522-8.

63. Rocha, L.M. 2001, Evolution with material symbol systems, Biosystems, 60, 95-121.

64. Abel, D.L. The GS (Genetic Selection) Principle [Scirus Topic Page].
http://www.scitopics.com/The_GS_Principle_The_Genetic_Selection_Principle.html (Last accessed May, 2011).

65. Abel, D.L.; Trevors, J.T. 2006, More than metaphor: Genomes are objective sign systems, Journal of BioSemiotics, 1, (2) 253-267.

66. Abel, D.L. 2010, Constraints vs. Controls, Open Cybernetics and Systemics Journal, 4, 14-27 Open Access at http://www.bentham.org/open/tocsj/articles/V004/14TOCSJ.pdf.

67. Origin of Life Science Foundation, I. Origin of Life Prize. http://www.lifeorigin.org

68. Shannon, C. 1948, Part I and II: A mathematical theory of communication, The Bell System Technical Journal, XXVII, (3 July) 379-423.

69. McMillan 1953, The basic theorems of information theory, Ann. Math. Stat., 24, 196-219.

70. Breiman, L. 1957, The individual ergodic theorem of information theory, Ann. Math. Stat., 28, 808-811 [Correction in **31**:809-810].

71. Kinchin, I. 1958, The concept of entropy in probabililty theory. Also, On the foundamental theorems of information theory. In *Mathematical Foundations of Information Theory*, Dover Publications, Inc. : New York.

72. Yockey, H.P. 1992, *Information Theory and Molecular Biology*. Cambridge University Press: Cambridge.
73. Allen, A.N. 2000, *Astrophysical Quantities*. Springer-Verlag: New York.

The First Gene, David L. Abel, Editor 2011, pp 325-356 ISBN: 978-0-9657988-9-1

Chapter 12: The Formalism > Physicality (F > P) Principle [*]

David L. Abel

Department of ProtoBioCybernetics/ProtoBioSemiotics
Director, The Gene Emergence Project
The Origin-of-Life Science Foundation, Inc.
113 Hedgewood Dr. Greenbelt, MD 20770-1610 USA

ABSTRACT: The F > P Principle states that "Formalism not only describes, but preceded, prescribed, organized, and continues to govern and predict Physicality." The F > P Principle is an axiom that defines the ontological primacy of formalism in a presumed objective reality that transcends both human epistemology, our sensation of physicality, and physicality itself. The F > P Principle works hand in hand with the Law of Physicodynamic Incompleteness, which states that physicochemical interactions are inadequate to explain the mathematical and formal nature of physical law relationships. Physicodynamics cannot generate formal processes and procedures leading to nontrivial function. Chance, necessity and mere constraints cannot steer, program or optimize algorithmic/computational success to provide desired nontrivial utility. As a major corollary, physicodynamics cannot explain or generate life. Life is invariably cybernetic. The F > P Principle denies the notion of unity of Prescriptive Information (PI) with mass/energy. The F > P Principle distinguishes instantiation of formal choices into physicality from physicality itself. The arbitrary setting of configurable switches and the selection of symbols in any Material Symbol System (MSS) is physicodynamically indeterminate—decoupled from physicochemical determinism.

Correspondence/Reprint request: Dr. David L. Abel, Department of ProtoBioCybernetics/ProtoBioSemiotics
The Origin-of-Life Science Foundation, Inc., 113 Hedgewood Dr. Greenbelt, MD 20770-1610 USA   E-mail:
life@us.net
*Sections from previously published peer-reviewed science journal papers [1-9] have been incorporated with permission into this chapter.

Introduction: The reality of nonphysical formalism

Both the physicodynamic force relationships of classical physics and quantum statistical reality conform to mathematical description. The prescriptive mathematical formulae known as "natural laws" are formal, not physical. Why do these mathematical expressions work so well not only to describe, but to predict future physicodynamic interactions?   Eugene Wigner [10], Hamming [11], Steiner [12], and Einstein [13] all published on the "unreasonable" effectiveness of formal mathematics to describe and predict physical interactions. Einstein asked, "How is it possible that mathematics, a product of human thought that is independent of experience, fits so excellently the objects of physical reality?" [13] Mathematics is the ultimate expression of formal logic. Numerical representation and quantification are highly prized in science. Quantification permits by far the best modeling of physicality. But quantification is formal, not physical. The rational rules of mathematics, logic theory, and the scientific method are also all formal, not physical. Together they provide for reliable prediction of physical events.

Relationships in nature tend to stay constant despite varying local initial conditions. This constancy is defined by numerical constants. We value laws and the constants they employ because they are invariant in nature (excepting quantum decoherence, for the moment). Invariance is the key to prediction. Despite the variables, universal mathematical relationships exist that tell us how forces and physical objects will interact. The preciseness of quantification in force relationships minimizes subjective factors, objectifying our understanding of physical reality. Most advances in science have resulted from the formal manipulation of these numerical representations. In short, nonphysical formalism is the glue that holds all forms of scientific investigation together.

Other formalisms include logic theory, language, and cybernetics. None of these formalisms can be explained by physicality alone within a materialistic, physicalistic, naturalistic worldview. Naturalism looks for derivation of everything though mass/energy interactions and through chance- and-necessity causation. But chance contingency does not explain computational programming, or any other form of nontrivial utility. Logic gates cannot be set to open-or-closed functionality by redundant fixed law, either. If logic gates were set by law, they would all be set to the same position. Logic would be impossible. Binary programs would consist either of all "1's," or of all "0's." No uncertainty would exist, and therefore no Prescriptive Information potential. There would be no freedom of purposeful choice from among real options. Programming of any kind requires choice contingency, not forced law, and not mere chance contingency.

In the case of evolution, we refer to choice contingency as "selection pressure." But as has been covered many times in this anthology, selection pressure cannot steer events towards *eventual* utility. Evolution cannot pursue *potential* function at the decision-node programming level where organization originates. Evolution cannot work at the genetic/genomic/epigenetic/ epigenomic programming level where the phenomenon of regulation and control originates (The Genetic Selection [GS] Principle [5]).

At the level of consciousness, choice contingency in the intuitive sense is undeniably real. Certainly at the level of human cognition, no one doubts the reality of purposeful choice. In addition, one cannot even argue for the worldview of a strict physicalistic determinism without employing formal choice contingency in the logical argument itself. Any defense of physicalism is therefore self-contradictory. There will be no escape in logical or scientific debate from the reality of choice contingency. 95% of the practice of physics, (the study of physicodynamics), consists of formal nonphysical mathematics and logic theory. The other 5% consists of observation which cannot be reduced to mass/energy either. Why do mathematics and logic theory serve us so well in the pursuit of scientific knowledge? And why should a fundamentally chaotic, irrational and non-formal physicality be so wonderfully able to be modeled by rational and mathematical formalism? Could it be that physicality has its roots in, and arose out of, a formalism even more fundamental and causative than physicodynamics?

1. Is physicality chaotic, or organized?

This question may pose a false dichotomy, but naturalistic science tends to just metaphysically presuppose initial disorganization in its various cosmologies and cosmogonies. Mass/Energy is allowed, but not formal organization. Sometimes initial order is considered in the notion of a cosmic egg. But not bona fide organization as the primal force. How was it determined that reality was initially chaotic and only physical? Certainly not scientifically. The pre-assumption of ultimate chaos is not only purely metaphysical; it is antithetical to repeated observations of current reality, and to abundant formal prediction fulfillments of an underlying organization. It is contrary to the logic theory upon which math and science are based. Overwhelming empirical evidence exists that reality is *not* fundamentally chaotic. Not only repeated observation, but innumerable fulfilled predictions of physical interactions based solely on mathematical models is far more suggestive that physicality unfolds according to formalism's ultimate integration, organization and control of physicality.

The effectiveness of mathematics in science is only "*un*reasonable" if we are foolish enough to begin our thinking with purely physicalistic metaphysical pre-assumptions.   The Formalism > Physicality (F > P) Principle explains away this supposedly "unreasonable" effectiveness of mathematics in the natural sciences.   What is unreasonable is a materialistic belief system that ignores or tries to deny the supremacy of formalism.   Materialism has never been empirically or logically established to be absolute Truth.   This physicalistic faith system is inherently self-contradictory.   No "ism" is physical.   Naturalism is illegitimately incorporated into the very definition of science.   The term, "Naturalistic science" is an oxymoron.   Science is a formal enterprise from beginning to end.   "Naturalistic science" exists only in name.   Science itself is an abstract, cognitive, epistemological quest of mind.   There is nothing "natural" about it, at least as philosophic naturalism would define "natural."   All of the formalisms inherent in scientific method cannot be reduced to cause-and-effect physicodynamic determinism.   Mathematics and science cannot be practiced within a consistently held materialistic and naturalistic metaphysical worldview.   Neither can cybernetic pursuits—activities involving any form of control.

Science is an epistemological system.   Science from the beginning was the pursuit of an ever-increasing collective knowledge of the way things actually are. The abandonment of realism led only to the decline of quality science.

The acquisition of information "about" reality is a purely formal enterprise.   Whatever qualitative aspects of science that cannot be quantified are still dealt with logically.   Linguistic logic theory, like mathematics, is also formal.   Science also depends upon categorization.   Categorization in turn depends upon drawing conceptual conclusions about distinctions between classes of objects and events.   Categorization is formal, not physical.

The collection, categorization and organization of data, the reporting of results using representational symbols (e.g. in tables), and the drawing of conclusions are formal enterprises, not physicodynamic interactions.   Physicality doesn't govern science.   Nonphysical formalism governs science.   In short, *formalism predominates,* not physicodynamics.

As Pattee has pointed out many times [14-18], even initial physical conditions must be formally represented with numbers within the laws of physics.   Physical conditions themselves cannot be plugged into the nonphysical mathematical equalities and inequalities that we call "the laws of physics." We insert numerical *representations* of initial conditions.   Initial conditions cannot measure or symbolically represent themselves.   Without formal representations of initial conditions and formal manipulations using equations, no physicist could predict any physical outcome.   The physics practiced by physi-

calists is not physical. It is nonphysical and formal. In order to practice physics, the materialist must violate his own metaphysical imperative; he must violate his or her own contention that physicality is all there is. Thus a dichotomy
exists that categorizes physicodynamic reality from its formal representation
and prediction. Physics and chemistry as sciences are dead without formalism.
The scientific method cannot be practiced without abiding by the formal rules
(not laws) of logic, mathematics and scientific ethics. In view of the historical
existence of mathematical orderliness in nature (presumed by naturalism itself)
prior to Homo sapiens, how should the scientific community respond to such a
ridiculous, totally metaphysical pontification as, "The physical cosmos is all
there is, ever was, or ever will be!"? [19] Professional philosophers of science
typically respond to such dogma with, "SEZ WHO?" or at least, "How was
this statement of absolute Truth established?"

2. Formalism is more than mathematics

We usually think of scientific formalism in terms of numerical axioms,
quantifications, and manipulations of mathematical equations. But the essence
of formalism is not just the use of number systems. Formal systems also involve choosing with intent characters from a finite alphabet of symbols, syntax, grammar, and assigned meanings to those symbols and symbol combinations.   Additional formal linguistic rules also apply. In the broadest sense,
formal enterprises include language itself, the rules of syllogistic deduction,
abduction, induction, semantics, derived theorems and corollaries, and cybernetic steering of events and calculations towards pragmatic benefit. The pursuit of utility is a rational and formal pursuit that cannot be reduced to mass
and energy.

3. The two subsets of contingency revisited

As covered in Chapter 2, *contingent* in a past tense context means that an
event could have occurred other than how it happened. In a present and future
context, contingency means that events can unfold in multiple ways despite
both local and seemingly universal law-like constraints. Contingency is not
forced by physicodynamic necessity. Contingency embodies an aspect of
freedom from physicochemical determinism. Refusal to acknowledge the reality of contingency tends to make the practice of even weighted statistics rather
difficult. No law of motion renders absolutely reliable predictions. Even the
most dogmatic adherents to "hard determinism" can easily be cornered into
admitting the reality of some degree of contingency.

But there are two kinds of contingency. The first is *chance contingency*. Chance contingency gives rise to random variation—"noise." The Brownian movement caused by the heat agitation of molecules is a seeming example of chance contingency. "Just how random is randomness?" remains an open question. Many have argued that seemingly random events are actually the result of unknown causes and complex interactions between multiple known physicodynamic causes [20-22]. But the bottom line of chance contingency is a non-willful, non-steered independence from straightforward cause-and-effect determinism. Possibilities and options are not purposefully chosen, but result from "the roll of dice."

The second type of contingency is *choice contingency*. Choice contingency is a purposeful selection from among real options. Choice contingency is exercised with directionality for a reason and purpose. The goal of choice contingency is almost always some form of utility that is valued by the chooser.

4. The essence of any formalism is purposeful choice contingency

Contingency—freedom from determinism—alone is not adequate to generate nontrivial function. No computationally successful program was ever written by a random number generator. Nontrivial programs can only be written by purposeful, wise choices at bona fide decision nodes. What is the garbage in the programmer's phrase, "Garbage in, garbage out!"? Usually it's bad data, but garbage can be bad programming choices too—something less than wise choice contingency—something moving either in the direction of chance contingency or physicodynamic self-ordering, neither of which can program formal function.

Randomness is contingent, but not formally determinative. To contingency must be added "choice with intent." Randomly occurring events have never been shown to generate optimized algorithms, nontrivial conceptual instructions, or sophisticated programming [8, 23, 24]. All formal systems, including mathematics, require *purposeful choice contingency*. Equation manipulations are a form of choice-contingent behavior.

Neither the rules of computation nor the computation itself are physical. What is computation? More than any other factor, the bottom line of *any* formalism is the exercise of *expedient choice with intent at bona fide decision nodes*. "Natural process" experiments that purport to have generated spontaneously occurring new programming, computational success, or non- trivial formal function can be shown invariably to be guilty of "investigator involvement" in experimental design and methodology [9]. Artificial rather than natural selection has been introduced.

Choice contingency has been thoroughly distinguished from chance contingency and law-like necessity in prior publications [2, 8, 9, 24]. Choice contingency cannot be derived from a combination of the chance contingency and necessity of physicodynamics. Any attempt to extirpate purposeful choice contingency from the explanation of sophisticated function invariably results in the rapid deterioration of that function. Noise replaces meaningful communication with gibberish. "Bugs" and "blue screens" replace programming. Failure to halt replaces successful computation. Nonsense replaces sound reason. No escape exists from choice contingency in any rational explanation of sophisticated function. Sophisticated utility is realized only at the behest of wise purposeful choices—the essence of formalism.

5. Formalism not only describes reality, it prescribes and controls reality.

The ability of mathematics to *predict* future physical interactions is a far more daunting problem for the physicalist than explaining how mathematics is able to *describe* so well those same interactions. Thus Wigner, Hamming, Steiner and Einstein, if anything, understated the problem. To the degree that mathematical formulae and their logical manipulations predict future physical interactions, it could be argued that they not only describe, but prescribe and control physical world unfoldings.

Statistical predictions are a special case in science. Assuming a theoretical total independence from any law-like orderliness, descriptions of future quantum outcomes are purely probabilistic. We would not consider statistical predictions to be determinative or controlling in any sense. Chance doesn't cause or even influence any physical event to happen. Chance is only a descriptive mental construct, not a physical cause of effects.

Most macroscopic "chance" events, however, do not conform to this theoretical total independence from law-like orderliness. Coin flips, for example, are not absolutely random because they are not absolutely "fair." The heads side may weigh ever so slightly more than the tails side. Thus a "fair coin flip" is theoretical rather than real unless a coin can be manufactured that has absolutely no variance of one side from the other. To whatever degree the coin is not "fair," law-like influence must be incorporated into "weighted means." In the quantum mechanical world, decoherence from expected events becomes an issue. But in the macroscopic world, mathematical formulae and rules govern physicodynamic unfoldings with amazing accuracy.

6. The derivation of formalism

How could purely formal mathematics and biological Prescriptive Information (PI) [6] utilizing linear digital programming choices be derived naturalistically from physicality alone? Much has been accomplished in science through reductionism. Let us therefore attempt to reduce the problem at hand to a maximally reduced and simplified query: Could inanimate physicodynamics have generated cybernetics, the mathematics of physics, the highly fine-tuned numerical force constants, and the linear digital programming upon which all life depends [25-27]?

Physical explosions (e.g., the Big Bang) do not create mathematical constructs and computational algorithms. The physical laws may have become *apparent* at 10^{-43} seconds. But that does not establish that they didn't exist prior to becoming physically instantiated and actualized. In addition, it does not establish that physicality produced those nonphysical formalisms. Indeed, as one of the reviewers of this paper pointed out, circular logic is involved in arguing that Physicality produced formalism which then produced physicality. It is much more likely that the nonphysical formal laws pre-existed the cosmic egg "explosion," and only became apparent at 10^{-43} seconds within the time-space physical medium. If true, the Big Bang was not a chaotic explosion, but a controlled unfolding of prescribed physical organization and reality.

7. The valuation and pursuit of utility in "applied science" is formal

The pursuit of functionality arises first out of a desire for and valuation of "usefulness." Inanimate nature (e.g., a prebiotic environment) possesses none of these formal attributes or motives. The environment does not value and does not pursue organization over disorganization. Physicodynamics *can* self-order spontaneously (e.g., Prigogine's dissipative structures: hurricanes, tornadoes, candle flames, falling drops of water forming spheres, etc.). But inanimate nature cannot self-organize itself into formal step-wise processes/procedures (e.g., algorithms) in pursuit of utility. A prebiotic environment had no sense of pragmatism. It exerted no pressure towards function over non-function. Only our minds imagine an environmental preference for function over nonfunction in order to make our molecular evolution models "work for us." Rationalization prevails rather than progressive communal discovery of what the objective world is actually like.

The postmodern concept of something "working for us" boils down to providing psychological, sociological and seemingly rational and empirical support for our already presupposed beliefs. Naturalism is already committed to the metaphysical presupposition that "physicality is sufficient to explain everything." Most of us bring with us this axiomatic pre-assumption *to* science. We were told from an early age on that science requires it. So most of us have cooperated fully with the incorporation of philosophic materialism and naturalism into our very definition of science.

If anyone dares to raise an eyebrow of healthy scientific skepticism about the all-sufficiency of mass and energy at any stage of our education, we are immediately pounced upon, ridiculed, shouted down by peers, and flunked out by professors. If we wait to raise any questions about the all-sufficiency of materialism until after we hold a degree, we are silenced by the peer review of true believers in physicalism. If we are fortunate enough to get a few open-minded peer reviewers, we are still stifled by a concerted effort of physicalists not to cite any paper that dares to challenge the all-sufficiency of physicodynamics to explain the whole of observational reality.

The subject of this paper is nothing more than a statement of what should have been obvious to every scientist all along. Mass and energy cannot represent meaning or programming choices using arbitrary symbol assignments. Mass and energy cannot state or manipulate mathematical equations. Physicality cannot organize data or draw abstracted conclusions. It cannot predict outcomes or practice any aspect of the scientific method.

Applied science values and pursues useful applications of academic scientific principles, data, results and conclusions in each specific field of study. To ascribe value to something is a formal function. To pursue utility is a formal undertaking. Cause-and-effect determinism knows nothing of value or function. It cannot identify or pursue "usefulness." In a materialistic world, whatever effects are caused are just "the way it is." Benefit is irrelevant.

Grant money is a lot easier to come by when academic interests are applied to solving everyday practical problems. The value of science is often judged by its practical usefulness to humans. NASA received a lot more funding when the general public and their political representatives started seeing the practical every day devices and benefits that arose directly out of the space program. Seeking knowledge for knowledge sake is noble, but rarely generates much grant money or pays anyone's salary. Thus a forensic scientist who is able to generate reliable methods of identifying serial rapists and murderers tends to get more attention and grant money than the scientist who first figured out how to sequence DNA for purely academic reasons. The forensic science wouldn't have been possible without the academics. Both the scientific academics and the pragmatic application of those academics are abstract, conceptual and formal, not merely mass/energy cause-and-effect determined interactions.

8. Controls and rules, not constraints and laws, achieve pragmatism

Science must follow certain rules. Rules are not laws [7]. Rules are agreed-upon conventions that govern voluntary behavior. Rules exist to guide choices. Rules can be broken at will. Rules govern procedures, competing interests, and ethical behavior. Rules are formal. The rules of the scientific method require honesty in the reporting of results, for example. There is nothing physical about the expectation of and demand for honesty. Science would collapse without adherence to certain ethical standards. We castigate scientists who falsify results or who plagiarize the work of others. Yet it is widely acknowledged that such moral "shoulds" and "oughts" are not derivable from a purely material world. Yet without these metaphysical and ethical demands, science could not be trusted as a source of reliable knowledge. Thus, science depends upon formal values, rules and honest behavior. It cannot be reduced to the chance and necessity of physicality.

Obedience to rules is not constrained. It is voluntary. But for any formalism to proceed, choices must be voluntarily made according to arbitrary rules with the intent of achieving formal function. This includes any mathematical or logical pursuit in science. It includes language. And it includes cybernetic programming. Loss of formal utility usually accompanies the disobedience of those rules unless a pragmatically superior rule system is being explored.

Most of what is really interesting in life was produced by choice contingency, not chance contingency or law. Our most fundamental problem in naturalistic science lies in explaining how physicodynamic determinism could have produced the bona fide choice contingency that we all observe and practice on a daily basis. The most fundamental question of biology is, "How did law-constrained physicochemical interactions along with "random" heat agitation generate a formally prescriptive linear digital genetic system?"

Language and any other form of sign/symbol/token system require deliberately choosing alphanumeric symbols from an alphabet of multiple options. Linguistic *rules* of language convention also must be arbitrarily chosen and adhered to. By arbitrary, we mean choice contingent, not chance contingent. Arbitrary does not mean that the chooser flips a coin to decide, or that the chooser does not care what is chosen. In addition to being choice contingent, "arbitrary" also means "unconstrained by natural law." Arbitrariness excludes determinism by law-like self-ordering. Self-ordering phenomena are extremely low in information [9]. High uncertainty and freedom are needed as a pretext to programming. No linguistic or cybernetic system has ever been organized by chance contingency or physicochemical determinism.

All forms of cybernetic programming in computer science are formal. Computational success can only be prescribed through formal choices with intent. The same is true of algorithmic optimization, the engineering of sophisticated function, and organization of any kind. Such formal utility cannot be achieved through after-the-fact selection of the best algorithms. A pool of "potential solutions" first has to exist before optimization is pursued. These stepwise discrete procedures ("potential solutions" are algorithms) must be programmed *at the decision node level*. A mere stochastic ensemble of symbols is not a potential solution. When Scrabble tokens are dumped out of the box onto the board and lined up upside down in strings, they sometimes contain happenstantial "words" when turned over. But this is only because our minds pick out those random sequences of letters by prior association. They are in reality just as random as any other letters in the string. Similarly, a random pool of supposed "potential solutions" are not the problem solutions they are claimed to be. Only our minds select them in pursuit of the solution and optimization we are pursuing. Consciousness is always smuggled in subconsciously in successful Markov processes. Strings of symbols have to be processed to function as programmed computational solutions. This requires either the selection of logic gate settings according to arbitrary conventions prior to the existence of any function [5], or the reading and processing of these instructions according to previously agreed upon rules, or both. Optimization requires motivation, the declaration of value, and the pursuit of a desired ever-improving utility. All of these factors are formal, not physicodynamic.

What empirical evidence and prediction fulfillment support do we have for the metaphysical belief that physicality generated formalism (e.g., that physical brain generated mind)? Has anyone ever observed a single instance of chance and necessity generating nontrivial computational function? Has anyone ever observed constraints generating bona fide controls that specifically steer events toward formal nontrivial utility? Do the laws of physics and chemistry ever generate creative new Prescriptive Information (PI)?

Says Howard Pattee:

> "The concept of control does not enter physical theory because it is the fundamental condition for physical laws that they describe only those relations between events which are invariant with respect to different observers, and consequently those relations between events over which the observer has no control. At the least, control requires, in addition to the laws, some form of local, structural constraint on the lawful dynamics. Pragmatic control also requires some measure of utility. To say that the river bed controls the flow of the river is a gratuitous use of

control since there is no utility, and the simpler term 'constraint' serves just as well." [21, pg. 69]

Without exception every sophisticated pragmatic tool, machine or mechanistic procedure known to humanity required decision-node programming or integrative configurable switch setting to achieve. No bona fide nontrivial organization has ever arisen without purposeful steering, controlling and regulating the process. Constraints and invariant laws cannot perceive or pursue utility. Constraints and laws could not have generated a single complex machine, let alone life.

9. The Law of Organizational and Cybernetic Decline (The OCD Law)

The OCD Law states that, absent the intervention of formal agency, any nontrivial organization or cybernetic/computational function instantiated into physicality (e.g., integrated circuits) will invariably deteriorate and fail through time. This deterioration may not be continual. But it will be continuous (off and on, but overall consistently downhill). Computers, robots, all forms of Artificial Intelligence and Artificial Life, messages instantiated into material symbol systems or electronic impulses, will invariably progress toward dysfunction and fail.

The OCD Law is not to be confused with the Second Law of Thermodynamics. The OCD Law is not concerned with the entropy of statistical mechanics or the "entropy" or "mutual entropy" of Shannon's probabilistic combinatorial uncertainty. Heat exchange, heat dissipation, phase changes, order and disorder are not at issue. The OCD Law addresses only the formal organization and utility already instantiated into physical media and environments. Only purposeful choice contingency at bona fide decision nodes can rescue from deterioration the organization and function previously programmed into physicodynamics.

Thermodynamicists differ widely in opinion as to whether entropy is physical. Most materialists find themselves seriously trying to argue that the negatives of log functions of probabilities are physical! Even if they were, entropy tells us nothing about organization or achieving nontrivial formal function.

The OCD Law, of course, raises the question of how organization arose in the first place. *The Organization (O) Principle* states that nontrivial formal organization can only be produced by Choice Contingency Causation and Control (CCCC). The O Principle, like the OCD Law, can still be treated as a mere null hypothesis if desired by skeptics and critics. The firm prediction is made that neither the OCD Law nor the O Principle will ever be falsified by empirical evidence or prediction fulfillment data. It will never be overturned by sound Aristotelian logic, either. A single legitimate exception to either generalization would serve as falsification. It is incumbent upon those who religiously believe in spontaneous self-organization of mass/energy into non-trivial formal utility to provide empirical evidence or prediction fulfillment support for their blind belief. Thus far, any logical defense of belief in self-organization has also been sorely lacking [9]. In the absence of scientific support, informationless self-organization hypotheses and models such as Ganti's [28] remain little more than superstition.

10. Is entropy physical?

Many thermodynamicists are uncertain as to whether "entropy" is physical. "Energy unavailable for work" is one of several common definitions of entropy. "Energy" would certainly have to be considered physical. But "unavailable for work" is a formal characterization, not a physical entity. "Work" as used in this context obviously does not refer to mere heat exchange between bodies. It refers to formal utilitarian potential. Can the energy be used to achieve function? When taken as a whole, "energy unavailable for work" is a formal construct that cannot be reduced to physicality. In addition, nonphysical formal mathematics is required to define entropy and measure it in scientific terms. Nevertheless, as mentioned above, most metaphysical naturalists find themselves seriously trying to argue that the negative of a log function is physical. This is especially true of those who insist that statistical mechanical entropy is one-in-the-same with Shannon entropy. Shannon uncertainty is a probabilistic measure. Reduced uncertainty (R) is still a mathematical subtraction based on "before" minus "after" uncertainty. Reduced uncertainty is equated with gained knowledge. But even reduced uncertainty is formal.

Neurophysiology has never had much success trying to reduce epistemology to physicodynamics. But even if entropy were physical, entropy tells us nothing about organization or achieving nontrivial formal function.

Many try to define 'entropy" in terms of "increasing disorder." But clearly many forms of crystallization simultaneously increase order while moving towards greater entropy within the system. This confusion was caused by the initial confusion of order with organization, and the confusion of constraints with controls. Self-ordering phenomena and constraints are physicodynamic properties. Organization and controls are formal properties. Physicality cannot generate nonphysical formalisms. They lie in different categories. Self-ordering phenomena and constraints arise from the near side of The Cybernetic Cut (Chap 3). Organization and controls arise only from the far side of The Cybernetic Cut. The one-way Configurable Switch (CS) Bridge allows controls to travel into the physicodynamic world from the formal non-physical world. Under no circumstances do physicodynamic phenomena ever traverse the CS Bridge from the near physicodynamic side to the far formal side. What makes reality especially interesting is not order, but organization. What generates utilitarian work is choice contingency and controls, not constraints and laws. We must learn to get order and disorder out of the discussion of organization.

11. Formalism's instantiation can alone temporarily and locally circumvent The 2nd Law

James Clerk Maxwell first stated his well known "demon paradox" in a letter to Peter Tait in 1867. A controllable trap door separating two compartments allows an imaginary demon to separate warmer and cooler ideal gas molecules on opposite sides of the door separating the two compartments. The temperature differential between compartments was to provide an energy potential needed to drive a potential heat engine.

There are good reasons why naturalism is forced to view Maxwell's demon as only a "thought experiment" [29]. Abstract concept and volition are required for the demon to *selectively* open and close the trap door. He must choose with intent to concentrate the fastest-moving particles on one side of the partition. No energy is required in this thought experiment for Maxwell's demon's *mind* to choose whether and when to open the trap door. No accounting is provided for the demon's brain or muscle energy requirements, either, to operate the trap door. The demon has no brain or physical reality. He and his purposeful choices are transcendent to physicality. The hotter faster-moving particles cannot be concentrated on one side of the partition without his purposeful choices. It is true that the actual opening of the trap door would require a physical force and energy. But the *vectors of door pull up or push down are not physically determined. They are formally chosen.* And the all-important choice of *when* to open or close the trap door is also purely formal.

The demon has always been prominent in physics and thermodynamics precisely because *he provides the energy-free formal agency that alone can explain temporary and local circumvention of the Second Law.* Take away the demon's formal purposeful choices—his agency—and equilibration of heat in both compartments is inevitable according to the 2nd Law of Thermodynamics. What is the natural-process equivalent of such a mystical demon? None exists in the naturalist's materialistic metaphysical world. The demon's persistence in physics texts is nothing less than a classic demonstration of naturalistic rational inconsistency. The physical cosmos clearly cannot be "all there is, ever was, or ever will be" [19]. Seemingly local and temporary circumvention of the Second Law is too evident and common; but only because the demon lives and purposefully chooses. All energy transduction mechanisms making non-trivial function possible can be traced back to the same formal controls. Mere constraints, laws and phase changes do not produce functional "work" and sophisticated utility.

But why couldn't some yet-to-be discovered natural-process law operate the trap door? The answer is that laws always preclude freedom of programming choice and control. The trap door would always be held open, or always locked closed, *by law*!

It could be argued that not even life violates the 2nd Law, at least the physical manifestation of life. But life's formal controls and regulation are nonphysical. Formalisms are not subject to the 2nd law because they are nonphysical. The instantiation of formalisms into physicality is alone what makes possible the seeming temporary and local circumvention of the 2nd law.

In the microscopic world, circumvention of the 2nd Law is considered a given by many. But the quantum world is highly laced with human epistemological factors. Some might argue that the microscopic world may be more of a subjective human mental construction than ontologic reality. Others might point to the role of mathematics, probabilism and imagination in quantum theory as further evidence of formalism being the most fundamental level and ultimate cause of overall reality, including physicality.

We might be quick to deny "vitalism;" but we will not succeed in denying the reality of life's formal programming, regulation and control. We will not be able to sweep under the rug the prokaryotes' representational symbol systems, cybernetic programming, tens of thousands of nanocomputers, firmware, operating systems, various application softwares, semiosis (messaging), coding and decoding, translation, and its orientation around the pragmatic goal of staying alive. All of these are formalisms, not mere physicochemical interactions. This leaves us with the uncomfortable question, "What exactly is the difference between the undeniable transcendence of all these formalisms that program and regulate life, and the vitalism we so vociferously decry?

In Rolf Landauer's review [30] of *Maxwell's Demon: Why Warmth Disperses and Time Passes* by Hans Christian von Baeyer [31], Landauer points out, "It is impossible to sort molecules without expending more energy than the work that can be extracted from the sorted molecules. The second law of thermodynamics does indeed hold true."

Szilard rightly argued that Maxwell's Demon must be "informed" in order to know when to open and close the trap door [32]. Uninformed and undirected constraints cannot operate the trap door so as to deliberately separate hot and cold particles. Only choice-based control can.

Gilbert Lewis wrote: "Gain in entropy always means loss of information, and nothing more." [4, 33, pg. 573).   Conversely, to reduce entropy requires increased information—not only increased Shannon information, but increased prescriptive information {Abel, 2007 #6367]. It takes prescriptive information for the demon to know how to achieve heat engine potential. But the problem is far greater than one of knowledge. It is one of deliberate steering, control, management and goal.   The Demon must *decide* when to open and close the door *for some useful reason.* The Demon must have desire and motive. With every approaching particle he must make a purposeful binary choice of whether to open or to close the door so as to create a future energy potential. By what naturalistic physical mechanism is this choice accomplished? The Cybernetic Cut {Abel, 2008 #6969;Abel, 2008 #8037} and The F > P Principle declares that no natural mechanism exists that can choose with intent to deliberately design, engineer and maintain a *Sustained Functional System (SFS)* [34] such as a thermal engine for pragmatic reasons. Prigogine's dissipative structures in chaos theory have little in common with SFSs. Falsification of the assertion that nontrivial SFSs do not spontaneously form in nature is simple: cite a single exception. Such falsification is invited to promote further discovery and to test axiomatic principles such as The F > P Principle "in the real world."

12. The source of Prescriptive Information (PI) is formal

Prescriptive information (PI) either instructs or directly produces nontrivial function [2, 8]. PI usually accomplishes this through programmed algorithmic processing. "Prescriptive Information either tells us what choices to make, or it is a recordation of wise choices already made" [6]. Prescription requires formal selective steering at successive decision nodes. The purpose of PI is to generate pragmatic results. Such utility is valued and pursued by agents. Inanimate nature cannot value or pursue a formal goal. Not even evolution has a goal. Expedient choice commitments must be made prior to the realization of function at each successive decision node in any program. Bifurcation points can be traversed randomly; but no significant computational halting success can be expected at the end of a random path. Decision nodes require true decisions, not "coin flips" or "dice rolls," to generate PI and sophisticated function.

The definition of PI centers on selection *for potential (not yet existent) function.* What exactly do we mean by function? "A function is a goal-oriented property of an entity" [35]. Says Voie, "Functional parts are only meaningful under a whole, in other words it is the whole that gives meaning to its parts" [35].

The road to utility is paved with algorithmic intent [36]. A goal-based algorithm is a step-wise, usually discrete process or procedure leading to *future* utility. Natural selection cannot generate such procedures. Evolution is blind to potential function and the future. It can only eliminate inferior formal programs (highly integrated computational haltings manifested as already-computed phenotypic organisms [The GS Principle] [5]). Goal-based algorithms control events and behavior, steering them toward organized, predictable usefulness. But such steering requires free and purposeful choices at bona fide decision nodes. Neither chance nor necessity can generate or optimize algorithms. These programming decisions must be made wisely with the intent to achieve computational halting. The only known source of conceptually integrated function is formally-generated PI.

Given the right processing algorithms, PI not only instructs, but can actually produce sophisticated function. But, to accomplish this pragmatism, at the very least constraints must be purposefully chosen through the selection of particular initial conditions in order to influence physical interactions to move towards Aristotelian "final function." Constraints are blind to function. Constraints and laws have no pragmatic goal. Constraints cannot generate the symbol systems used by semiosis. It is only the purposeful choice of certain constraints (e.g., the choice of initial conditions in designing an experiment) that creates bona fide controls. The F > P Principle states that the fundamental ingredients of any semiotic system are *representationalism* and *choice contingency*, not chance contingency or necessity. Meaning is always formal, never physicodynamic.

13. Naturalistic "efficient causation" (Aristotle) is grossly inadequate

Physicodynamic cause-and-effect was classified by Aristotle as "efficient causation." Naturalistic science attempts to explain seemingly teleological (teleonomic) phenomena solely in terms of efficient causation. Naturalistic biologists universally just presuppose functionality in scientific literature without any explanation of its derivation: "The purpose of the kidneys is to excrete waste products from the blood stream." "Mitochondria function as the powerhouse of the cell." "Each amino acyl tRNA synthetase is present *in order to* bind the appropriate amino acid to its own tRNA." Naturalistic science has to be able to explain all of these purely formal "in order to's" with nothing but mass/energy interactions. It fails miserably.

How can we refine evolutionary explanations to incorporate "in order to" into efficient causation? We point to selection pressure as the cause. But environmental selection favors only the best already-living phenotypic "effects." It does not explain the cause—the programming, algorithmic processing, preservation schemes and optimization that produced those effects (The GS Principle) [5].

In evolutionary theory, the chaperone proteins cannot come into existence "in order to" fill the need of helping other proteins fold correctly. They, too, have to be folded. For the consistent naturalist, "Folding correctly" must ultimately be purely accidental prior to secondary selection for after-the-fact fitness. Virtually every player in homeostatic metabolism participates actively in pursuing and eventually achieving cooperative holistic integration. Evolution theory provides no mechanism for anticipation or pursuit of goals. In addition, the probability of thousands of needed players all coming together at the same place and time, all to contribute their role in achieving the final function of homeostatic metabolism, is statistically prohibitive for any purely materialistic conglomerate. The notion that physicodynamics alone can accomplish even a protometabolism can be definitively falsified by the Universal Plausibility Metric and Principle [37].

14. The genomic symbol system's prescription, control and regulation are formal

Küppers [38, pg 166] makes the same point as Jacques Monod [39], Ernst Mayr [40, 41], and Hubert Yockey [42, 43], that physics and chemistry do not explain life. Niels Bohr argued that "Life is consistent with, but undecidable from physics and chemistry."[44] "Undecidable" means that life cannot be explained by mere physical interactions alone. What exactly is the missing ingredient that renders life unique from inanimate physics and chemistry? The answer lies in the fact that life, unlike inanimacy, traverses the Cybernetic Cut (See Chapter 3) [4] . The Cybernetic Cut dichotomizes reality into two fundamental categories. The dynamics of physicality ("chance and necessity") lie on one side of the great divide. On the other side lies the ability to choose with intent what aspects of ontological being will be preferred, pursued, selected, rearranged, integrated, organized, preserved, and used (formalism). Algorithmic programs and their optimization require traversing the Cybernetic Cut. Life is further differentiated from non-life by its linear digital Prescriptive Information that uses a material symbol system (MSS) [45, 46]. Says Hubert Yockey, "The existence of a genome and the genetic code divides living organisms from non-living matter. . . . There is nothing in the physico-chemical world that remotely resembles reactions being determined by a sequence and codes between sequences." [26, pg. 54]

Linear digital programming occurs prior to any folding. The source of this programming lies in the selection and sequencing of rigidly bound nucleotide (token) "choices." Primary structure (sequencing) is the main determinant of tertiary structure (the globular molecular machine). Chaperones and other factors contribute to folding. But rigidly-bound monomeric sequencing largely determines what folding thermodynamic tendencies will be. And chaperones are themselves prescribed by the same linear digital symbol system. The far weaker H bonding of average folding is primarily mediated by primary structure. Thus true selection must take place at the point of polymerization of each additional monomer onto the forming positive strand. Since polymerization of the primary single strand is nearly dynamically inert in coding regions, physicodynamics plays no role in sequencing. Nothing is left but randomness with which to program in a naturalistic context. Yet coin flips have never been observed to program computational halting in any cybernetic system. There is no escaping the reality that all known organisms are prescribed and largely controlled by this linear digital programming. A representational MSS is clearly employed in the triplet codon table of amino acid prescription. Even most epigenetic factors are produced only through linear digital instruction and control (e.g. regulatory peptides, proteins and small RNAs) [47-49]. Even DNA methylation and protein binding to histone tails are at least indirectly prescribed by nucleotide sequencing. Non-coding regulatory RNA prescribed by DNA controls much of the genome [48-51].

Even more confounding is that all of these processes require sophisticated nanocomputers, firmware, "high tech" operating systems and software. Formal algorithms are required. Sequencing has no meaning or function independent of an overarching formal system of arbitrary assignment of specific amino acid correspondences. No physical force or law explains these arbitrary correspondence assignments. Their formal functions are not physicodynamically mandated. They are formally prescribed.

DNA genetic sequencing *seems* 99% "random" when considered only from a Shannon probabilistic and combinatorial perspective. But of course this perspective is blind to the meaning or function of any message or program. A string of 1s and 0s, as the result of compiled computer source code, can look random even though every logic gate position represented in that string was purposefully chosen for maximized utility. From a Shannon point of view, any truly random mutation in a genetic sequence would seem to have minimal effect on the already *seemingly* random frequency of the four nucleotides. Since 99% of genes already appear to be random, random mutations would tend to randomize only the 1% of *apparent* order within that gene's bit content. Random mutations would have a much more dramatic effect on the Shannon uncertainty found in redundant sequences (e.g., in introns rather than exons).

But genes are in reality programming strings. They are not analogous to programming strings. They ARE programming strings. It could be seriously argued that computer programs are analogous to genetic and genomic programming. Each nucleotide added to the string is an additional configurable switch setting added to the programming syntax. If genetic prescription is random, why are we spending billions of dollars on ascertaining reference sequences? Further, a mutation of a random sequence is more than bordering on a non-sequitur.

Mutations, whether random or ordered by varying degrees of physicodynamic determinism, corrupt existing programming "choices." Random mutations of PI strings will consistently result in noise pollution and degradation of the meaning and function of that PI. Mutations resulting from extremely low-informational cause-and-effect determinism will also reduce any programming efficiency of an existing gene. The fixed orderliness of nature described by laws cannot program formal function because it freezes up logic gates. Switches must be freely configurable to program formal function.

The GS Principle (See Chapter 7) [5, 52] states that genetic determinism's strong contribution to life requires selection at its *formal* configurable-switch level, not just at the post-computational phenotypic level. Nucleotides must be selected and covalently bound into primary structures (sequence strings) prior to the realization of selectable function. Environmental selection cannot occur until final function and the fittest already-living organisms exist.

Replicative function is often confused with information prescription in the literature. These two functions have nothing to do with each other. Templating and complementary base-pairing are purely physicodynamic. They are both highly ordered with high probability and very low uncertainty. There is no formal component to templating or base pairing. It is largely "forced" via physicochemical constraints ("laws" and local initial conditions). Templating and base-pairing, therefore, are unrelated to Prescriptive Information (PI) generation. The only exception to this is the prescriptive sequencing of the template itself. Naturalistic templates are all low informational (e.g., clay adsorption produces homopolymers, not informational strands). Yet templating and self-replication are often erroneously appealed to as an explanation for the source of biological information. Point mutations and wobbles are noise producers, not programmers. Pointing to a template does not explain the origin of PI in the initial linear digital sequence of the template itself. No explanation is ever provided by naturalism for the *source* of PI in any template or biopolymer. The *sequencing* of nucleotides in a single, positive, prescriptively informational strand is formal, not physical. Untemplated, merely physicochemical polymerizations of over 100 mers at higher temperatures produce homopolymers, not PI polymers [53].

Once the functional sequencing is established in a positive informational strand, base-pairing is purely physicochemical. In our naïveté, we would expect that replication would merely copy the existing PI in reverse direction. The discovery that the complementary negative strand of DNA is simultaneously prescriptive of entirely different regulatory function only bespeaks the added dimensions of formal causality instantiated into molecular biology that totally defy all physicodynamic explanation. Mere physicodynamic base-pairing will never answer how each complementary strand is able to prescribe a different formal function.

Technically, duplication yields no new information even in the Shannon sense of "information." Duplication plus variation does yield new Shannon uncertainty. But duplication plus variation has never been demonstrated to produce new nontrivial Prescriptive Information (PI) [6].

How can nonphysical formal mathematics and formal biological cybernetics be derived naturalistically from physicality alone? Admits Weinberger, " . . . a theory such as ours must explicitly acknowledge purposeful action, or 'agency', in such diverse fields as evolutionary theory . . ." [54, pg. 105] Yet the whole point of evolutionary theory was to obviate the need for purposeful action and "agency."

15. Formal biocybernetics predates *Homo sapiens* and our cognition

All known life is cybernetic. If one assumes that humans evolved from previous lesser life forms in only the last one thousandth of life's history on earth, it follows that cybernetics predates humans. The simplest known life forms all display undeniable evidence of linear digital prescription using a representational Material Symbol System (MSS) [45, 46] and cybernetic regulation [4]. The biosemiosis that produced life, humans and their minds included, is formal. Even at a primordial life level, each ribonucleotide selection in a polymer is a configurable switch-setting [2, 8]. It is a memory token in a material symbol system [55]. In a theoretical RNA World, each linear digital symbol sequence (syntax) prescribes a certain three-dimensional configuration space of potential ribozyme function [5, 9, 56].

Pre-metazoan life utilizes the same representational symbol systems, linear digital programming, coding/decoding/translation between language/operating systems, and redundancy block-coding for noise reduction. They cannot be attributed to human mentation or heuristics. Neither chance nor necessity can explain these phenomena. Linear, digital, genetic algorithmic programming requires ontologically real selection contingency. Life could have arisen only through selection operating at the covalently-bound level of primary structure formation. Environmental selection of the fittest already-computed phenotypes is irrelevant to the question of how initial genes were programmed. Formally functional configurable switch settings could not possibly have been programmed by physicodynamics.

The destination of any message must have knowledge of the cipher and possess the ability to use it. Deciphering is a formal function—as formal as mathematics and the rules of inference. Deciphering of the source's code and prescriptive intent at the destination cannot be done by the chance and necessity of physicodynamics. An abstract and conceptual handshake must occur between source and destination. Shared lexicographical meaning must exist between source and destination. Source and destination must be in sync regarding pragmatic significance of the arbitrarily chosen language system in order to create a protocol in a communication sense.

Natural selection is always *post-computational*. Natural selection is *after-the-fact* of relatively bug-free program halting. Environmental selection does not explain how the program got "written." Genetic digital selections must be distinguished from analog dynamic folding and from environmental phenotypic selection. Molecular evolution models of the spontaneous generation of life must be able to demonstrate selection at the covalently-bound decision-node level. No such theory or model currently exists in naturalistic scientific literature. No empirical evidence or rational support exists for attributing genetic programming to stochastic ensembles. This would be like attributing a Ph.D. thesis to nothing but a secretary's typographical errors. Although a stochastic ensemble could happen to match a reference sequence, no operational context would exist for that particular sequence to mean anything metabolically. An entire formal operating system (or several), power plant, and manufacturing factory would have to simultaneously arise from sequence space at the same time and place. Cybernetics is required to generate homeostatic metabolic utility in the face of thermodynamic decline. Since cybernetics is a formalism, and since life at all levels is cybernetic, formalism predates not only *Homo sapiens*, but even invertebrates. Cybernetics cannot be reduced to human mentation. Cybernetics is not just a heuristic tool or metaphorical epistemology generated by our minds [55]. Molecular biological cybernetics produced our consciousness, not the other way around.

16. The F > P Principle

The Formalism > Physicality (F > P) Principle states that Formalism not only describes, but preceded, prescribed, organized, and continues to control, regulate, govern and predict physicodynamic reality and its interactions. The F > P Principle is an axiom that defines the ontological primacy of formalism. Formalism is the source of all aspects of reality, both nonphysical and physical. Formalism organized physicality before the fact of physicality's existence. Formalism gave rise to the equations, structure and orderliness of physicality rather than to chaos. This alone explains why the scientific method must be conducted in a rational manner, why the applicability of mathematics to physical interactions is reasonable rather than unreasonable, and why such formalism can predict physical interactions.

The quest for a mathematical unified field of knowledge presupposes the F > P Principle. The F > P Principle further states that reality is fundamentally arbitrary—rule and choice-contingency-based, not indiscriminately forced by an infinite regress of cause-and-effect determinism. Physicality cannot even spawn a study of itself—physics—because physics is a formal enterprise. Nothing within the "chance and necessity" of physicality itself is capable of generating formal logic, computation, mathematical relationships, or cybernetic control. Only formalisms can measure, steer, manage, and predict physicality. Physicodynamics constrains; formalism controls.

In this paper, we have defined critical terms, presented fundamental concepts related to emergence, and reviewed repeated and predictable observations that collectively demand acknowledgement of the F > P Principle as the most fundamental axiom of science. Reality is first and foremost formal; physicality is realized only secondarily. Formalism can be instantiated into physicality through the use of configurable switches, material symbol systems, and through the integration of components into a holistic functional system.

Physicality cannot merge with formalism. Physicality can *be used by* logical formalism, but physicality cannot merge with or control formalism. Only formalism can measure, steer, organize, manage, and predict physicality. The F > P Principle explains why and how design and engineering principles can be incorporated into physicality to render it uniquely functional and/or computational. Physicality cannot do this on its own.

A corollary of the F > P Principle is acknowledgement that humans did not create the formal physical laws; our minds just discovered them. Before our minds existed, physicality obeyed these mathematical rules of physical interaction. Their prescription and control are in no way dependent upon human consciousness. $F = ma$ governed physicality long before human mentation arrived on the scene to recognize such formal relationships.

While the initial formal rules were arbitrary (choice-contingent), once instantiated into physicality they became physical fixed "laws." Their formal prescription and control became translated into fixed invariant directives of physicodynamic determinism. Cause-and-effect chains became "ordered" or forced into regularities. The fundamentally formal rules became physical laws. From the physicality side of The Cybernetic Cut [4], the choice contingency of the initial rule-writing and instantiation can seem imperceptible. We see only the forced regularities described by the laws of nature. But the prescription of these regularities prior to instantiation into physicality was free, choice-contingent, and purely formal.

This formal rationality extends even to the roles of heat agitation, undetermined degrees of freedom in nature, and stochastic quantum events. Even randomness, chaos and dissipative structures can be formally and mathematically described, defined and predicted. The only thing that Einstein got wrong in his statement "How is it possible that mathematics, a product of human thought that is independent of experience, fits so excellently the objects of physical reality?" [13] was that mathematics is "a product of human thought." Human thought did not create mathematics. Human thought is just progressively discovering it and its role in cosmic organization. As we have learned throughout this anthology, it is a logical impossibility for order to have produced PI or organization. The orderliness of nature could not have produced mathematics, cybernetics, language capacity, the scientific method, scientific ethics, and all the other non-material formalisms; rather, it's the other way around.

The F > P Principle states that the flow of control and organization is unidirectional from formalism to physicality. No reversibility exists between the law-based necessity of physicality and the rule-based choice contingency of formalism. Physicality cannot generate formalism. Phase changes at the edge of chaos, fitness landscapes, so-called evolutionary algorithms, neural networks, cellular automata, and the infodynamics perspective cannot circumvent the F > P Principle. In every case, nontrivial function requires formal, choice-based, behind-the-scenes, artificial selection in experimental design in order to produce nontrivial utility. The fundamental modus operandi of all uphill climbs to optimize the "fitness functions" of evolutionary algorithms is subtle choice contingency. Markov processes ("Drunken walks") are not devoid of experimenter steering. Optimization of fitness functions is formal, not physicodynamic. Genetic algorithms start with a pool of potential *formal solutions* to a problem. The preferred choices can be instantiated secondarily into material tokens and into Material Symbol Systems (MSS) [57]. Once instantiated into physicality, MSS's then can cause physicodynamic effects. But their utility was formally, not physico-dynamically, programmed.

The F > P Principle is a far more contemporary and less metaphysical axiom than Plato's original notion of Form [58, 59]. The F > P Principle adds to Plato's and Aristotle's early metaphysical explorations many benefits of the Enlightenment, modern and postmodern scientific thought and empirical experience. This axiom should be considered the most foundational principle of science. Without it, no basis exists for demanding science's subjection to logic theory. It explains science's demand for quantification (formal representation with numbers followed by numerical manipulations). The axiom provides a basis for trust in repeated observations and demand for prediction fulfillment. It explains why falsifiability is a valid test of scientific objectivity. Apart from the F > P Principle, the requirement of mathematical quantification in science makes little sense. The sciences of physics, chemistry, and biology, along with applied mathematics, computer science, and engineering, all *demand* formalism's dominion and control over physicality.

Belief in "self-organization" and "emergence" in the absence of choice contingency is blind belief bordering on superstition. It completely lacks empirical confirmation, prediction fulfillment, and rational justification. The hypotheses of "self-organization" and "emergence" are not even falsifiable. What is potentially falsifiable is the null hypothesis that nontrivial *"self-organization does not happen absent choice contingency."* This null hypothesis was first published quite succinctly in peer-reviewed literature around the turn of the millennium [23, 60] and many times thereafter [1-9, 34, 52, 61-64]. The scientific community has been rigorously invited to provide such falsification. After a decade, no falsification has been provided. The firm scientific prediction is hereby made that no falsification of this null hypothesis will ever be provided without behind-the-scenes investigator involvement in experimental design (artificial selection rather than natural selection). After another decade or two with no worldwide success at falsification, the above formal scientific prediction should become a mature generalized theory or theorem, if not a tentative law of science. This proposed tentative law states that inanimate physicodynamics is completely inadequate to generate, or even explain, formal processes and procedures leading to sophisticated function (The Law of Physicodynamic Incompleteness). Chance and necessity alone cannot steer, program or optimize algorithmic/computational success to provide desired nontrivial utility.

The time has come to extend this null hypothesis into a formal scientific prediction"No nontrivial algorithmic/computational utility will *ever* arise from chance and/or necessity alone."

How can such a bold, dogmatic prediction possibly be made by any reputable scientist? The answer lies first in the fact that it is just a null hypothesis designed for open-minded testing. The author of the hypothesis himself actively pursues falsification. Its deliberately absolutist tone begs falsification all the more in the challenging spirit of quality science. Second, the hypothesis itself arises from logical inference in addition to seemingly universal empirical observation. The statement is not just a product of inductive reasoning. The latter would be subject to overturning with minimal new data that could arise around the next blind empirical corner. The prediction is rather a logically valid inference enjoying deductive absoluteness within its own axiomatic system. Baring fallacious inference, the only possibility of falsehood would be that the logic flows from a faulty axiom. If a presupposition (pre-assumption about the nature of reality) is "out of touch with reality (ontologic, objective being)" then the prediction might not be "helpful." Unhelpfulness would be realized in the form of a prediction failure. Since no axiom is ever proven, science tends to proceed by assuming an axiomatic system to be tentatively valid, and testing it from many different directions through time. In this sense, all laws of science are considered best-thus-far generalizations subject to continuing experiment falsification.

After another decade or two with no worldwide success at falsification, the above formal scientific prediction should become a mature generalized theory, if not a tentative law of science, which Abel has named in advance *"The Law of Physicodynamic Incompleteness."* This proposed tentative law states that inanimate physicodynamics is completely inadequate to generate, or even explain, the mathematical nature of physical interactions (the laws of physics and chemistry). The Law further states that physicodynamic factors cannot institute formal processes and procedures leading to sophisticated function. Chance and necessity alone cannot steer, program or optimize algorithmic/computational success to provide desired nontrivial utility.

When we see sophisticated function of any kind, we have strong evidence suggesting that the Cybernetic Cut has been traversed across the one-way-only CS Bridge [4]. Nonphysical formalisms are the product of purposeful choice contingency [4, 7]. Choice contingency is instantiated into physicality via logic gates, configurable switch-settings, the purposeful selection of tokens from an alphabet of tokens, or cooperative integration of physical components into formal systems or conceptually complex machines [1-9, 23, 34, 55, 61, 62]. Mere physicodynamic constraints can accomplish none of the above examples of formal organization. Organization and sophisticated function in the physical world are all the products of formalisms instantiated into physicality. Physicality cannot generate nonphysical formalisms.

Physicality can self-order. But it cannot organize itself into or optimize formal algorithmic systems [9]. Physicodynamics cannot integrate parts into holistic, cooperative, functional metasystems. Inanimate physicality is incapable of producing organization because it cannot generate choice from among options or pursue the goal of function. The environment has no pragmatic preferences or values. It cannot generate nonphysical Prescriptive Information (PI) [6]. It cannot program logic gates or configurable switches [1]. Physicodynamics *does* include spontaneous non-linear phenomena; but it cannot practice the formal applied-science/math known as "non-linear dynamics." The latter is produced only by agents, not by inanimate nature.

But what is the utility of the F > P Principle? What does it *do* for us? The principle tells us to stop wasting time and hundreds of millions of research dollars trying to explain algorithmic optimization from physicodynamics alone. The Principle states that formal computational function cannot be generated by chance and necessity. Organization cannot be produced by physicodynamic self-ordering phenomena. Organization can only be generated through educated, expedient "choice with intent" at successive decision nodes. Organization arises out of the formal pursuit of desired utility.

Philosophical and metaphysical considerations are minimized in accord with Einstein's tenet of exercising a "minimum metaphysic" in scientific thought. Science, however, simply cannot be practiced competently without presupposing The F > P Principle. We already do this without realizing it. We just need to name and acknowledge the axiom we already subconsciously presuppose, and scrap the one we consciously incorporate erroneously into the very definition of science.

17. The axiomatic nature of all laws and principles

The axiom of ontological primacy of Formalism and its governance of Physicality flows from a combination of repeated observation and rational plausibility. It is still axiomatic, of course, as are all laws and principles of science and mathematics. But human experience and reason are far more consistent with the axiom of formalism's primacy than the pre-assumption of chaos and/or physicality's primacy.

It is easy to demand proof of The F > P Principle, and in the absence of proof immediately discount it. This is true of all axiomatic principles. It is not so easy to falsify it, or to find the slightest bit of evidence inconsistent with the Principle. Metaphysical naturalism's rejection of the Principle is purely philosophic, not scientific. The dogmatic pontification that physicality is everything is easily falsified. The bottom line of reality repeatedly traces back to formalism's choice contingency and organization (e.g., the periodic table; the Anthropic Principle, the reliability of mathematical laws to predict future physical interactions).

Like all axioms and "universal" laws, absolute proof of such principles is unattainable. Whether hypothetico-deductive or empirico-inductive, universal principles and laws must be viewed tentatively. At best, they represent "best-thus-far" knowledge. We accept them primarily because they are internally consistent and because they seem to work for us across a broad array of disciplines. Note that both of these criteria are formal requirements.

Principles should support a metanarrative (an over-arching story) of our experience of the whole of reality. We typically have a large sample space of observational data which conform to the principle. Fulfilled predictions made by the principle are especially convincing when they occur in unrelated and unexpected areas of science. But the principle nonetheless must be potentially falsifiable to be considered scientific [65, 66]. The F > P Principle is indeed potentially falsifiable. Only one example of physicodynamic causation of a single formalism is required.

Theorems are deduced from unproven axiomatic commitments. We choose to tentatively believe these axioms, and we choose to abide by the rules of logic theory within the deductive systems that flow from those axioms. We presuppose that self-contradiction cannot lead to progressive discovery of an objectivity outside our minds. We obey the rules of inference believing it will lead to pragmatic benefit or some computational utility. Obeying the rules seems to "work for us."

The reason Einstein advocated a "minimum metaphysic" in science rather than banning metaphysics from science was his realization of the inseparability of science from philosophy. He appreciated the axiomatic nature of mathematics and the presuppositional starting point of all scientific logic. The nature of the human condition is such that even scientific knowledge is inescapably finite, perspectival, and tentative. Some ideas must be pre-assumed to be true without absolute certainty. It is a non-sequitur to fallaciously conclude from our epistemological problem that objective reality is relative. Objective reality is exactly what it IS. We can only validly conclude that *our knowledge of* objectivity is subjective and relative, not reality itself.

Short-term usefulness can be provided even by ill-founded axiomatic systems. But long-term usefulness in many unrelated areas strongly suggests that an axiomatic system *corresponds to* objective reality—to the way things actually are. This is the realist's interpretation, at least. For the anti-realist, the centrality of choice with intent is all the more true. The solipsist's dreams of reality are not forced by external constraints and laws. The dream is a formal one, free and unconstrained by physicality or any inescapable objectivity outside of the solipsist's mind. Thus reality for the realist and anti-realist, for the modernist and the post-modernist, is ultimately formal, not physical. The F > P Principle holds either way.

The F > P Principle is nothing new. But it does need parsimonious expression using a formal term, and it needs to take its place as the most fundamental principle of science. It should not be surprising or controversial to presuppose that formalism preceded and controlled the very birth of physicality and physicodynamic relationships (Figure 3). Only dogmatic metaphysical imperatives and a long-standing Kuhnian paradigm rut preclude our admission of the obvious. Physics flows from formalism, not from physicality (its object of study). Physicality cannot explain physicality.

The F > P Principle is fully falsifiable through documentation of a single observed incident of nontrivial spontaneous physicodynamic enlightenment of any formalism. The firm scientific prediction is made that no exceptions to the F > P Principle will ever be observed.

References

1. Abel, D.L. 2009, The capabilities of chaos and complexity, Int. J. Mol. Sci., 10, (Special Issue on Life Origin) 247-291 Open access at http://mdpi.com/1422-0067/10/1/247
2. Abel, D.L.; Trevors, J.T. 2006, More than metaphor: Genomes are objective sign systems, Journal of BioSemiotics, 1, (2) 253-267.
3. Abel, D.L. 2007, Complexity, self-organization, and emergence at the edge of chaos in life-origin models, Journal of the Washington Academy of Sciences, 93, (4) 1-20.
4. Abel, D.L. 2008, 'The Cybernetic Cut': Progressing from description to prescription in systems theory, The Open Cybernetics and Systemics Journal, 2, 234-244 Open access at www.bentham.org/open/tocsj/articles/V002/252TOCSJ.pdf
5. Abel, D.L. 2009, The GS (Genetic Selection) Principle, Frontiers in Bioscience, 14, (January 1) 2959-2969 Open access at http://www.bioscience.org/2009/v14/af/3426/fulltext.htm.
6. Abel, D.L. 2009, The biosemiosis of prescriptive information, Semiotica, 2009, (174) 1-19.
7. Abel, D.L. 2010, Constraints vs. Controls, Open Cybernetics and Systemics Journal, 4, 14-27 Open Access at http://www.bentham.org/open/tocsj/articles/V004/14TOCSJ.pdf.
8. Abel, D.L.; Trevors, J.T. 2005, Three subsets of sequence complexity and their relevance to biopolymeric information., Theoretical Biology and Medical Modeling, 2, 29 Open access at http://www.tbiomed.com/content/2/1/29.
9. Abel, D.L.; Trevors, J.T. 2006, Self-Organization vs. Self-Ordering events in life-origin models, Physics of Life Reviews, 3, 211-228.
10. Wigner, E.P. 1960, The unreasonable effectiveness of mathematics in the natural sciences, Comm. Pure Appl., 13 (Feb).
11. Hamming, R.W. 1980, The unreasonable effectiveness of mathematics, The American Mathematical Monthly, 87, (2 February) 81-90.
12. Steiner, M. 1998, The Applicability of Mathematics as a Philosophical Problem. Harvard University Press: Cambridge, MA.
13. Einstein, A. 1920, Sidelights on Relativity. Dover: Mineola, N.Y.
14. Pattee, H.H. 1972, Laws and constraints, symbols and languages. In Towards a Theoretical Biology, Waddington, C. H., Ed. University of Edinburgh Press: Edinburgh, Vol. 4, pp 248-258.
15. Pattee, H.H. 1973, Physical problems of the origin of natural controls. In Biogenesis, Evolution, and Homeostasis, Locker, A., Ed. Springer-Verlag: Heidelberg, pp 41-49.
16. Pattee, H.H. 1995, Evolving Self-Reference: Matter, Symbols, and Semantic Closure, Communication and Cognition-Artificial Intelligence, 12, 9-28.
17. Pattee, H.H. 1995, Artificial Life Needs a Real Epistemology. In Advances in Artificial Life Moran, F., Ed. Springer: Berlin, pp 23-38.
18. Pattee, H.H. 2001, The physics of symbols: bridging the epistemic cut, Biosystems, 60, (1-3) 5-21.
19. Sagan, C. 2000, Cosmos, PBS TV series.
20. Koons, R.C. 2000, Realism Regained: An Exact Theory of Causation, Teleology, and the Mind. Oxford University Press: Oxford.
21. Pattee, H.H. 2000, Causation, Control, and the Evolution of Complexity. In Downward Causation: Minds, Bodies, and Matter, Andersen, P. B.; Emmeche, C.; Finnemann, N. O.Christiansen, P. V., Eds. Aarhus University Press: Aarhus, DK, pp 63-77.
22. Pearle, J. 2000, Causation. Cambridge University Press: Cambridge.
23. Abel, D.L. 2002, Is Life Reducible to Complexity? In Fundamentals of Life, Palyi, G.; Zucchi, C.Caglioti, L., Eds. Elsevier: Paris, pp 57-72.
24. Trevors, J.T.; Abel, D.L. 2004, Chance and necessity do not explain the origin of life, Cell Biology International, 28, 729-739.
25. Yockey, H.P. 2000, Origin of life on earth and Shannon's theory of communication, Comput Chem, 24, (1) 105-123.

26. Yockey, H.P. 2002, Information theory, evolution, and the origin of life. In Fundamentals of Life, Palyi, G.; Zucchi, C.Caglioti, L., Eds. Elsevier: Paris, pp 335-348.

27. Yockey, H.P. 2002, Information theory, evolution and the origin of life, Information Sciences, 141, 219-225.

28. Gánti, T. 2003, The Principles of Life. Oxford University Press: Oxford, UK.

29. Leff, H.S.; Rex, A.F. 1990, Maxwell's Demon, Entropy, Information, Computing. Princeton Univer. Press: Princeton, N.J.

30. Landauer, R. 1991, How molecules defy the demon. Book Review of Maxwell's Demon: Why Warmth Disperses and Time Passes by Hans Christian von Baeyer, reprinted 1998 by Random House, New York, last accessed in Oct of 2006 at http://physicsweb.org/articles/review/12/1/1. In Physics World

31. von Baeyer, H.C. 1998, Maxwell's Demon: Why Warmth Disperses and Time Passes. Random House New York.

32. Szilard, L. 1964, On the decrease of entropy in a thermodynamic system by the intervention of intelligent beings, Behav Sci, 9, (4) 301-10.

33. Lewis, G.N. 1930, The symmetry of time in physics, Science, 71, 569-576.

34. Abel, D.L. 2011, Moving 'far from equilibrium' in a prebiotic environment: The role of Maxwell's Demon in life origin. In Genesis - In the Beginning: Precursors of Life, Chemical Models and Early Biological Evolution Seckbach, J.Gordon, R., Eds. Springer: Dordrecht.

35. Voie, A. 2006, Biological function and the genetic code are interdependent, Chaos, Solitons & Fractals, 28, (4) 1000-1004.

36. Berlinski, D. 2000, The Advent of the Algorithm: The Idea that Rules the World. Harcourt, Inc.: New York.

37. Abel, D.L. 2009, The Universal Plausibility Metric (UPM) & Principle (UPP), Theor Biol Med Model, 6, (1) 27 Open access at http://www.tbiomed.com/content/6/1/27.

38. Küppers, B.-O. 1990, Information and the Origin of Life. MIT Press: Cambridge, MA.

39. Monod, J. 1972, Chance and Necessity. Knopf: New York.

40. Mayr, E. 1988, Introduction, pp 1-7; Is biology an autonomous science? pp 8-23. In Toward a New Philosophy of Biology, Part 1, Mayr, E., Ed. Harvard University Press: Cambridge, MA.

41. Mayr, E. 1982, The place of biology in the sciences and its conceptional structure. In The Growth of Biological Thought: Diversity, Evolution, and Inheritance Mayr, E., Ed. Harvard University Press: Cambridge, MA, pp 21-82.

42. Yockey, H.P. 1992, Information Theory and Molecular Biology. Cambridge University Press: Cambridge.

43. Yockey, H.P. 2005, Information Theory, Evolution, and the Origin of Life. Second ed., Cambridge University Press: Cambridge.

44. Bohr, N. 1933, Light and life, Nature, 131, 421.

45. Rocha, L.M. 1997, Evidence Sets and Contextual Genetic Algorithms: Exploring uncertainty, context, and embodiment in cognitive and biological systems. . State University of New York, Binghamton.

46. Rocha, L.M. 2000, Syntactic autonomy: or why there is no autonomy without symbols and how self-organizing systems might evolve them, Annals of the New York Academy of Sciences, 207-223.

47. Ledford, H. 2010, Mystery RNA spawns gene-activating peptides: Short peptides that regulate fruitfly development are produced from 'junk' RNA. In NATURE, Vol. Published online 15 July.

48. Craig, J.M.; Wong, N.C. 2011, Epigenetics: A Reference Manual | Book Caister Academic Press: p 450.

49. Robertson, M. 2010, The evolution of gene regulation, the RNA universe, and the vexed questions of artefact and noise, BMC Biology, 8, (1) 97.

50. Beiter, T.; Reich, E.; Williams, R.; Simon, P. 2009, Antisense transcription: A critical look in both directions, Cellular and Molecular Life Sciences (CMLS).

51. Royo, H.; Cavaille, J. 2008, Non-coding RNAs in imprinted gene clusters, Biol Cell, 100, (3) 149-66.

52. Abel, D.L. The GS (Genetic Selection) Principle [Scirus Topic Page]. http://www.scitopics.com/The_GS_Principle_The_Genetic_Selection_Principle.html (Last accessed September, 2011).

53. Costanzo, G.; Pino, S.; Ciciriello, F.; Di Mauro, E. 2009, RNA: Processing and Catalysis: Generation of Long RNA Chains in Water, J. Biol. Chem., 284, 33206-33216.

54. Weinberger, E.D. 2002, A theory of pragmatic information and its application to the quasi-species model of biological evolution, Biosystems, 66, (3) 105-19.

55. Abel, D.L.; Trevors, J.T. 2007, More than Metaphor: Genomes are Objective Sign Systems. In BioSemiotic Research Trends, Barbieri, M., Ed. Nova Science Publishers: New York, pp 1-15

56. Durston, K.K.; Chiu, D.K.; Abel, D.L.; Trevors, J.T. 2007, Measuring the functional sequence complexity of proteins, Theor Biol Med Model, 4, 47 Free on-line access at http://www.tbiomed.com/content/4/1/47.

57. Rocha, L.M. 2001, Evolution with material symbol systems, Biosystems, 60, 95-121.

58. Plato 1996, Parmenides In Parmenides' Lesson, Sayre, K. M., Ed. University of Notre Dame Press: Notre Dame.

59. Plato 1999, Allegory of the Cave, Book 7 of the Republic (514A–520A) 360BC. In Great Dialogues of Plato: Complete Texts of the Republic, Apology, Crito Phaido, Ion, and Meno, WarmingtonRouse, Eds. Signet Classics: New York, Vol. 1, p 316.

60. Abel, D.L. 2000 Is Life Reducible to Complexity?, Workshop on Life: a satellite meeting before the Millennial World Meeting of University Professors, Modena, Italy,

61. Abel, D.L. 2006 Life origin: The role of complexity at the edge of chaos, Washington Science 2006, Headquarters of the National Science Foundation, Arlington, VA

62. Abel, D.L. 2008 The capabilities of chaos and complexity, Society for Chaos Theory: Society for Complexity in Psychology and the Life Sciences, International Conference at Virginia Commonwealth University, Richmond, VA., Aug 8-10.

63. Abel, D.L. The Cybernetic Cut [Scirus Topic Page]. http://www.scitopics.com/The_Cybernetic_Cut.html (Last accessed Sept, 2011).

64. Abel, D.L. Prescriptive Information (PI) [Scirus Topic Page]. http://www.scitopics.com/Prescriptive_Information_PI.html (Last accessed September, 2011).

65. Popper, K. 1963, Conjectures and Refutations. Harper: New York.

66. Popper, K.R. 1972, The logic of scientific discovery. 6th impression revised. ed., Hutchinson: London.

Glossary

Abiogenesis—the belief that life emerged spontaneously from non-life through natural process.

"Adjacent other"—the wonderfully inviting, mystical, poetic notion of Stuart Kauffman describing his belief in a spontaneously arising formal capability of physicodynamics (the inanimate mass/energy interactions, forces and laws of motion that are the subject of physics). Unfortunately, such imagination is purely metaphysical, never once observed, unfalsifiable, and has never logged a single prediction fulfillment. It can best be described as superstition or fairy tale—certainly not science.

Agency—the ability to choose from among real options and to voluntarily pursue goals such as formal utility. Agents are able to program logic gates, steer courses of action through long strings of decision nodes, and assemble and organize objects and events to create potential function—function not yet existent at the time choices must be made. Agency is invariably associated with life. Life itself is utterly dependent upon cybernetic programming—a phenomenon never observed independent of agency.

Algorithm—a step-wise, discrete process or procedure—often computational—leading to future utility. Algorithms require wise choices at decision nodes, logic gates and configurable switches prior to the realization of any function. Algorithms cannot be generated by after-the-fact natural selection of the fittest computational result or already-programmed species.

Animate—living.

Arbitrary—unconstrained by initial conditions or cause-and-effect determinism. As used in the context of cybernetics, arbitrary means more specifically choice-contingent, not chance-contingent. Arbitrary does not mean that the chooser flips a coin to decide, or that the chooser does not care what is chosen. Symbol systems, for example, require purposeful, choice-contingent assignment of certain symbolic "strokes of pen" to represent specific meaning. By convention, arbitrary rules of interpretation are followed that allow sender and receiver to communicate the same meaning and function from those symbols and symbol syntax.

Artificial Selection—change brought about by the purposeful choice contingency of agents selecting from among real options at bona fide decision nodes.

Change induced by choice-contingent causation and control (CCCC). Selection FOR POTENTIAL fitness—something that natural selection cannot do.

Axiom—A deductively underivable, and empirically and logically unprovable, propositional statement that is tentatively assumed to be true, or self-evident, and which serves as the basis for a whole deductive system of thought and inference.

Bijection—a mapping, correspondence, or translation, usually one to one, of one symbol system to another. When Hamming redundancy block-coding is used to reduce noise pollution in the Shannon channel, mapping can be many to one (e.g. triplet codons prescribing each amino acid).

Blueprint—a two-dimensional picture, or composite of signs, representing the plans of a building or other physical structure. The term blueprint is often misapplied to genetic and genomic instruction. Genomics does not employ signs or blueprints. Codons serve only as block codes of symbols in a formal linear digital material symbol system (MSS). No direct physicochemical interaction is involved in the polycodonic prescription of polyamino acid sequencing that determines which protein is produced in ribosomes.

Chance contingency—non-willful, non-steered independence from apparent "necessity" (cause-and-effect determinism). Possibilities and options are not purposefully chosen, but result from "the roll of the dice." Chance contingency gives rise to random variation—"noise." The Brownian motion caused by the heat agitation of molecules is an example of seeming chance contingency. "Just how random is randomness?" remains an open question. Many have argued that seemingly random events are actually the result of thus-far unknown causes, and highly complex interactions between multiple known physicodynamic causes.

Chaos—disorganization, not disorder! Abundant highly-ordered dissipative structures of Prigogine's chaos theory form momentarily out of chaos in nature. No spontaneous dissipative structure shows any evidence of formal organization. In fact, most self-ordered dissipative structures such as hurricanes and tornadoes only destroy organization.

Chemoton—Tibor Ganti's abstract model of the simplest all-or-none unit of life. It consists of three non-living, autocatalytic chemical components: a motor, boundary, and prescriptive information system. The stable motor is capable of self-reproduction and synthesis of everything needed for the other two subsystems. The chemical boundary is envisioned to be semipermeable and to allow transport in of needed nutrients and the excretion of wastes. The prescriptive information must be capable of self-replication and must control, not just constrain, metabolism, growth, and reproduction. The chemoton model lacks enzymes and genetic code. The problems with Ganti's model are many, starting with the fact that no one has ever observed such a minimal unit of life short of the cell itself. The mechanisms provided in the model are entirely inadequate to explain the derivation of most of this unit's attributes and capabilities.

Choice contingency—freedom from determinism involving a purposeful selection from among real options. Choice contingency is exercised by agents with intent for a reason and purpose. The goal of choice contingency is almost always some form of utility that is valued by the chooser.

Choice-Contingent Causation and Control (CCCC)—the steering of physical events and the organizing of physical entities into potential usefulness. CCCC can generate extraordinary degrees of unique functionality that have never been observed to arise from randomness or law-described necessity. Neither physicodynamics nor evolution can pursue potential utility (e.g., the programming of computational success prior to its realization). CCCC does. CCCC is the only known cause and governor of formalisms.

Code—a representational symbol system used to assign associations (e.g. via a codon table) or to convey meaningful messages (e.g., messenger molecules). In an everyday connotation, coding signs and symbols are usually substituted for letters or words. Most codes (e.g., ASCII, Zip code) are "open," (non-encrypted) with arbitrary meaning to communicate between two independent worlds. The codon/amino acid code is the most widely known code in life, but more than 20 other biosemiotic codes have been discovered in the past decade, each with no known physicochemical "cause." In molecular biology, genetic code is specifically used for:

1. instantiation of formal, immaterial programming choices into physicality
2. efficiency in translation between two different material symbol systems where molecules serve as "physical symbol vehicles" (tokens) in two different material symbol systems (MSS) rather than being mere physico-chemical interactants/reactants
3. ease-of-transmission
4. noise pollution prevention in the Shannon channel (e.g., redundancy block coding)

5. proof reading and error correction (e.g. the processing of parity bit coding to detect noise pollution)

Complexity—the opposite of regularity, order, redundancy, and pattern. Complexity does not lend itself to algorithmic compressibility. Maximum complexity corresponds to randomness which contains no order, pattern or compressibility. Complexity is at opposite extremes with order on a bidirectional vector. Combinatorial complexity itself has nothing to do with functionality or the choice-contingent causation and control (CCCC) that generates nontrivial utility. The only relation of complexity to positive formalism is the mathematical probabilism used to measure complexity's negative uncertainty.

Composome—a hypothesized "metabolism-first" model referred to as an "ensemble replicator" or "compositional genome." The model imagines a self-reproducing assembly of different molecular species that manifests protometabolic "networks." The model was advanced because of serious problems with 1) template replication, 2) non-enzymatic biopolymer synthesis, and 3) a lack of Prescriptive Information (PI) source to program functional sequencing in RNA-World related models. No explanation has ever been provided for how protometabolic cybernetic networks could have spontaneously organized from physicodynamics alone, or how an ensemble of molecular species could have reliably reproduced themselves. Recent work by well-known and respected investigators has shown that the replication of compositional "information" is so inaccurate that fitter composomes could not possibly have evolved into metabolism-first life forms.

Configurable Switch—a purely physical device designed specifically to record (instantiate) nonphysical, formal choices into physical reality without any influence of physicodynamic forces, laws and constraints. Configurable switch settings are physicodynamically indeterminate (inert, decoupled, incoherent).

Configurable Switch (CS) Bridge—the one-way bridge that spans The Cybernetic Cut. Choice contingency causation and control (CCCC) traverses the vast ravine known as The Cybernetic Cut allowing traffic only from the formal far side to the physicodynamic near side. All formal meaning, function and bona fide organization enters the physical realm via this one-way bridge. Through "configurable" switch settings, formal choice contingency can become a source of physical causation. The setting of these configurable switches and logic gates constitutes the building of the CS Bridge. Nonphysical formalism itself can never be physical. In addition, the chance and necessity of physicality cannot steer objects and events towards formal utility. Chance and necessity cannot compute or make programming choices. Mere constraints cannot control or regulate. The inanimate environment does not desire or pursue function over nonfunction. So how does physicality ever get organized into usefulness of any kind? How does stone and mortar ever become a building? The answer lies in our ability to build a CS Bridge from the far side of The Cybernetic Cut—the formal side of reality—to the near side—the physicodynamic (physical) side of the ravine. The scaffolding needed to build this bridge consists of devices that allow instantiation of formal choices into physical recordations of those choices. This is accomplished through the construction of physical logic gates—the equivalent of Maxwell's demon's trap door. The gate can be opened or closed by agent choice at different times and in difference contextual circumstances. The open or shut gate corresponds to "yes" vs. "no," "1" vs. "0." Because the gate can be opened or closed by the operator at will, we call it a "configurable" switch. Another means of crossing the one-way CS Bridge across The Cybernetic Cut is to select physical symbol vehicles (tokens) from an alphabet of tokens available in a material symbol system. Assembling components into a holistic Sustained Functional System (SFS) or machine is another example of the one-way traffic flow across the CS Bridge from formalism to physicality.

Computational halting—a program finishes running rather than going on forever. Computational "success" is usually implied with the term halting, meaning that the program does what it is supposed to do within a finite period of time.

Constraints—a restriction or limitation of possibilities caused by initial (starting) conditions or by the regularities of nature described by physical law. Constraints themselves play no role in steering, controlling or regulating events to achieve formal function. Constraints are blind to formalisms. However, constraints can constitute barricades and bottlenecks for agent-pursued goals.

Contingency—in a past-tense context, contingency means that an event could have occurred other than how it happened. In a present and future context, contingency means that events can unfold in multiple ways despite both local and seemingly universal law-like constraints. Contingent behavior is not forced by physicodynamic necessity. Contingency embodies an aspect of freedom from physicochemical determinism.

Control—to purposefully steer toward the goal of formal function and pragmatic success. To regulate. To select for potential usefulness.

Cybernetic Cut—the most fundamental dichotomy of reality. The dynamics of physicality ("chance and necessity") lie on one side of a great divide. On the other side lies the ability to choose with intent what aspects of ontological being will be preferred, pursued, selected, rearranged, integrated, organized, preserved, and used (formalism). Life is unique from inanimate physics and chemistry in that life's control and regulation arise from the far side of The Cybernetic Cut.

Cybernetics—the study of control and of various means of programming, organizing, steering, and regulating physicality. Mere physicodynamic constraints are blind and indifferent to formal success. Only controls, not constraints, steer events toward pragmatic goals such as being alive and staying alive.

Decision nodes—bifurcation points which cannot be traversed by a mere "flip of the coin," at least not if one expects pragmatic results or reliable escape from danger. Decision nodes, as the name implies, require wise purposeful choices to achieve goals. A classic example is the purposeful setting of a "logic gate" in computing in order to integrate circuits or achieve computational success.

Decision theory—the study of various outcomes resulting from purposeful decisions at bona fide decision nodes. Decision nodes are more than mere "bifurcation points," which could be traversed using a fair coin flip to determine which way to go at each "fork in the road." When decision nodes are replaced with mere bifurcation points, universal experience shows a rapid deterioration of formal function potential.

Decode—to decipher the meaning of a message through mapping representational symbols to meaningful language or computation. The interpretation of symbols and symbol syntax in a symbol system.

Decrypt—to decode, but with the connotation that the original encoding was not "open," but written with the intent to make decoding very difficult by an enemy at war, for example.

Descriptive Information (DI)—positive background semantic information coming from an external source that serves to reduce uncertainty and to educate one's knowledge. DI provides valued common-sense knowledge to human beings about the way things already are. Thus, being can be described to provide one form of Functional Information (FI: intuitive and semantic information). However, the DI subset of FI is very limited and grossly inadequate to address many forms of instruction (Prescriptive Information (PI) and "how to" information for design, creativity, engineering, control and regulation.

Dissipative Structures of Chaos Theory—spontaneously self-ordered, momentary phenomena usually occurring in rapid succession so as to give the impression of a sustained structure (e.g., a candle flame; a tornado). Dissipative structures occur naturally out of mass/energy interactions alone. They require no choice-contingent causation and control (CCCC). Dissipative structures are often mistakenly viewed as evidence of self-organization in nature when in fact they example nothing more than spontaneous self-ordering with no formal components and no attention to the goal of functionality of any kind.

Edge of Chaos— the wonderfully inviting and mystical notion of complexity pursued by Christopher Langton, Doyne Farmer, J.P. Crutchfield, Melanie Mitchell, Stuart Kauffman and others that loosely describes a state of spontaneously realizable formal capability and self-organization arising out of physicodynamics alone. Melanie Mitchell has since questioned the validity of this notion. Such imagination is purely metaphysical, unobserved in inanimate nature, unfalsifiable, and no record exists of a single prediction fulfillment. It can best be described only as superstition or fairy tale, except where formalism is smuggled in through the back door to illegitimately redefine such terms as "phase transitions" and "constraints" (e.g., using the word "constraints" to mean formal "controls," where the constraints of inanimate cause-and-effect determinism are illegitimately granted the ability to purposefully steer events toward formal functionality or pragmatic success).

Emergence—the spontaneous occurrence in nature of more complex patterns arising from multiple simpler interactions. The spontaneous formation of symmetrical patterns in snowflakes during atmospheric precipitation is an example of emergence arising from purely physicodynamic self-ordering. Candle flame shapes, vortices of swirling water at bathtub drains, tornadoes and hurricanes all self-order spontaneously into rapid successions of momentary dissipative structures (the subject of chaos theory). Poorly understood is that no known cases of emergent self-ordering have anything to do with organization, and especially not "self-organization." Organization is formal and always arises through choice contingent causation and control (CCCC) from the far side of The Cybernetic Cut. No instance of bona fide "self-organization" has ever been observed; only unimaginative, redundant, lo-informational, self-ordering occurs spontaneously in inanimate nature out of chaos (which means disorganization, not disorder!).

Encode—To use a symbol system to represent, record and communicate meaningful messages. Molecular biology stores and passes along into progeny Prescriptive Information (PI, of which linear digital cybernetic programming is a major component) needed for organization and metabolic function. Encoding involves conversionary algorithms that biject or translate one symbol system into another.

Encrypt—to encode using a symbol system not easily deciphered and purposefully inaccessible to unwanted decoders.

Entropy—energy not available for formally useful work; the progressing formal disorganization observed in nature that is so often erroneously confused with increasing "disorder." Evidence of the 2nd Law is regularly observed with simultaneous increases in order, as with crystallization. Clearly, increasing entropy is not synonymous with increasing disorder. Physicodynamic entropy is not the same as informational entropy, which is a measure of epistemological uncertainty associated with a random variable. Informational entropy is a purely formal concept which, being nonphysical, has nothing to do with mass or energy, and everything to do with mathematical probabilism.

Epigenetic—the study of variation in heritable gene expression that is not caused by variation in nucleotide sequence of the genes. Histone deacetylation and DNA methylation are classic examples of gene suppression that does not affect nucleotide sequencing. Such alterations continue to alter gene expression throughout multiple future generations. Differentiation of the zygote (fertilized egg) into different cell types during development involves still other aspects of epigenetic control.

Epigenomics—the study of factors such as epigenetic DNA methylation, histone protein modifications, and chromatin structure on overall genomics and upper-level DNA structural (three-dimensional) Prescriptive Information (PI).

Falsifiability—the possibility that a claim, particularly a universal assertion, can be evaluated and potentially refuted by empirical testing showing results incongruous with that claim. The capability of disproving a proposition, hypothesis or theory by showing logical contradiction, or by finding, through experimentation, repeatable contradictory exceptions.

Fits—functional bits. The measurement of Functional Sequence Complexity, denoted as ζ, is defined as the change in functional uncertainty from the ground state $H(X_g(t_i))$ to the functional state $H(X_f(t_i))$, or $\zeta = \Delta H (X_g(t_i), X_f(t_j))$. The resulting unit of measure is defined on the joint data and functionality variable. The unit Fit thus defined is related to the intuitive concept of functional information, including genetic instruction and, thus, provides an important distinction between functional information and Shannon information.

Formal—relating to Plato's forms and Aristotle's appreciation of general classes of form and function that transcend particular physical structure and shape. Formal behavior is abstract, mental, arbitrary, nonphysical, and choice-contingent. The cognitive behavior of agents is typically goal- and function-oriented.

Formalism—a system of rules of thought or action typically involving symbol systems and requiring choices to be made at decision nodes, logic gates or configurable switch settings. Formalisms employ conceptual representationalism, mathematics, language, and/or categorical groupings of related ideas. Formalisms arise out of uncoerced choices in the pursuit of function and utility. Formalisms are typically computationally successful, integrated-circuit producing, and/or algorithmically optimizing. Formalisms require bona fide decision nodes, not just "bifurcation points. Language, mathematics, programming, and logic theory are all formalisms. Formalisms are governed by arbitrary rules, not laws. Listed below are aspects of reality that are all formalisms. None of these formalisms can be encompassed by a consistently held naturalistic worldview that seeks to reduce all things to physicodynamics:

1. Mathematics
2. Language
3. Inferential and deductive logic theory
4. The sign/symbol/token systems of semiosis
5. Decision theory
6. Cybernetics (including computer science)
7. Computation
8. Integrated circuits
9. Bona fide organization (as opposed to mere self-ordering in chaos theory)
10. Semantics (meaning)
11. Pursuits of goals
12. Pragmatic procedures and processes
13. Art, literature, theatre, ethics, aesthetics
14. The personhood of scientists themselves

All of the above formalisms depend upon choice contingency rather than chance contingency or necessity. Formalism also entails choices made in pursuit of potential function. Natural selection (NS) cannot select for potential function. NS can only favor the fittest already-programmed, already-existing, already living phenotypic organisms.

Formalism > Physicality (F > P) Principle—the most fundamental axiom of science states that Formalism not only describes, but preceded, prescribed, organized, and continues to control, regulate, govern and predict physicodynamic reality and its inter-actions. The F > P Principle is an axiom that defines the ontological primacy of formalism. Formalism is the source of all aspects of reality, both nonphysical and physical. Formalism organized physicality before the fact of physicality's existence. Formalism gave rise to the equations, structure and orderliness of physicality rather than to chaos (disorganization, not disorder!). This alone explains why the scientific method must be conducted in a rational manner, why the applicability of mathematics to physical interactions is reasonable rather than unreasonable, and why formalism can reliably predict physical interactions. The quest for a mathematical unified field of knowledge presupposes the F > P Principle. The F > P Principle further states that reality is fundamentally arbitrary—rule and choice-contingency based, not indiscriminately forced by an infinite regress of cause-and-effect determinism. Physicality cannot even spawn a study of itself—physics—because physics is a formal enterprise. Nothing within the "chance and necessity" of physicality itself is capable of generating formal logic, computation, mathematical relationships, or cybernetic control. Only formalisms can measure, steer, manage, and predict physicality. Physicodynamics constrains; formalism controls.

Function—usefulness; utility; contributing to productivity and efficiency. "A function is a goal-oriented property of an entity," Says Voie. "Functional parts are only meaningful under a whole, in other words it is the whole that gives meaning to its parts" [35].

Functional Sequence Complexity (FSC)—a sequence of subunits that produces utility in some larger context, as a string of amino acids performing a protein function of importance and value in a larger metabolic scheme. Also, a linear, digital, cybernetic string of symbols representing syntactic, semantic and pragmatic prescription; each successive symbol in the string is a representation of a decision-node configurable switch setting---a specific selection for potential function. FSC prescribes or produces usefulness, usually via algorithmic processing.

Functional Information (FI)—Intuitive semantic information that serves some purpose such as educating prior uncertainty, or instructing how to accomplish some goal. FI technically has two subsets: Descriptive (DI) and Prescriptive (PI), each discussed in this glossary.

Genetic Code—the arbitrary representational symbol system used by life to assign associations (e.g. via a codon table) or to convey meaningful messages (e.g., messenger molecules). In an everyday connotation, coding signs and symbols are usually substituted for letters or words. The codon/amino acid code is the most widely known code in life, but more than 20 other biological semiotic codes have been discovered in the past decade, each with no known physicochemical "cause." In molecular biology, genetic code is specifically used for:

1. instantiation of formal, immaterial programming choices into physicality
2. efficiency in translation between two different material symbol systems where molecules serve as "physical symbol vehicles" (tokens) in two different material symbol system (MSS) rather than being mere physicochemical interactants/reactants
3. ease-of-transmission
4. noise pollution prevention in the Shannon channel (e.g., redundancy block coding)
5. proof reading and error correction (e.g. the processing of parity bit coding to detect noise pollution)

Genetic Selection (GS) Principle—states that biological selection must occur at the point when the sequencing of monomers is established. Nucleotides must be selected at the molecular-genetic level of 3'5' phosphodiester bond formation. After-the-fact differential survival and reproduction of already-programmed, already-living phenotypic organisms (natural selection) does not explain polynucleotide sequence prescription and coding.

Genetics—the study of the prescription of form, function and metabolic contribution by the arbitrarily programmed material symbol system of polynucleotide sequencing in DNA. Triplet codon sequence in coding regions is translated into amino acid sequence in ribosomes which in turn determines minimum Gibbs-free-energy folding into three-dimensional protein globular structure. Genetics includes not only the study of coded genetic control through the inheritance of discrete units called genes, but variation through mutations, environmental factors, and the effects of many non-coding regulatory RNAs and epigenetic elements that affect biomolecular structure, function, metabolism and phenotypic expression.

Genomics—a more holistic study than genetics that investigates the interactions of all of the various networks of the entire genome, mRNA transcriptome, and proteome. Genetics tends to focus more on the effects of individual gene knock-outs. Genomics includes a study of pleiotropy (where one gene affects multiple phenotypic traits), epistasis (where additional modifier genes affect a single main gene), and heterosis (where outbreeding leads to hybrid vigor).

Hamming Block Code—an error-correcting redundancy code using a fixed or constant number of multiple loci comprising each "block" of a linear string of symbols to represent each prescribed unit of instruction. Triplet codons in coding regions of DNA, for example, always consist of a block of three nucleotides in a row to prescribe each amino acid. Discounting the stop codons, 61 ways exist to prescribe formally 20 amino acid options in the ribosomes. Catastrophic "frame shift" errors can result if decoding is not begun at the correct starting locus in the string, or if the number of loci in each block does not remain constant, or if additional amino acids are added to the code through time (each of which needing a new triplet codon block of representational symbols). The latter realities make the notion of gradual evolution of the genetic code from purely physicodynamic factors fraught with seemingly insurmountable problems.

Hypercycle—an autocatalytic cycle induced by circular constraints that lead to redundant self-replication. Hypercycles are envisioned to generate formal self-organization and progressively higher levels of formal organization. The model suffers from the confusion of formal programming and organizational controls with mere circular physicodynamic constraints. In the real world, these self-reinforcing loops lead only to the consumption of all resources in the production of the same few redundant products. The result is the depletion of the tremendous phase space that would be needed for any other theoretically contributing players to "evolve" into a legitimate protometabolism. Like all molecular evolution models of life origin, it suffers from a lack of organizational directionality and pursuit of formally useful interactive products. Empirical support for Eigen and Schuster's original notion of spontaneous hypercycles and their ever-increasing protometabolic competence has never accumulated.

Inanimate—non-living.

Instantiate—to insert or infuse aspects of one category into another normally separate and distinct category. In the context of cybernetics, the term is used to denote incorporating programming choices into physical computational devices. Nonphysical formalisms can only be instantiated into physical reality through the setting of configurable switches, the selection of "physical symbol vehicles" (tokens) from an alphabet of tokens, or though the design and engineering of physical devices (e.g., sophisticated machines, robots). In object-oriented analysis, design and programming, creating an object from a class is called instantiating the class. A class has certain aspects that are "infused", or become aspects of the object. Therefore, the word "instantiate" in this context involves not a "separate and distinct" category, but an "instance" of the category (class).

Law of Organizational and Cybernetic Deterioration/Decline (OCD Law)—The OCD Law states that, absent the intervention of formal agency, any nontrivial organization or cybernetic/computational function instantiated into physicality (e.g., integrated circuits; programmed computational success) will invariably deteriorate and fail through time. This deterioration may not be continual. However, it will be continuous (off and on, but overall consistently downhill). Computers, robots, all forms of Artificial Intelligence and Artificial Life, messages instantiated into material symbol systems or electronic impulses, will invariably progress toward dysfunction and fail. The OCD Law is not to be confused with the Second Law of Thermodynamics. The OCD Law is not concerned with the entropy of statistical mechanics or the "entropy" or "mutual entropy" of Shannon's probabilistic combinatorial uncertainty. Heat exchange, heat dissipation, phase changes, order and disorder are not at issue. The OCD Law addresses only the formal organization and utility already instantiated into physical media and environments. Only purposeful choice contingency at bona fide decision nodes can rescue from deterioration the organization and function previously programmed into physicality.

Law of Physicodynamic Incompleteness—an axiomatic proposition stating that physicochemical interactions are inadequate to explain the mathematical and formal nature of physical law relationships. Physicodynamics cannot generate formal processes and procedures leading to nontrivial function. Chance, necessity and mere constraints cannot steer, program or optimize algorithmic/computational success to provide desired nontrivial utility. nanimate physicodynamics is completely inadequate to generate, or even explain, the mathematical nature of physical interactions (the laws of physics and chemistry). The Law further states that physicodynamic factors cannot institute formal processes and procedures leading to sophisticated function. Chance and necessity alone cannot steer, program or optimize algorithmic/computational success to provide desired nontrivial utility. As a major corollary, physicodynamics cannot explain or generate life. Life is invariably cybernetic. Inanimate physics and chemistry are inadequate to explain the spontaneous self-organization of even a protometabolism, let alone the generation of life from non-life (abiogenesis.)**Laws**—generalized reduction algorithms, extracted and derived from observed regularities in reams of data, describing and predicting different aspects of regular physical interactions in nature despite varying initial conditions.

Linear digital symbol system—A system of recordation, transmission, and communication of messages between sender and receiver made possible by both following the same set of arbitrarily assigned rules of formal symbolization. Messages consist of a succession of discrete symbols and symbol syntax having arbitrarily assigned meaning and communicative function. Language, computer programs consisting of a succession of 0's and 1's, and polycodonic prescription of amino acid sequence in proteins by coding DNA are examples of linear digital symbol systems.**Liposomes**—artificially produced vesicles designed to deliver drugs and other agents to various locations within living cells, and used to mimic hypothesized protocells in life-origin studies.

Logic gates—a type of cybernetic configurable switch that can be set to either open or closed in a binary programming mode. Logic gates allow formal purposeful choices to be instantiated into physical computational systems and integrated circuits.

Machine— a physical device, often a relatively independent functioning contrivance, that utilizes mass and energy to accomplish a nonphysical formal function. The classical definition of machine involved the forces of motion and power to accomplish some desired task referred to as "work." Such "work" is far more than the mere transfer of energy. Even the "simple machines" are used by agents to transform the direction or magnitude of a force in order to accomplish a desired goal. Physicodynamics do not pursue goals. The advent of electronics and computers broadened our definition of "machine" no longer to require moving parts. Molecular biology has opened our eyes further to a vast array and coordinated interplay of the most sophisticated machines of all—molecular machines.

Macroevolution—the belief that evolution can spontaneously give rise to ever more sophisticated genetic and genomic PI programming, and to increasing conceptual complexity in organisms, giving rise to "higher" families, orders, classes, and phyla. No observations or prediction fulfillments exist in support of macroevolution. Falsification is not possible, raising the question of whether the notion of macroevolution is a scientifically respectable theory.

Material Symbol System (MSS)—A symbol system that formally assigns representational meaning to physical objects (tokens, physical symbol vehicles). The Game of Scrabble employs physical symbol vehicles, wood block tokens with inscribed symbols, that can be resorted to spell meaningful words and messages.

Meaning—Aboutness; function; the sense, importance, significance, implication, value, consequence, import or purpose of a message; the reason for sending a communication. In molecular biology, "meaning" is usually defined in terms of contribution to biofunction and holistic metabolism.

Mechanism—a means, directed process, programmed procedure, technique, system, or component of a machine that achieves some pragmatic goal. "Mechanism" is a formal term, not a physicodynamic term. "Mechanism," like the term "useful work," has no place in a consistently held naturalistic physics and chemistry context. The etiology of "mechanism" from both Latin and Greek derives from the word "machine." Metaphysical naturalism has never demonstrated the ability of physicodynamics and so-called "natural process" to produce nontrivial machines or sophisticated pragmatic mechanisms.

Message—a signal that contains interpretable meaning, and that manifests or fosters functionality at its destination. A signal that conveys Descriptive (DI) and/or Prescriptive Information (PI), both of which are subsets of Functional Information (FI).

Metabolism-First World—A model of life-origin that proposes that a protometabolism spontaneously self-organized, probably in a vesicle, without the aid of any Prescriptive Information contained in a material symbol system (such as DNA nucleotide or codon sequence) or RNA memory or catalysis. Variations include the Garbage-First model, Clay Life and other Mineral First models, Chemoton World, Peptide World, Lipid World, and Protein world.

Micelle— A spherical aggregate of surfactant molecules containing often containing a liquid colloid. In water, the surfactant molecules spontaneously self-order (NOT formally organize) with the hydrophilic (water-loving) "heads" aimed outward towards the aqueous solvent, and the hydrophobic (water-hating) tails aimed into the center of the sphere. A micelle is a crudely self-ordered structure similar to an oil-in-water droplet.

Microevolution—the universally acknowledged, spontaneously acquired, change in heritable phenotypic traits within a species, possibly within a family, but never extending to evolutionary transition to a more conceptually complex ("higher") order, class or phyla.

Molecular evolution—as used in this volume, molecular evolution pertains mostly to prebiotic evolution from inanimate molecules into a living state—abiogenesis. Of prime interest is how ordinary molecules could have self-organized, in a formal sense, under the influence only of physicochemical forces and attractions, to produce so many integrated biochemical pathways, cycles, highly tailored "parts" or components, and such goal-oriented holistic metabolism. All of these are needed to organize and sustain even the simplest conceivable life form.

Multiverse—the purely metaphysical rather than scientific notion that this Universe is only one of countless universes.

Mutations—alterations in genomic nucleotide sequencing, including the ribo-nucleotide sequencing of RNA viruses. A special case of mutation is when protein structure "mutates" (misfields) in prions in a way that affects the folding of other protein molecules in that family. Prion misfoldings are contagious and are subject to natural selection. Replication errors, mutagenic chemicals, radiation, transposons and deliberate hypermutation in immune cells are common causes of mutations. Mutations can be neutral (having no selective advantage, and no immediate apparent deleterious phenotypic effect), deleterious (most mutations), or in extremely rare instances, beneficial, at least in some very indirect way (e.g., sickle cell anemia rendering erythrocytes more resistant to the malaria parasite). The very recent discovery of vast new areas of functionality performed by non-coding DNA and non-mRNAs raises the question of whether most supposedly neutral mutations are really neutral. Far more likely is the progressive accumulation of noise pollution of what were highly refined regulatory instructions, the effects of which will only become apparent through time as our knowledge of molecular biology and microRNA regulation grows.

Natural selection—differential survivability and reproduction of the best already-programmed, already-living phenotypic organisms. Natural selection (NS) has no creative programming ability at the genetic or genomic level (See The GS Principle). NS is purely eliminative of less fit phenotypes. It cannot program genomes or other material symbol systems at the molecular level. Natural selection results only in the differential preservation and reproduction of the fittest already-existing organisms.

Necessity—a term often used almost synonymously with Law, as in Monod's Chance and Necessity, referring to the physicodynamic cause-and-effect determinism of inanimate nature. Necessity refers to regular physical interactions in nature that are so dependable, despite varying initial conditions, that the outcomes seem unavoidable, completely predictable, or "necessary."

Neural net—originally, the central nervous system consisting of circuits of neurons and their interconnections. Artificial neural networks are mathematical and computational models of the central nervous system and are used to model information processing and artificial intelligence. Neural networks are formal cybernetic constructs, not just physicodynamic "buttons and strings."

Noise—chance-contingent, meaningless, non-functional, unwanted disturbances or perturbations that corrupt meaningful, functional, desired, choice-contingent messages and Prescriptive Information (PI) commands.

Order—regularity, recurring pattern, redundancy, algorithmic compressibility. Order is antithetical to complexity and at opposite extremes with complexity on a bidirectional vector. Maximum complexity corresponds to randomness, which contains no order or compressibility. Order contains very little information, whereas organization typically contains high Prescriptive Information (PI) content from instantiated choice contingent causation and control (CCCC).

Ordered Sequence Complexity (OSC)—a linear string oflinked units, the sequencing of which is patterned either by the naturalregularities described by physical laws (necessity) or by statistically weightedmeans (e.g., unequal availability of units), but which is not patterned bydeliberate choice contingency (agency). OSC is marked by repetition or redundancy, or recurring pattern in its sequence. Reuse of programing modules or structures needed for construction can create the illusion of OSC when in fact the recurring pattern is generated by choice contingency (FSC). . The more highly ordered (patterned) a sequence, the more highly compressible that sequence becomes, the less Shannon uncertainty, and the less potential prescriptive information that can be instantiated into that sequence.

Organization—the choice-contingent association, categorization, configuring, steering, controlling, arranging or integrating of ideas or physical parts into a productive scheme, system or device that accomplishes formally useful work. Organization should never be confused with low-informational "order" or "pattern." Organization typically arises only out of high Prescriptive Information (PI) and sophisticated choice-contingent causation and control (CCCC).

Organization (O) Principle—Nontrivial formal Organization can be produced only by Choice-Contingent Causation and Control (CCCC). See Chap 12, Sec 9.

Panspermia— the belief that life originated elsewhere in the Universe and was spread to earth, probably by meteoroids or asteroids. This same definition applies to exogenesis. Panspermia suggests that life is more generalized throughout the Cosmos, whereas exogenesis does not necessarily make this claim. The notion of panspermia does nothing to help explain how life could have spontaneously self-organized out of nothing but physicodynamics. It does little to extend the time available for molecular evolution since the Big Bang, since the age of the cosmos is believed to be only three times that of the earth.

Pattern—predictable, regular or repetitive form. A recurring, compressible order that reduces Shannon uncertainty and the ability to instantiate functional choices (semantic information) into that medium. Patterns can arise, however, in meaningful messages and programs from deliberate reuse of linguistic elements and programming modules.

Peptide World hypothesis—the belief that life arose as a metabolism-first self-organization from interactions between short peptides and polypeptides. Adherents to this model point to the near impossibility of spontaneous ribonucleotide formation in a prebiotic environment, activation problems of ribonucleotides, difficulties of polymerization bond formation in water, short half-lives, etc.

Phenotype—the already-programmed, already-organized, already-living, holistic physical organism.

Physical symbol vehicle—a token; a physical object employed as a formal representational symbol. Meaning is consciously assigned arbitrarily to each physical object, thereby making possible the instantiation of choice contingency into the physical world. The physical token then functions as a formal meaningful and functional symbol in a material symbol system rather than as a physical interactant. The blocks of wood with inscribed letters in a Scrabble game, or the nucleotides in genes serve as physical symbol vehicles.

Physicodynamic determinism—cause-and-effect physicochemical interactions that lead back in an infinite regress of determinism to some physical first cause. Physicodynamic determinism, often referred to as "necessity," does not explain the reality of choice contingency—the freedom to choose from among real options to achieve choice-contingent causation and control (CCCC).   It also does not explain the rational, mathematical and formal nature of reality.

Physicodynamically indeterminate—Contingent; undetermined by cause-and-effect determinism; could have happened other than it did; having multiple possible options despite initial constraints and the laws of physics and chemistry.

Physicodynamically inert—physicodynamically indeterminate; contingent; undetermined by cause-and-effect determinism; could have happened other than it did; having multiple possibilities or options of occurrence despite initial constraints under the laws of physics and chemistry.

Physicodynamically incoherent— physicodynamically indeterminate; contingent; undetermined by cause-and-effect determinism; could have happened other than it did; having multiple possibilities or options of occurrence despite initial constraints under the laws of physics and chemistry.

Physicodynamic discontinuity— physicodynamically indeterminate; contingent; undetermined by cause-and-effect determinism; could have happened other than it did; having multiple possibilities or options of occurrence despite initial constraints under the laws of physics and chemistry.

Potential function—Formal function not yet existent, which, when nontrivial, only comes into existence through advanced planning, assembling of component parts or processes, programming and engineering choices.   Physicodynamics alone is incapable of producing sophisticated formal function. Natural selection (NS) cannot select for potential function at the genetic programming level (The GS Principle). NS can only prefer existing fittest phenotypic organisms.

Pragmatic—functional, useful, helpful, utilitarian, productive, contributory to a larger or higher organization or goal.

Prebiotic—referring to the inanimate physical environment (nature) that existed prior to the origin of life.

Prescriptive Information (PI)—a subset of Functional Information (FI) that either instructs or indirectly produces nontrivial formal function. PI is semantic "how to" information. PI provides the instructions required to organize and program sophisticated utility.   Potential formal function and computational success must be prescribed in advance by PI programming prior to halting, not just described after the fact. PI requires anticipation and "choice with intent" at bona fide decision nodes. PI either tells us what choices to make, or it is a recordation of wise choices already made. PI is positive, as opposed to negative uncertainty. Prescriptive information (PI) does far more than merely describe (Descriptive Information [DI])).   We can thoroughly describe a new Mercedes automobile, providing a great deal of DI in the process.   However, this functional DI might tell us almost nothing about how to design, engineer and build that Mercedes. PI provides the instructions required to organize and program sophisticated utility.   PI designs, creates, engineers, controls and regulates. The inanimate physical environment is incapable of participating in such formal pursuits.   So-called "natural" physicodynamics cannot generate nonphysical PI. PI can perform nonphysical "formal work."   PI can then be instantiated into physicality to marshal physical work out of nonphysical formal work. Cybernetic programming is only one of many forms of PI.   Ordinary language itself, various communicative symbol systems, logic theory, mathematics, rules of any kind, and all types of controlling and computational algorithms are forms of PI. Neither chance nor necessity has been shown to generate PI. Choice contingency, not chance contingency, prescribes nontrivial function. PI typically is recorded into a linear digital symbol system format. Symbols represent purposeful choices from an alphabet of symbol options. Symbol selection is made at bona fide decision nodes.

ProtoBioCybernetics—the study of the derivation of control and regulation in the first life forms. Cybernetics incorporates Prescriptive Information (PI) into various means of steering, programming, communication, instruction, integration, organization, optimization, computation and regulation to achieve formal function. "Bio" refers to life. "Proto" refers to "first." Thus, the scientific discipline of ProtoBioCybernetics specifically explores the often-neglected derivation through "natural process" of initial control mechanisms in the very first theoretical protocell.

Protobiont—a hypothesized initial precursor of living organisms, usually thought to have been a protocell with some semblance of a vesicular-like phospholipid or bilayer "membrane." Contained within this vesicle is believed to have been the minimal unit of protolife or life. Tibor Ganti's minimal unit of life, the chemoton, includes the vesicular or membrane-like barrier.

ProtoBioSemiotics—the study of meaningful or functional messaging and how it arose within and between the first protobionts.

Protocell—a hypothesized initial "cell" with a vesicular-like phospholipid or bilipid "membrane" in which life is imagined to have spontaneously self-organized.

Protometabolism—the hypothesized first semblance of integration of bio-chemical pathways and cycles into a holistic, organized, functional metabolic system.

Random Sequence Complexity (RSC)—a linear string of stochastically linked units, the sequencing of which is dynamically inert, statistically un-weighted, and is unchosen by agents; a random sequence of independent and equiprobable unit occurrence. RSC is the most complex of the three kinds of sequence complexity, the reason being that a random sequence contains no al-gorithmically compressible order. Its sequence cannot be enumerated using any representational string shorter than itself. RSC manifests the absence of any order or pattern. RSC represents maximum uncertainty, and therefore con-tains the maximum number of Shannon bits. Although maximally complex, RSC does nothing functional, emphasizing that complexity is not an explana-tion for utility or pragmatic worth.

Regulation—the choice-contingent steering, controlling, adjusting and fine-tuning of some formal process, procedure, or reaction sequence. To regulate presupposes freedom from law sufficient to manage events by formal choice-contingent causation and control (CCCC).

RNA analogues— Molecules similar in structure to RNA, but having the phosphate, ribose or nucleobase replaced with some alternative. Alternate nu-cleobase Molecules similar in structure to RNA, but having the phosphate, ri-bose or nucleobase replaced with some alternative. Altering nucleobases (e.g. fluorophores) typically result in altered base pairing and stacking properties. Peptide nucleic acid (PNA) is a phosphate-sugar backbone analogue. Other backbone analogues include threose nucleic acid (TNA), glycol nucleic acid (GNA), Morpholino or locked nucleic acid (LNA). Originally, it was hoped that RNA analogues might solve the many problems of prebiotic RNA chemis-try that threatened the RNA World hypothesis. However, the Pre-RNA World hypothesis has encountered many roadblocks of its own.

RNA World hypothesis—the belief that initial life consisted primarily of RNA rather than the DNA and protein necessary for current life. RNA can potentially retain nonphysical information in its physical matrix and self-replicate. RNA can act as a crude catalyst compared to proteins. Numerous biochemical hurdles in a prebiotic environment have rendered the RNA World hypothesis highly suspect. The PreRNA, RNA analog, and RNA World models probably remain the most favored models in life origin theory today. Ribonucleoprotein enzymes such as ribosomes are thought to have arisen from molecular evolution prior to DNA-protein life.

Rules—Choice-contingent guidelines intended to guide procedures, competing interests, and ethical behavior. Rules are nonphysical, formal, mental constructions. Rules are not laws. Laws describe and predict deterministic physicodynamic interactions. Loss of formal utility usually accompanies the disobedience of rules unless a pragmatically superior rule system is being explored. Rules can also be arbitrarily agreed-upon conventions that govern language and voluntary behavior. Rules exist to guide choices. Rules can be broken at will, often at the expense of efficiency or efficaciousness in accomplishing some pragmatic goal.

Semantic—meaningful or functional.

Semiotics—the study of symbolization using sign and symbol systems, meaningful message generation, language, programming, and the communication methods employed. The three main branches of semiotics are 1) semantics—the meaning generated by how symbols are arbitrarily assigned to represent objects and ideas, 2) Syntactics—the sequencing and relation of symbols to one another to create higher meaning, and 3) Pragmatics—the usefulness of symbol system applications and their communication.

Sign—a two-dimensional picture or drawing conveying representational meaning to one's senses. The picture or drawing is self-explanatory because we recognize by sight physical objects that are being depicted from our every-day empirical world. A visual image of real world objects is delivered by the sign. Our consciousness links the two-dimensional picture with our experience of and with that object. A picture of an automobile with two wavy lines emanating from behind its rear tires is a street sign conveying the message of slippery road conditions.

Signal—a transmission of mass/energy from one location to another, as a pulsating emission of light from a distant star. A signal need not have any meaning or function, and should be carefully distinguished from "message." Messages always contain formal meaning, and can only be instantiated into physicality through choice contingent causation and control (CCCC) from the far side of The Cybernetic Cut. Signals, on the other hand, can be entirely physicodynamic.

Stoichiometry—the branch of chemistry dealing with the relative quantities of reactants and products. Whole numbers usually represent the ratio of reactants to products.

Structure—a recognizable framework of categorization, pattern or order in an entity or relationship between entities. The manner in which the parts of a whole are assembled. Primary structure refers to the sequencing of monomers in a linear polymer. Secondary structure refers to the two-dimensional representation, at least, of alpha helices and beta strands (in proteins) and helices and stem-loops (in nucleic acids) due to base pairing and base stacking. Tertiary structure refers to the three-dimensional globular shape of folded proteins, ribozymes, and chromatin.

Sustained Functional Systems (SFS)—Any device, machine, network or system that both 1) continues on in time (is a non-dissipative structure in the sense of Prigogine's chaos theory) and that 2) generates sustained non trivial functionality. Prescriptive Information (PI) and Organization alone make Sustained Functional Systems (SFS) far from equilibrium possible. Maxwell's Demon's choice contingency of when to open and close the trap door so as to accomplish the goal of a sustained energy potential represents the very first true decision-node instantiation into physicality. The Demon's first choice is the birth of engineering and the artificial intelligence movement. Deciding when to open and close the trap door is the very first logic gate—the very first configurable switch-setting. The Demon's voluntary (arbitrary) trap-door operation represents the birth of integrated circuits, computational cybernetics, and life's regulatory mechanisms. No natural mechanism exists that can choose with intent to deliberately design, engineer and maintain a SFS. Yet without SFS's, life is impossible. SFS's predate and produced Homo sapiens. They therefore cannot be attributed solely to human mentation and creativity.

Symbol—an arbitrarily-shaped/generated character representing some assigned meaning by definition. The meaning of these "strokes of pen" is just arbitrary assigned by the sender and agreed to by the recipient. Otherwise, the message will not have meaning or function at its destination. A symbol, unlike a sign, conjures no meaning from one's sight memory of physical objects. The letters of most language alphabets are not signs, but symbols. Strings of such symbol characters spell words leading to lexicons of words. Hierarchies of phrases, clauses, sentences, and paragraphs can be constructed from the lexicon of words according to syntactical rules. Sometimes only one letter symbol, such as "H" or "C" on a faucet handle, conveys meaning. Mathematical symbols such as $\pi, \Omega, \xi, \Delta, =$, and $\neq$ are symbols, not signs. We cannot ascertain the meaning of these symbols from the symbol itself, except that we sometimes become so familiar with a certain symbol's assigned meaning that it begins to take on a function similar to a picture or drawing, thereby having a sign-effect from our sight memory (e.g., the symbol " $=$ " begins to be recognized visually as the a physical sign of equality). Codons function as symbols in molecular biology, not as direct physicochemical reactants or pictorial signs. Genes are not blueprints (two-dimensional pictures).

Symbol Systems—a means of recordation or communication that employs symbols to represent and encode meaning. Symbol systems allow recordation of deliberate choices and the transmission of linear digital prescriptive information. Formal symbol selection can be instantiated into physicality using physical symbol vehicles (tokens). Material symbol systems (MSS) formally assign representational meaning to physical objects. Even the analog perturbations of verbal semiosis can be symbolized with numerical representations in voice recognition software.

Token—a physical symbol vehicle. A physical object on which a symbol has been inscribed or to which symbolic meaning has been ascribed.

Transcribe—in molecular biology, to synthesize meaningful/functional RNA sequences containing Prescriptive Information (PI) using RNA polymerase enzymes from a DNA template.

Translate—to map one symbol system onto another in an effort to decode the initial system.

Turing machine and tape—a thought experiment imagining a device that can algorithmically process a string of successive symbols on a linear tape according to a table of rules.   An infinite memory is afforded by an infinite tape. Each symbol represents not only meaning, but also arbitrary choice contingency rather than chance and/or necessity.   The rules are also choice- contingent. The thought experiment can simulate the function of modern computers and their computational limits.

Undecidable— a decision problem that is impossible to always answer with a "Yes" or "No" using a single algorithm.   The term is most applicable to computational complexity theory. Alan Turing, for example, proved that the halting problem is undecidable for Turing machines. A verbal statement can also be considered "undecidable" with relation to Gödel's incompleteness theorems when that statement is neither provable nor refutable within a certain deductive axiomatic system.

Universal Probability Bound (UPB)—A quantifiable limit to an extremely low probability resulting from the limitation of probabilistic resources in that context. Statistical prohibitiveness cannot be established by an exceedingly low probability alone. Rejection regions and probability bounds need to be established independent of (preferably prior to) experimentation in any experimental design.

Universal Plausibility Metric—a numerical value measuring the plausibility (not probability) of extremely low probability events in view of the probabilistic resources in each context. The UPM employs the symbol ξ (Xi, pronounced zai in American English, sai in UK English, ksi in modern Greek) to represent the computed UPM according to the following equation:

$$\xi = \frac{f^{L}\Omega_{A}}{\omega}$$

where f represents the number of functional objects/events/scenarios that are known to occur out of all possible combinations (lower case omega, ω) (e.g., the number [f] of functional protein family members of varying sequence known to occur out of sequence space [ω]), and $^{L}\Omega_{A}$ (upper case Omega, Ω) represents the total probabilistic resources for any particular probabilistic context.   The "L" superscript context of Ω describes which perspective of analysis, whether quantum (q) or a classical (c), and the "A" subscript context of Ω enumerates which subset of astronomical phase space is being evaluated: "u" for universe, "g" for our galaxy, "s" for our solar system, and "e" for earth. Note that the basic generic UPM (ξ) equation's form remains constant despite changes in the variables of levels of perspective (L: whether q or c) and astronomic subsets (A: whether u, g, s, or e).

Universal Plausibility Principle—states that definitive operational falsification of any chance hypothesis is provided by the inequality of:

$$\xi < 1$$

where ξ is the measured UPM for that context. This definitive operational falsification holds for hypotheses, theories, models, or scenarios at any level of perspective (quantum or classical) and for any astronomical subset (Universe, galaxy, solar system, and earth). The UPP inequality's falsification is valid whether the hypothesized event is singular or compound, independent or conditional. Both UPM and UPP pre-exist and are independent of any experimental design and data set. No low-probability hypothetical plausibility assertion should survive peer-review without subjection to the UPP inequality standard of formal falsification ($\xi < 1$).

Utility—formal usefulness or functionality, usually as decided or evaluated by agents with reference to their desires and goals. A more objective concept of "utility" might be found in the biofunctionality of molecular machines, for example, with reference to the holistic metabolic goals of cells and organisms.

Vesicles—a complex version of the micelle containing one or more phospholipid bilayers that can enclose, transport and digest other substances. Cellular vacuoles, lysosomes, transport and secretory vesicles in living organisms have attracted much attention as models of possible protobionts (protocells) with crude "membranes." Phospholipids can form bilipid layer walls of artificially prepared liposomes.

Index

225, 226, 227, 246, 249, 273, 279, 280, 281, 286, 288, 292, 293, 295, 297, 300, 304, 322, 323, 353, 354, 358, 361, 367, 369, Glossary

composome, 14, 189, 239, 240, 259, 267, 272, 292, Glossary

computational halting, 26, 37, 57, 91, 92, 95, 97, 98, 101, 165, 178, 182, 213, 216, 298, 340, 341, 343, Glossary

configurable switch, 1, 3, 6, 11, 15, 19, 26, 27, 31, 33, 38, 41, 42, 43, 44, 45, 47, 48, 49, 57, 59, 62, 63, 64, 66, 67, 81, 83, 92, 94, 95, 100, 102, 104, 119, 130, 138, 140, 147, 150, 161, 170, 171, 172, 177, 178, 182, 183, 201, 211, 215, 219, 220, 245, 319, 336, 344, 345, 346, 363, 366, Glossary

Configurable Switch (CS) Bridge, 55, 58, 177, 219, 338, 359, Glossary

constraints, 1, 2, 6, 9, 11, 22, 24, 25, 33, 34, 35, 36, 37, 38, 40, 41, 42, 43, 44, 45, 50, 57, 58, 60, 61, 68, 70, 73, 83, 89, 110, 114, 118, 119, 121, 126, 129, 130, 131, 140, 147, 148, 149, 163, 170, 171, 176, 182, 189, 190, 191, 193, 203, 210, 211, 215, 217, 219, 230, 241, 242, 247, 248, 249, 251, 252, 256, 262, 269, 272, 276, 280, 299, 301, 318, 319, 325, 329, 334, 335, 337, 339, 340, 341, 344, 351, 353, 359, 360, 362, 365, 371, Glossary

control, 1, 2, 3, 6, 10, 11, 12, 13, 14, 24, 29, 33, 34, 38, 40, 42, 48, 49, 50, 51, 53, 55, 56, 57, 58, 60, 61, 62, 65, 69, 70, 93, 105, 107, 108, 110, 115, 119, 135, 142, 147, 148, 150, 152, 154, 157, 158, 160, 161, 162, 163, 164, 165, 170, 177, 179, 180, 182, 186, 190, 193, 194, 201, 203, 206, 208, 209, 210, 211, 212, 214, 219, 222, 227, 243, 244, 246, 248, 249, 252, 254, 259, 269, 270, 271, 272, 282, 283, 285, 287, 288, 289, 292, 294, 295, 296, 299, 300, 319, 327, 328, 331, 335, 339, 340, 341, 342, 343, 347, 348, 349, 356, 357, 358, 359, 360, 361, 362, 364, 369, 370, 372, 373, 374, Glossary

cybernetic, 7, 10, 15, 17, 29, 37, 48, 49, 52, 55, 56, 57, 58, 59, 60, 61, 63, 65, 68, 69, 71, 72, 82, 83, 94, 108, 112, 140, 141, 146, 148, 152, 159, 160, 177, 184, 211, 219, 221, 228, 229, 279, 280, 298, 300, 302, 310, 319, 322, 336, 338, 340, 342, 348, 353, 355, 359, 360, 362, 366, 372, 374

Cybernetic Cut, 55, 56, 57, 60, 69, 71, 148, 338, 342, 359, 360, Glossary

cybernetic string, 119

decision node, 2, 9, 10, 26, 27, 28, 40, 44, 48, 67, 104, 105, 130, 142, 153, 179, 183, 201, 220, 248, 253, 269, 310, 335, 340, Glossary

decision theory, 21, 56, 71, 245, Glossary

decode, 146, 158, 361, 375, Glossary

Descriptive Information (DI), 10, 15, 105, 142, 146, 216, 233, 361, Glossary

Dissipative Structures of Chaos Theory, 361, Glossary

edge of chaos, 91, 93, 103, 104, Glossary

Emergence, 1, 19, 55, 75, 135, 161, 189, 191, 206, 223, 224, 227, 228, 231, 278, 279, 286, 302, 305, 325, 362, Glossary

encode, 13, 73, 206, 225, 250, 362, 374, Glossary

encoding, 96, 119, 130, 145, 220, 250, 361, Glossary

entropy, 8, 34, 45, 53, 54, 81, 93, 113, 133, 177, 227, 276, 323, 336, 337, 340, 354, 362, 366, Glossary

epigenetic, 29, 57, 63, 67, 107, 151, 156, 158, 160, 164, 165, 168, 181, 182, 189, 199, 204, 288, 327, 343, Glossary

epigenomics, 16, 108, 146, 156, 165, 168, 327, Glossary

evolvability, 228, 256, 267, 280, 287, 293

falsification, 62, 71, 96, 98, 99, 108, 180, 198, 199, 201, 211, 212, 287, 297, 305, 306, 308, 311, 315, 317, 318, 320, 321, 337, 340, 349, 375, Glossary

Fits (Functional bits), 68, 89, 151, 174, 236, 326, 348, Glossary

formal, 1, 2, 3, 5, 6, 7, 8, 10, 11, 13, 14, 15, 19, 20, 21, 23, 24, 25, 26, 27, 28, 29, 30, 31, 32, 34, 35, 36, 37, 38, 40, 41, 42, 43, 44, 45, 46, 48, 49, 50, 55, 56, 57, 58, 59, 60, 61, 62, 63, 65, 66, 67, 68, 69, 70,

71, 75, 77, 80, 81, 83, 85, 86, 88, 91, 92, 93, 94, 97, 98, 99, 100, 101, 102, 103, 104, 105, 106, 107,
108, 110, 130, 135, 136, 137, 139, 140, 141, 143, 144, 145, 146, 147, 148, 149, 150, 151, 152, 153,
154, 155, 156, 157, 158, 161, 163, 165, 166, 167, 169, 170, 171, 172, 173, 175, 176, 177, 178, 180,
181, 182, 189, 190, 191, 192, 193, 194, 195, 197, 198, 199, 200, 201, 202, 203, 204, 205, 206, 207,
208, 210, 211, 212, 213, 214, 215, 216, 218, 219, 220, 221, 236, 237, 241, 242, 243, 247, 248, 249,
251, 252, 253, 255, 256, 262, 268, 269, 271, 272, 287, 294, 297, 299, 300, 301, 305, 307, 310, 319,
320, 325, 326, 327, 328, 329, 330, 331, 332, 333, 334, 335, 336, 337, 338, 339, 340, 341, 342, 343,
344, 345, 346, 347, 348, 349, 350, 351, 356, 357, 359, 360, 361, 362, 364, 365, 366, 367, 368, 369,
370, 371, 372, 373, 374, 376, Glossary

formalism, 3, 4, 5, 6, 25, 27, 31, 37, 40, 41, 42, 45, 50, 55, 56, 58, 63, 68, 69, 71, 83, 148, 155, 193, 203,
215, 247, 287, 295, 300, 325, 326, 327, 328, 329, 330, 331, 332, 334, 335, 339, 342, 346, 347, 348,
349, 350, 351, 352, 358, 359, 360, 362, 364, Glossary

Formalism > Physicality (F > P) Principle, 69, 247, 325, 328, 347, 364, Glossary

function, 1, 2, 3, 5, 6, 7, 8, 10, 11, 12, 13, 14, 15, 19, 20, 21, 23, 24, 25, 26, 27, 28, 29, 31, 32, 33, 35, 36,
37, 38, 39, 40, 41, 42, 43, 44, 47, 48, 49, 50, 55, 56, 57, 58, 59, 60, 61, 62, 63, 64, 65, 67, 70, 72, 76,
77, 79, 80, 81, 82, 83, 85, 86, 88, 89, 91, 92, 93, 94, 95, 97, 98, 99, 100, 101, 103, 104, 105, 106, 107,
118, 119, 120, 121, 123, 124, 128, 129, 130, 131, 132, 136, 137, 138, 139, 140, 141, 142, 144, 145,
146, 147, 148, 149, 150, 152, 153, 154, 155, 156, 157, 161, 162, 163, 164, 165, 166, 167,168, 169,
170, 171, 172, 173, 174, 175, 176, 178, 179, 180, 181, 182, 183, 186, 189, 190, 192, 193, 194, 195,
201, 203, 204, 205, 206, 211, 212, 213, 214, 215, 216, 218, 220, 232, 235, 238, 243, 245, 246, 247,
248, 249, 250, 251, 253, 254, 255, 258, 259, 260, 262, 266, 267, 268, 269, 270, 271, 272, 273, 276,
280, 295, 296, 297, 310, 316, 319, 322, 325, 327, 330, 331, 332, 333, 334, 335, 336, 337, 339, 340,
341, 342, 343, 344, 345, 346, 348, 350, 354, 356, 359, 360, 361, 362, 363, 364, 366, 367, 371, 372,
374, 375, Glossary

Functional Information (FI), 8, 142, 150, 216, 233, 266, 367, 371, Glossary

Functional Sequence Complexity (FSC), 15, 64, 76, 82, 88, 89, 117, 119, 174, Glossary

functional sequences, 117, 120, 123, 124, 126, 127, 131

functional state, 2, 89, 121, 123, 125, 363

functional uncertainty, 88, 89, 120, 121, 124, 126, 363

functionality, 11, 24, 25, 45, 81, 83, 85, 88, 92, 106, 108, 117, 118, 119, 120, 121, 123, 131, 132, 136,
140, 153, 156, 165, 167, 174, 176, 194, 231, 252, 253, 289, 299, 326, 332, 341, 358, 361, 362, 363,
367, 369, 376, Glossary

genetic code, 110, 122, 157, 264, 265, 284, 358, 364, 365, Glossary

Genetic Selection (GS) Principle, 67, 161, 162, Glossary

genetics, 8, 17, 32, 141, 157, 164, 177, 210, 225, 276, 287, 298, 364

genome, 5, 12, 15, 65, 66, 113, 152, 160, 162, 181, 184, 186, 188, 203, 208, 218, 229, 230, 241, 260,
266, 267, 270, 284, 287, 288, 289, 290, 292, 295, 296, 298, 299, 300, 343, 358, 364, Glossary

ground state, 89, 121, 122, 123, 126, 129, 130, 363

Hamming block code, 33, 172, 176, 213, 365, Glossary

highly ordered, 46, 76, 77, 78, 89, 91, 92, 95, 98, 110, 119, 129, 142, 149, 179, 196, 233, 244, 247, 344

hypercycle, 96, 114, 223, 242, 272, 279, Glossary

inanimacy, 50, 59, 65, 94, 203, 205, 342, Glossary

information theory, 36, 81, 83, 90, 112, 133, 186, 295, 323

instantiate, 34, 38, 41, 57, 79, 94, 95, 107, 140, 148, 220, 246, 359, 366, 370, Glossary

Law of Organizational and Cybernetic Deterioration/Decline (The OCD Law), 336, 366, Glossary

Law of Physicodynamic Incompleteness, 325, Glossary

structure, 72, 90, 114, 186, 226, 229, 285, 322, 323, Glossary

Sustained Functional Systems (SFS), 91- 92, 340, 360, Glossary

symbol, 3, 5, 10, 11, 12, 13, 18, 20, 29, 32, 34, 40, 44, 46, 48, 49, 50, 52, 60, 64, 67, 72, 73, 83, 85, 86, 90, 94, 101, 106, 108, 113, 118, 119, 130, 135, 136, 137, 138, 139, 140, 141, 142, 143, 144, 147, 148, 149, 150, 151, 152, 153, 155, 156, 157, 158, 159,162, 163, 166, 168, 170, 171, 173, 175, 177, 178, 180, 182, 186, 189, 200, 201, 203, 206, 207, 208, 211, 213, 214, 215, 217, 218, 220, 229, 230, 245, 248, 258, 259, 260, 269, 280, 285, 312, 319, 323, 329, 333, 334, 336, 339, 341, 342, 343, 345, 346, 347, 354, 356, 357, 358, 360, 361, 362, 363, 364, 365, 366, 367, 368, 369, 370, 372, 373, 374, 375, Glossary

symbol systems, 15, 135, 136, 349, 374, Glossary

syntactical, 12, 67, 138, 142, 374

syntax, 11, 13, 23, 29, 33, 63, 67, 136, 144, 145, 150, 152, 154, 162, 173, 204, 211, 218, 244, 246, 300, 319, 329, 344, 346, 356, 361

token, 5, 11, 20, 32, 60, 83, 94, 136, 140, 141, 145, 150, 154, 155, 172, 176, 214, 218, 220, 258, 259, 334, 343, 345, 370, Glossary

translate, 100, 146, 362, Glossary

translation, 13, 20, 64, 66, 67, 90, 113, 143, 144, 145, 150, 151, 152, 162, 172, 212, 213, 220, 277, 292, 319, 339, 346, 357, Glossary

Turing machine and tape, 181, 209, 227, 233, Glossary

undecidable, 65, 178, 342, Glossary

Universal Plausibility Metric (UPM), 18, 52, 74, 134, 159, 184, 198, 223, 302, 305, 306, 308, 311, 312, 321, 354, Glossary

Universal Plausibility Principle (UPP), 198, 308, 311, 315, 320, 321, Glossary

universal proteins, 126

upper probability bound or limit, 127, 308-10, 313

utility, 2, 3, 5, 6, 7, 10, 13, 14, 15, 18, 19, 23, 24, 27, 28, 29, 35, 36, 37, 38, 43, 48, 50, 55, 56, 57, 58, 59, 61, 62, 75, 77, 83, 85, 86, 88, 89, 91, 95, 100, 102, 103, 106, 110, 136, 137, 138, 146, 153, 164, 166, 167, 170, 172, 176, 177, 179, 191, 193, 202, 204, 214, 215, 218, 248, 249, 253, 254, 269, 281, 310, 325, 326, 327, 329, 330, 331, 332, 333, 334, 335, 336, 337, 339, 340, 341, 343, 346, 348, 350, 351, 356, 358, 359, 363, 364, 366, 371, 373, 376, Glossary

vesicles, 196 232, 236, 239,278, 279, 356, 366, Glossary

2-D complexity space, 149

3-D complexity space, 149

www.ingramcontent.com/pod-product-compliance
Lightning Source LLC
Chambersburg PA
CBHW082128210326
41599CB00031B/5905